RECURSIVE FUNCTIONS AND METAMATHEMATICS

SYNTHESE LIBRARY

STUDIES IN EPISTEMOLOGY,

LOGIC, METHODOLOGY, AND PHILOSOPHY OF SCIENCE

VOLUME 286

ROMAN MURAWSKI
Adam Mickiewicz University,
Poznań, Poland

RECURSIVE FUNCTIONS AND METAMATHEMATICS

Problems of Completeness and Decidability, Gödel's Theorems

KLUWER ACADEMIC PUBLISHERS
DORDRECHT / BOSTON / LONDON

A C.I.P. Catalogue record for this book is available from the Library of Congress.

ISBN 0-7923-5904-6

Published by Kluwer Academic Publishers,
P.O. Box 17, 3300 AA Dordrecht, The Netherlands.

Sold and distributed in North, Central and South America
by Kluwer Academic Publishers,
101 Philip Drive, Norwell, MA 02061, U.S.A.

In all other countries, sold and distributed
by Kluwer Academic Publishers,
P.O. Box 322, 3300 AH Dordrecht, The Netherlands.

Printed on acid-free paper

Printed in the Netherlands.

To Hania and Zosia

One can be sure that if God does actually exist then Gödel is in direct contact with him.

A. Mostowski

Preface

Gödel's incompleteness theorems and the decidability theory constitute one of the most significant parts of mathematical logic and the foundations of mathematics. Being very important and meaningful for mathematics they also have far-reaching epistemological and methodological consequences. They arose in the atmosphere of the late twenties and early thirties of the twentieth century marked by Hilbert's program of validation and justification of (infinite) classical mathematics. This program was one of the attempts to find a way out of the difficulties revealed by the discovery of antinomies connected with the notion of the actual infinity (on the turn of the nineteenth century).

Gödel's theorems and the results on undecidability of theories have shown that Hilbert's program cannot be realized in the proposed form. They also forced the revision of certain views on the nature and structure of mathematical knowledge.

The indicated results are not a closed chapter of logic and the foundations of mathematics. On the contrary, new interesting theorems being generalizations and strengthenings of Gödel's theorems have been obtained in the seventies. Quite recently the so-called reverse mathematics opened a new perspective showing that Hilbert's program can be partially realized. Problems of (un)decidability of theories got a new dimension and new impulses by the development of computer science and computers themselves (cf. problems of practical decidability). All those issues are still a subject of interesting philosophical discussions and investigations.

The aim of this book is to present Gödel's incompleteness theorems and the decidability theory in their historical and philosophical context. This means that the book contains both exposition of technical details of the mentioned results as well as the discussion of their historical and philosophical background and consequences. Hence the book can be classified somewhere between mathematical logic and the foundations of mathematics on the one hand and history and philosophy of mathematics and mathematical logic on the other. We hope that it fullfils the gap in the literature where one can find books and papers devoted either to mathematical exposition of Gödel's theorems and related topics without sufficient emphasis on their philosophical aspects (treating Gödel's theorems as mathematical results only and abstracting from their deep philosophical meaning) or, *vice versa*, discussions of the philosophical meaning of those results without sufficient mathematical background.

The book is organized in the following way. The introduction is devoted to the discussion of the historical and philosophical background of the studies presented in the book. In Chapter 1 elements of the recursion theory are sketched. The aim of this chapter is not to give a comprehensive course of this theory but to develop the tools which will be necessary in the sequel. Hence the following items are discussed there: recursive functions and their basic properties (Section 1.2), Markov algorithms and Turing machines (being other formalizations of the notion of effectiveness) (Section 1.3), primitive and elementary recursive functions (Section 1.4), the arithmetical hierarchy of numbertheoretic functions and relations (Section 1.5) as well as Church's thesis (Section 1.6).

Chapter 2 is devoted to Gödel's incompleteness theorems, their strengthenings and generalizations as well as to Tarski's theorem and related topics. We start by describing a standard formal system of the arithmetic of natural numbers, i.e., Peano arithmetic. In Section 2.2 the Representability Theorem is proved indicating connections between recursive functions and relations on the one hand and Peano arithmetic on the other. The next section is devoted to another main tool, i.e., the arithmetization of language and syntax. Gödel's incompleteness theorems and various related topics (such as theorems of Rosser, Löb and Tarski) are considered in Section 2.4. In Section 2.5 generalizations of the incompleteness theorems due to Paris, Harrington and Kirby are discussed. Section 2.6 is devoted to the study of connections between satisfaction and consistency. In particular it is shown how and where the notion of satisfaction for Peano arithmetic can be defined and Mostowski's theorem on the reflexivity of arithmetic is given.

In Chapter 3 one finds elements of the decidability theory. Having introduced the notions of decidability and undecidability in Section 3.1 we consider basic methods of proving the decidability (Section 3.2) and undecidability (Section 3.3) of theories. The considerations are illustrated by many examples.

The final Chapter 4 consists of philosophical and methodological comments. Various philosophical and methodological consequences of Gödel's incompleteness theorems and their generalizations as well as of theorems on undecidability of theories are considered here. We discuss direct implications of Gödel's results (Section 4.1), the problem of connections between incompleteness theorems and Hilbert's program (Section 4.2), generalizations of the program of Hilbert, i.e., generalized and relativized Hilbert's program, and the influence of results of the so-called reverse mathematics on the latter (Sections 4.3 and 4.4), Hilbert's tenth problem and Matiyasevich's theorem (Section 4.5) as well as connections between Gödel's theorems and computer science (Section 4.6) and general philosophy (Section 4.7).

Each chapter (except the final one) contains a section entitled "Historical Remarks". The aim of this section is to give the historical background of results discussed in the chapter, their further development, generalizations and strengthenings as well as to give some bibliographical information. The book contains also an extensive bibliography.

The book is self-contained and presupposes only some knowledge of elementary mathematical logic.

I would like to thank Professor Jan Woleński from Jagiellonian University (Cracow) who encouraged me to write this book. I thank also my wife Hania and my daughter Zosia for their enormous patience and understanding they showed during my work on this book.

Roman Murawski

Poznań

Contents

Introduction

It is commonly believed that mathematics is a pattern and an ideal of certainty and precision, and that within mathematics itself the role of an ideal is played by geometry. The sources of such opinions could be seen in the fact that just geometry has been developed (and taught) in the axiomatic-deductive form almost since its beginning. The origins of the axiomatic method should be detected in ancient Greece — it seems that Plato was the first who invented it and considered it to be the proper method of mathematics. This was connected with his assertion that objects of mathematics belong to the world of ideas and that mathematics which discovers interrelations between those objects and describes them is just a science about ideas. This implies in particular that the mathematical knowledge should be based on reason and that the proper method of mathematics is the axiomatic method consisting of accepting certain theorems without proof (axioms and postulates) and next deriving from them all other theorems. As an ideal example (and pattern) of applying this method to mathematics were considered *Elements* by Euclid (about 300 B.C.). This work played an enormous role in the development of the methodology of mathematics. In fact the paradigm established by *Elements* was the working paradigm till the beginning of the twentieth century.

For a long time (till the end of the nineteenth century) geometry was in fact the only part of mathematics possessing a firm axiomatic basis. During the last two centuries attempts were made to apply the axiomatic method to other theories — various sets of axioms for them were proposed. It was tacitly assumed that any mathematical knowledge can be based on an appropriate set of axioms which suffice to derive from them infinitely many theorems true in a given domain.

The turn of the nineteenth century was characterised by the intense development of mathematics — but it also brought results of a new type. Thanks to the achievements of algebra, and in particular of the theory of algebraic equations, it was possible to show that all three classical problems of Greek mathematics are unsolvable. Recall that we mean here the problems formulated in the fifth century B.C. and concerning: (1) doubling the cube, i.e., finding the cube root of 2, (2) trisecting any given angle, and (3) squaring the circle, i.e., constructing a square equal in area to that of a given circle. They should be solved by constructions with ruler and compass only (this condition was laid down presumably in Plato's Academy). After centuries of futile attempts (though on the way various interesting "by-products" were obtained, as, e.g., the theory of conic sections) one proved in the nineteenth century that the problems cannot be solved by the admitted methods. Hence it was shown that certain constructions are impossible — this was in fact a result of a new type.

Another example of research of this type are studies on the fifth postulate of Euclid. Ever since Euclid's time attempts were made to show that the parallel postulate can be deduced from other postulates or from some logically equivalent set of axioms. And again all those attempts turned out to be futile. In the nineteenth century it was shown that the fifth postulate is independent of other postulates of geometry. This was done by constructing non-Euclidean geometries in which instead of the parallel postulate its negation is assumed. The idea of such a system of geometry arose almost simultaneously (i.e., about 1830) in three different places with three authors: N.I. Lobachevsky, professor of the University of Kasan, J. Bolyai, officer of the Austro-Hungarian army, and C.F. Gauss, professor of the University of Göttingen.

It slowly became clear that the essence of mathematics was the deduction of theorems from the adopted assumptions (axioms) and not, as assumed so far, answering the question whether those assumptions are true. In the course of time mathematics became an abstract and formal knowledge (axioms say something not about particular concrete objects but about any objects). This was aphoristically stated by Russell who wrote:

> ... mathematics may be defined as the subject in which we never know what we are talking about, nor whether what we are saying is true. (1901, p. 84)

In this way the direct reference and connection of mathematics and reality disappeared. Consequently also the basis of the conviction that mathematics is consistent disappeared — in fact, mathematical theorems were no longer a description of facts in the empirical reality. So there arose a problem of consistency of a given set of axioms for a given mathematical theory. In particular one should ask whether the non-Euclidean geometry is consistent. Since the axioms and postulates of Euclid's geometry seem to be fulfilled in the real space, the question about consistency of geometry was not asked until the nineteenth century. But then, when the non-Euclidean geometries were constructed, such a problem should be formulated the more so because their axioms are not true in the real space.

One of the methods by which one attempted to solve the problem of consistency was the construction of models, i.e., the method of interpretation. As an example here can serve a proof of the consistency of the Riemann's plane geometry. It is one of the non-Euclidean geometries (called 'elliptic geometry' in distinction to the geometry of Lobachevsky-Bolyai-Gauss which is called 'hyperbolic'). In this geometry through a point not on a line there is no line parallel to that line (in hyperbolic geometry there is more than one line parallel to that line). If one interprets the notion of plane in the Riemann's axioms as a 3-dimensional sphere in the Euclidean space, the notion of point as a point on that sphere and the notion of straight line as a big circle on the sphere then all the axioms of Riemann's geometry become theorems of Euclidean geometry. In this way one obtains a relative (i.e., non-absolute) consistency proof of one theory (in this case, Riemann's geometry) by using consistency of another theory (in our example, Euclidean geometry). Another example of a proof of this type is the proof of the consistency of Euclidean geometry via analytic geometry (i.e., by using the consistency of the theory of real numbers) proposed by Hilbert. In all such proofs one shows consistency of a given theory assuming the consistency of another theory. Hence the value of such proofs is not very high from a methodological and epistemological point of view.

A method of finite models consisting of constructing a finite model of a given theory seems to be better. Indeed, there is no need to assume consistency of other theories because finite models can be inspected in a finite number of steps and in this way the consistency of the considered theory can be established in an absolute way. Unfortunately finite models are not so interesting from the mathematical point of view, i.e., most mathematical theories do not have finite models.

The problem of consistency proofs became more important when some difficulties in the foundations of mathematics appeared on the turn of the nineteenth century. Main controversy centered around the problem of the legitimacy of abstract objects.

The works of Karl Weierstrass (1815–1897) have contributed to the clarification of the role of the infinite in calculus. Set theory founded and developed by Georg Cantor (1845–1918) promised new heights of generality, clarity and rigor within mathematics. Unfortunately paradoxes appeared. Some of them were known already to Cantor (e.g., the paradox of the set of all ordinals and the paradox of the set of all sets) and they could be removed by appropriate modifications of set theory (cf. Cantor's distinction between *absolut unendliche* or *inkonsistente Vielheiten* and *konsistente Vielheiten*, i.e., between classes and sets) — see Cantor's letters to Dedekind in (Cantor, 1932, pp. 443–447 and p. 448)). Frege's (1848–1925) attempt to realize the idea of the reduction of mathematics to logic (which was in fact a continuation of the idea of the arithmetization of analysis developed among others by Weierstrass) led to a really embarrasing contradiction discovered in his system, discovered by Bertrand Russell (1872–1970) in 1901 and known today as 'Russell's antinomy' or as 'the antinomy of non-reflexive classes'. This meant a crisis of the foundations of mathematics (called the 'second crisis' — the first being the crisis caused by the discovery of incommensurable segments in ancient Greek mathematics).[1]

Various ways of overcoming those difficulties and of securing the edifice of mathematics were proposed. Great mathematicains — e.g., Leopold Kronecker, Henri Poincaré, Luitzen Egbertus Jan Brouwer, Hermann Weyl — challenged the validity of all infinitistic reasonings and proposed to restrict methods of mathematics to secure finite ones. Kronecker (1823–1891) rejected any infinite objects restricting mathematics to integers only ("Die ganzen Zahlen hat der lieber Gott gemacht, alles andere ist Menschenwerk" — as he formulated his scientific and methodological credo during a meeting in Berlin in 1886). Poincaré (1854–1912) saw the source of antinomies in impredicativity of mathematics and demanded a restriction to predicative methods only.[2] The radical proposal of Brouwer (1881–1966), known today as 'intuitionism', was based on the idea that mathematics should be founded on the primitive intuition of natural number. He claimed that mathematics is a free activity of the human mind,

[1]It is not clear who introduced the name "crisis of the foundations of mathematics" (*Grundlagenkrise der Mathematik*) but it was Hermann Weyl (1885–1955) who popularized it by his lecture "Über die neue Grundlagenkrise in der Mathematik" held in Zurich.

[2]Impredicative definitions are definitions in which one defines an object N by referring to a certain totality of objects E (for example by quantification on E) to which N belongs as an element. An example of an impredicative definition is the following one: the set of natural numbers $\mathbf{N}$ is the smallest set containing zero and closed under the successor operation. One defines here the unknown object $\mathbf{N}$ by referring to the totality of sets having the indicated properties and such that the defined object $\mathbf{N}$ is one of its elements.

it can (and should) be developed independently of any language, one should restrict it only to constructive methods, hence in particular any non-constructive proofs of existential sentences should be rejected (Brouwer claimed that proofs of that type were the source of all antinomies). He only accepted countable infinity and rejected any uncountable one.

All those proposals meant in fact a restriction of mathematics and a rejection of a great part of it, especially that one dealing with infinite objects. David Hilbert (1862–1943) was definitely against it. He wrote:

> What Weyl and Brouwer do comes to the same thing as to follow in the footsteps of Kronecker! They seek to save mathematics by throwing overboard all that which is troublesome ... They should chop up and mangle the science. If we would follow such a reform as the one they suggest, we would run a risk of losing a great part of our most valuable treasure! (cf. Reid, 1970, p. 155)

And added:

> I believe that as little as Kronecker was able to abolish the irrational numbers ... just as little will Weyl and Brouwer today be able to succeed. Brouwer is not, as Weyl believes him to be, the Revolution — only the repetition of an attempted Putch. (cf. Reid, 1970, p. 157)

And he stressed firmly that:

> Aus dem Paradies, das Cantor uns geschaffen hat, soll uns niemand vertreiben können.
>
> (No one should be able to drive us from the paradise that Cantor created for us.) (cf. Hilbert, 1926)

Hilbert proposed a method of justification of (infinite) mathematics known today as 'Hilbert's program'. It was the core of a new doctrine in the philosophy of mathematics called 'formalism' (which became one of the main trends of the modern philosophy of mathematics beside Frege's and Russell's logicism and Brouwer's intuitionism).

Hilbert was first of all a mathematician and — as Smoryński writes in (1988) — "had little patience with philosophy, his own philosophy of mathematics being perhaps best described as naïve optimism — a faith in the mathematician's ability to solve any problem he might set for himself". His aim was to save the integrity of classical mathematics (dealing with actual infinity) by showing that it is secure.[3] The problem was first stated by him in his lecture at the Second International Congress of Mathematicians held in Paris in 1900 (Hilbert, 1901). Among twenty three main problems which should be solved he mentioned there as problem 2 the task of proving the consistency of axioms of arithmetic (by which he meant number theory and analysis). He has been returning to the problem of justification of mathematics in his lectures and papers (especially in the twenties) where he proposed a method of solving it.[4] One should mention here his lecture from 1901 held at the meeting of Göttingen Mathematical Society in which he spoke about the problem of completeness and decidability. Hilbert asked, in E. Husserl's formulation: "Would I have the right to say that every proposition dealing only with the positive integers must be either true or false on the basis of the axioms for positive integers?" (Husserl, 1891, p. 445). In a series of lectures in the twenties Hilbert continued to formulate the problems more precisely and simultane-

[3]Detlefsen writes that "Hilbert did want to preserve classical mathematics, but this was not for him an end in itself. What he valued in classical mathematics was its efficiency (including its psychological naturalness) as a means of locating the truths of real or finitary mathematics. Hence, any alternative to classical mathematics having the same benefits of efficiency would presumably have been equally welcome to Hilbert" (Detlefsen, 1990, p. 374).

[4]A good account of the development of Hilbert's views can be found in Smoryński (1988); see also Peckhaus (1990) where detailed analysis of Hilbert's scientific activity in the field of the foundations of mathematics in the period 1899—1917 can be found, as well as Detlefsen (1986) and Ketelsen (1994).

ously communicated partial results obtained by himself and his students and fellow researchers: Paul Bernays, Wilhelm Ackermann, Moses Schönfinkel, John von Neumann. One must mention here Hilbert's lectures in Zurich (1917), Hamburg (1922), Leipzig (1922), Münster (1925) (cf. Hilbert, 1926), second lecture in Hamburg (1927) (cf. Hilbert, 1927) and the lecture "Probleme der Grundlegung der Mathematik" at the International Congress of Mathematicians held in Bologna (1928) (cf. Hilbert, 1929). In the latter Hilbert set out four open problems connected with the justification of classical mathematics which should be solved: (1) to give a (finitist) consistency proof of the basic parts of analysis (or second-order functional calculus), (2) to extend the proof for higher-order functional calculi, (3) to prove the completeness of the axiom systems for number theory and analysis, (4) to solve the problem of completeness of the system of logical rules (i.e., the first-order logic) in the sense that all (universally) valid sentences are provable. In this lecture Hilbert claimed also — wrongly, as it turned out — that the consistency of number theory had already been proved.[5]

One should add that Hilbert saw the supramathematical significance of the whole issue. He wrote:

> The definite clarification of the nature of the infinite has become necessary, not merely for the special interests of the individual sciences but for the honor of human understanding itself. (Hilbert, 1926)

Hilbert's program of clarification and justification of mathematics was Kantian in character.[6] One can see here a turn in the direction of idealism (or, as Detlefsen (1986) calls it, to in-

[5]Only after Gödel published his incompleteness theorems in 1931 did Hilbert come to realize that Ackermann's proof, which he meant here, did not establish the consistency of all of number theory. In fact, Ackermann showed (1924–25) only the consistency of a fragment of number theory. Cf. also Ackermann (1940). Other results of that type were also obtained by Von Neumann (1927) and Herbrand (1931).

[6]Cf. Detlefsen (1993); see also Murawski (199?a).

strumentalism). It was taken already in Hilbert (1918) and is evident in Hilbert (1926). At the end of the latter he wrote:

> Finally, we should recall our true theme and draw the net result of our reflections for the infinite. That net result is this: we find that the infinite is nowhere realized. It is neither present in nature nor admissible as a foundation in that part of our thought having to do with the understanding (in unserem verstandesmäßigen Denken) — a remarkable harmony between Being and Thought ...
>
> The role which remains for the infinite is rather that of an idea — if, following Kant's terminology, one understands as an idea a concept of reason which transcends all experience and by means of which the concrete is to be completed into a totality ...[7]

In Kant's philosophy, ideas of reason, or transcendental ideas, are concepts which transcend the possibility of experience but on the other hand are the answer to a need in us to form our judgements into systems that are complete and unified. Therefore, we form judgements concerning an external reality which are not uniquely determined by our cognition, judgements concerning things in themselves. To do that we need ideas of reason.

In comparing the infinite to a Kantian idea, Hilbert suggests that it is to be understood as a regulative rather than a descriptive device. Therefore, sentences concerning the infinite, and generally expressions which Hilbert called ideal propositions, should not be taken as sentences describing externally existing entities. In fact they mean nothing in themselves, they have no truth-value and they cannot be the content of any genuine judgement. Their role is rather regulative than descriptive. But on the other hand they are necessary in our thinking. Hence their similarity to Kant's ideas of reason is evident — they play the similar cognitive role.

We use ideas of reason and ideal elements in our thinking because they allow us to retain the patterns of classical logic

---

[7] English translation after Detlefsen (1993).

in our reasoning. But the operations of the classical logic can no longer be employed semantically as operations on meaningful propositions (there is nothing in the externally existing reality that would correspond to ideal elements and ideas of reason) but should be understood only syntactically, as operations on signs and strings of sings. In 1926 Hilbert wrote:

> We have introduced the ideal propositions to ensure that the customary laws of logic again hold one and all. But since the ideal propositions, namely, the formulas, insofar as they do not express finitary assertions, do not mean anything in themselves, the logical operations cannot be applied to them in a contentual way, as they are to the finitary propositions. Hence, it is necessary to formalize the logical operations and also the mathematical proofs themselves; this requires a transcription of the logical relations into formulas, so that to the mathematical signs we must still adjoin some logical signs, say
>
> | & | ∨ | → | — |
> |---|---|---|---|
> | and | or | implies | not |

Hence the abstracting from meaning of expressions is connected with Hilbert's attempt to preserve the laws of classical logic as laws governing mathematical thinking and reasoning. It is also connected with the distinction between real and ideal propositions according to which real propositions play the role of Kant's judgements of understanding (Verstand) and the ideal propositions the part of his ideas of pure reason. In 1926 Hilbert wrote:

> Kant taught — and it is an integral part of his doctrine — that mathematics treats a subject matter which is given independently of logic. Mathematics, therefore, can never be grounded solely on logic. Consequently, Frege's and Dedekind's attempts to so ground it were doomed to failure.
>
> As a further precondition for using logical deduction and carring out logical operations, something must be given

> in conception, viz., certain extralogical concrete objects which are intuited as directly experienced prior to all thinking. For logical deduction to be certain, we must be able to see every aspect of these objects, and their properties, differences, sequences, and contiguities must be given, together with the objects themselves, as something which cannot be reduced to something else and which requires no reduction. This is the basic philosophy which I find necessary not just for mathematics, but for all scientific thinking, understanding and communicating. The subject matter of mathematics is, in accordance with this theory, the concrete symbols themselves whose structure is immediately clear and recognizable.

According to this Hilbert distinguished between the unproblematic, 'finitistic' part of mathematics and the 'infinitistic' part that needed justification. Finitistic mathematics deals with so-called real propositions, which are completely meaningful because they refer only to given concrete objects. Infinitistic mathematics, on the other hand, deals with so-called ideal propositions that contain reference to infinite totalities.

By Hilbert, analoguously as it was by Kant, ideal propositions (and ideal elements) played an auxilary role in our thinking, they were used to extend our system of real judgements. Hilbert believed that every true finitary proposition had a finitary proof. Infinitistic objects and methods enabled us to give easier, shorter and more elegant proofs but every such proof could be replaced by a finitary one. This is the reflection of Kant's views of the relationship between the ideas of reason and the judgements of the understanding (Kant, 1787, p. 383). Hilbert was also convinced that consistency implies existence and that every proof of existence not giving a construction of postulated objects is in fact a presage of such a construction.[8]

[8]Compare in connection with this Hilbert's solution in 1888 to Gordan's problem in the theory of invariants in which he proved, without construction, the existence of a finite base for any ideal in the polynomial ring $K[X_0, \ldots, X_{n-1}]$ over a field K.

Unfortunately Hilbert did not give a precise definition of finitism — one finds only some hints how to understand it.[9] Hence various interpretations are possible. Usually it is assumed that a finitist reasoning is essentially a primitive reasoning in the sense of Skolem (this will be explained in Chapters 1 and 4) — cf. Tait (1981) and Resnik (1974). But there are also other interpretations — cf., e.g., Detlefsen (1979) where it is suggested that even some variants of ω-rule can be regarded as finitistic or Smoryński (1988) where instead of a dichotomy real/ideal a trichotomy real/finitary general/ideal is proposed (cf. also the criticism of this proposal in Detlefsen (1990)). Prawitz (1981) argues that real sentences are the decidable ones (i.e., numerical equations and truth-functional compositions of them) and the ones of the form $\forall x A(x)$ where each instance $A(t)$ is decidable. The rest are considered to be ideal. This emphasizes the role of Π^0_1 sentences in Hilbert's program (this notion will be explained in Section 1.5) — cf. Kitcher (1976) and Tait (1981).

The infinitistic mathematics can be justified only by finitistic methods because only they can give it security (*Sicherheit*). Hilbert's proposal was to base mathematics on finitistic mathematics via proof theory (*Beweistheorie*).[10] Its main goal was to show that proofs which use ideal elements in order to prove

[9]This can possibly be explained in the following way: Hilbert cherished hopes that his program would be realized. In this situation no precise definition of finitism was needed — one could simply check the proposed methods of proving consistency and conservativeness (see below). Only in the case of a negative solution to the problem a precise definition of the finitist methods is needed.

[10]Later Hilbert named it metamathematics (*Metamathematik*). This name was used by him for the first time in his lecture "Neubegründung der Mathematik" (1922). It is worth noting that the very term "Metamathematik", though in another meaning, already appeared in the nineteenth century in connection with discussions on non-Euclidean geometries. It was constructed in the analogy to the word "Metaphysik" (metaphysics) and had a pejorative meaning. For instance, Schultze (1882) was talking about "die metamathematischen Spekulationen über den Raum" (metamathematical speculations about the space). B. Erdmann and H. von Helmholtz contributed to the change of the meaning of this term to a positive one.

results in the real part of mathematics always yield correct results. One can distinguish two aspects here: consistency problem and conservation problem. In some of Hilbert's publications (cf., e.g., Hilbert, 1926 and 1927) both aspects are stressed but usually (cf., e.g., Hilbert's last publication on this subject, namely the first volume of Hilbert and Bernays, 1934–39) the one-sided emphasis is put on the consistency problem only (cf. also Kreisel (1968 and 1976) where the author calls for a proper formulation taking into account both aspects).

The consistency problem consists in showing (by finitistic methods, of course) that the infinitistic mathematics is consistent; the conservation problem consists in showing by finitistic methods that any real sentence which can be proved in the infinitistic part of mathematics can be proved also in the finitistic part, i.e., that infinitistic mathematics is conservative over finitistic mathematics with respect to real sentences and, even more, that there is a finitistic method of translating infinitistic proofs of real sentences into finitistic ones. Both of those aspects are interconnected — indeed, if one identifies real sentences with Π^0_1 sentences (see above) then (as shown by Kreisel) a solution to the consistency problem yields a solution to the conservation problem.[11]

Hilbert's proposal to carry out this program consisted of two steps. To be able to seriously study mathematics and mathematical proofs one should first of all accurately define the notion of a proof. In fact, the concept of a proof used in mathematical practice is intuitive, loose and vague, it has clearly a subjective character. This does not cause much trouble in practice. On the other hand, if one wants to study mathematics as a science — as Hilbert did — then one needs a precise notion of proof.

[11]It can be shown that if φ is a Π^0_1 sentence, i.e., a sentence of the form $\forall x \psi(x)$, where ψ contains atomic formulas, logical connectives and maybe bounded quantifiers, and $T \vdash \varphi$ (T being an infinitistic theory) then $S + Con_T \vdash \varphi$ (S being a finitistic theory and Con_T is a sentences stating that the theory T is consistent). If one showed now that $S \vdash Con_T$ then any Π^0_1 sentence φ provable in T would be also provable in S. Kreisel's results are presented, e.g., in Smoryński (1977, pp. 858–860).

This was provided by mathematical logic. In works of Frege and Russell (who used ideas and achievements of Peano) one finds an idea (and its implementations) of a formalized system in which a mathematical proof is reduced to a series of very simple and elementary steps, each of which consists in performing a purely formal transformation on the sentences which were previously proved. In this way the concept of mathematical proof was subjected to a process of formalization. Therefore the first step proposed by Hilbert in the realization of his program was to formalize mathematics, i.e., to reconstitute infinitistic mathematics as a big, elaborate formal system (containing classical logic, infinite set theory, arithmetic of natural numbers, analysis). An artificial symbolic language and rules of building well-formed formulas should be fixed. Next axioms and rules of inference (referring only to the form, to the shape of formulas and not to their sense or meaning) ought to be introduced. In such a way theorems of mathematics become those formulas of the formal language which have a formal proof based on a given set of axioms and given rules of inference. One condition was put on the set of axioms (and rules of inference): they ought to be chosen in such a way that they suffice to solve any problem formulated in the language of the considered theory as a real sentence, i.e., they ought to form a complete set of axioms with respect to real sentences.

The second step of Hilbert's program was to give a proof of the consistency and conservativeness of mathematics. Such a proof should be carried out by finitistic methods. This was possible since the formulas of the system of formalized mathematics are strings of symbols and proofs are strings of formulas, i.e., strings of strings of symbols. As such they can be manipulated finitistically. To prove the consistency it suffices to show that there are not two sequences of formulas (two formal proofs) such that one of them has as its end element a formula φ and the other $\neg\varphi$ (the negation of the formula φ). To show conservativeness it should be proved that any proof of a real sentence can be transformed into a proof not referring to ideal objects.

One should note here that formalization was for Hilbert only an instrument used to prove the correctness of (infinitistic) mathematics. Hence the objections raised to him by Brouwer are mistaken. As indicated in Kreisel (1958) the real opposition between Brouwer's and Hilbert's approach to mathematics was between: (1) the conception of what constitutes a foundation, and (2) two informal ways of reasoning, namely finitist and intuitionist. Recall that Brouwer ignored non-constructive mathematics altogether.

Hilbert was professor at the University of Göttingen in Germany. This university was one of the main centers of research in logic and foundations of mathematics in the twenties and thirties. Beside it one should mention the Lvov-Warsaw school and the University in Vienna as well as the American Postulate Theorists. The most important from our point of view is Vienna. Here was the Vienna Circle active — a group of scientists of various specialities who founded a new trend in the philosophy called neopositivism or logical positivism or logical empiricism. The group was founded by a mathematician Hans Hahn (1879–1934), its senior members were Otto Neurath (1882–1945) and Moritz Schlick (1882–1936). They gathered many others around them: Rudolf Carnap (1891–1970), Richard von Mises, Felix Kaufmann, Viktor Kraft, Olga Hahn (-Neurath), Karl Menger and Kurt Gödel (1906–1978) who was then a student of Hans Hahn. It is just through his contacts with the Vienna Circle that Gödel became interested in mathematical logic and the foundations of mathematics.[12]

[12]One should add here that Gödel did not accept the positivistic philosophy of the Vienna Circle. As he wrote in a letter to Grandjean of 19th August 1975 (not sent): "It is true that my interest in the foundations of mathematics was aroused by the "Vienna Circle", but the philosophical consequences of my results, as well as the heuristic principles leading to them, are anything but positivistic or empiricistic" (cf. Wang Hao, 1987, p. 20). In the thirties Gödel's conntacts with the Vienna Circle were weaker. He was engaged instead in Menger's *Mathematisches Kolloquium* and helped to publish *Ergebnisse eines Mathematischen Kolloquiums.* For further infor-

Hans Hahn took part in the Congress of Mathematicians in Bologna in 1928 during which Hilbert gave his lecture "Probleme der Grundlegung der Mathematik" mentioned above. Hahn communicated Hilbert's view to the Vienna Circle and encouraged his student Kurt Gödel to set to work on problems formulated by Hilbert. Gödel made already an indepth study of the recently published book by Hilbert and Ackermann (1928) which, beside Hahn's suggestions, influenced his approach and choice of problems. In that book the notions of completeness and decidability were clarified (two other notions crucial to understand the work of Hilbert and Gödel, namely the notions of consistency and independence were already clear at the turn of the century). By mid-1929 Gödel had solved the fourth problem of Hilbert's Bologna address. Taking first-order logic as presented in Hilbert and Ackermann (1928), he showed that the first-order logic is complete, i.e., every true statement can be derived from the axioms. Moreover he proved that, in the first-order logic, every consistent axiom system has a model. Those results were included in Gödel's doctoral dissertation (1929).

It is worth paying attention here to some remark at the beginning of Gödel's dissertation. He noted there that to take consistency to be the sole criterion for the existence of a model (as Hilbert did — see above) presupposed that we cannot establish the unsolvability of any problem. He wrote:

> For, if the unsolvability of some problem (in the domain of real numbers, say) were proved, then, from the definition above, there would follow the existence of two non-isomorphic realizations of the axiom system for the real numbers, while on the other hand we can prove the isomorphism of any two realizations. We cannot at all exclude out of hand, however, a proof of the unsolvability of a problem if we observe that what is at issue here is only unsolvability by certain *precisely stated formal*

mation on Gödel's contacts with the Vienna Circle cf. Köhler (199?) and Sigmund (1995) as well as Wang Hao (1987).

> means of inference. For, all the notions that are considered here (provable, consistent, and so on) have an exact meaning. ... These reflectons, incidentally, are intended only to properly illuminate the difficulties that would be connected with such a definition of the notion of existence, without any definitive assertion being made about its possibility or impossibility.

Observe that Gödel carefully noted here that he had not shown the unsolvability of some problem. The quoted remarks were omitted when he published his thesis in 1930. Nevertheless it reveals that he considered the subject of incompleteness even while he was establishing the completeness of first-order logic.

Some months later, in 1930, Gödel solved three other problems posed by Hilbert in Bologna (and proved that the possibility mentioned above really holds) by showing that arithmetic of natural numbers and all richer theories (i.e., theories extending it) are essentially incomplete. This means that there are sentences undecidable in them, hence sentences φ such that neither φ nor $\neg\varphi$ can be proved in the given theory. This property cannot be changed by adding new axioms (provided that the set of axioms is recursive, i.e., it can be effectively recognized whether a given sentence is an axiom or not). Furthermore, Gödel also showed that no such theory can prove its own consistency.

So Gödel's answers to Hilbert's questions from Bologna were the opposite of what the latter had expected. In this way Hilbert's hopes and plans to justify classical mathematics were essentially weakened or even destroyed. The problem of the interrelations between Gödel's results and Hilbert's program will be discussed in Chapter 4.

Chapter 1

Recursive Functions

1.1 Computable Functions

The notion of effectiveness occurs in various domains. It can appear both in empirical and in theoretical sciences where one has often to do with a task of finding a method which would solve a given problem in an effective way. For instance, in analytic chemistry we want to have methods which make it possible to find out effectively the chemical composition of a substance by means of a finite number of successive tests. In mathematics we are looking for effective methods which enable us to find solutions of a given algebraical equation of a given type in a finite number of simple operations or to decide whether a given natural number is a prime or not. If we succeed, i.e., if such an effective method can be found, then one says that the considered problem is decidable.

But what does "effective method" mean in fact? It is impossible to answer this general question precisely. Therefore we shall restrict ourselves to a special class of objects, namely to functions and relations in the set $\mathbb{N}$ of natural numbers and try to find an answer in this case only. By applying the precise definition of effectiveness for this class and by using appropriate interpretations one can characterize effectiveness also in other cases.

So let f be a k-ary function whose arguments and values belong to the set $\mathbb{N}$ of natural numbers, i.e., $f : \mathbb{N}^k \to \mathbb{N}$. What does it mean that values of the function f can be calculated effectively? It means that there exists a mechanical method by which for any k-tuple $a_1, \ldots, a_k$ of arguments the value $f(a_1, \ldots, a_k)$ can be calculated in a finite number of steps. The adjective "mechanical" used here is imprecise, of course. The best way to elucidate it is to give some examples of methods which are excluded by it. So methods which involve chance procedures, magic or insight are excluded. Functions for which a mechanical method exists of calculating their values will be called *computable* and the class of all such functions will be denoted by $\mathcal{O}$. Addition and multiplication are examples of computable functions. Indeed, for any two given natural numbers we can find their sum and product in a finite number of simple mechanical steps (we were taught in school how to do it).

Now let R be a k-ary relation among natural numbers. What does it mean that the property described by R can be effectively decided for given natural numbers? It means that there exists a mechanical method by which for any given k-tuple $a_1, \ldots, a_k$ of natural numbers it can be said in a finite number of steps if the relation $R(a_1, \ldots, a_k)$ holds or not. Call such relations *computable*. An example of such a relation is the relation of being prime — indeed, using methods taught in school one can find all divisors of a given number and check if the only divisors are 1 and the given number.

The definitions of effectiveness indicated above cannot be satisfactory to a mathematician. The reason is that they are using imprecise and unclear colloquial notions such as "mechanical" or "simple". So one has to look for more precise definitions of the notion of effective calculability. This will be done in succeeding sections where a notion of recursive functions and recursive relations will be introduced and basic properties of them will be proved.[1] Those considerations belong to the domain of the

[1] For the discussion of historical, philosophical and methodological aspects of using the terms 'computable' and 'recursive' cf. Soare (1996).

foundations of mathematics called *recursion theory*. In this theory various properties of functions and relations in the set $\mathbb{N}$ of natural numbers connected with computability are studied. Hence it has much in common with the theory of mathematical machines, with automata theory and generally with computer science. It is an extensive and highly developed mathematical theory which has applications in various parts of mathematical logic and mathematics. The study of calculable functions and relations arose from the studies of the problems of completeness and decidability of formal systems (first of all of systems of arithmetic — cf. Section 1.7).

1.2 Recursive Functions

We shall consider functions with arguments and values in the set $\mathbb{N}$ of natural numbers and relations among natural numbers. Hence we are interested in functions $f : \mathbb{N}^k \to \mathbb{N}$ and relations $R \subseteq \mathbb{N}^l$. From now on the words "function" and "relation" denote such functions and relations and the word "number" means "natural number".

The aim of this section is to specify the intuitive notion of computability by giving a mathematical definition of a class of functions called recursive functions and denoted by $\mathcal{R}$. We intend this class to be equal to the class $\mathcal{O}$ of computable functions.

To simplify the notation we shall write $\overline{a}$ instead of $a_1, \ldots, a_n$. Hence the symbol $f(\overline{a})$ will denote $f(a_1, \ldots, a_n)$, the symbol $\forall\overline{a}\varphi$ will stand for $\forall a_1 \ldots \forall a_n \varphi$, where $\forall$ is the symbol of the universal quantifier, and the symbol $\exists\overline{a}\varphi$ for $\exists a_1 \ldots \exists a_n \varphi$, where $\exists$ denotes the existential quantifier.

For any n-ary relation R we define its characteristic function $K_R : \mathbb{N}^n \to \mathbb{N}$ as follows:

$$K_R = \begin{cases} 0, & \text{if } R(\overline{a}), \\ 1, & \text{if } \neg R(\overline{a}). \end{cases}$$

It can be easily seen that the relation R is computable if and only if the function K_R is computable.

Before we define the class of recursive functions, let us consider some examples of computable functions and of operations leading from computable functions to other computable functions. Observe first of all that the n-ary projection function I_i^n $(n \in \mathbb{N}, 1 \leqslant i \leqslant n)$, where $I_i^n(a_1, \dots a_n) = a_i$ is computable. The binary functions $+$ and $\cdot$ are also computable (we are taught how to calculate their values in school). The relation $<$ and hence its characteristic function $K_<$ are also computable.

If functions g, h_1, ... h_k are computable then their composition, i.e., the function

$$f(\overline{a}) = g(h_1(\overline{a}), \dots, h_k(\overline{a}))$$

is also computable. Indeed, to calculate the value $f(\overline{a})$ it is enough to calculate first the values $h_1(\overline{a})$, ..., $h_k(\overline{a})$ and then the value $g(h_1(\overline{a}), \dots, h_k(\overline{a}))$.

Consider one more operation leading from computable functions to computable functions. If φ is a formula with a free variable x then by $\mu x \varphi(x)$ we denote the least number x such that $\varphi(x)$. This operation will be called the minimization operation and μx — the μ-operation. Observe that the variable x occurs in $\mu x \varphi(x)$ as bounded — hence it can be replaced by any other variable which does not occur in φ and the value will not be changed. Let now a computable function g be given and assume that the effectiveness condition is satisfied, i.e., that for any $\overline{a}$ there exists an x such that

$$g(\overline{a}, x) = 0. \tag{1.1}$$

Then we can define a new function f as follows:

$$f(\overline{a}) = \mu x[g(\overline{a}, x) = 0].$$

The function f is also computable. Indeed, to calculate the value $f(\overline{a})$ it suffices to calculate successively the values $g(\overline{a}, 0), g(\overline{a}, 1)$, ... till we get $g(\overline{a}, x_0) = 0$, and the effectiveness conditions (1.1) guarantees that such x_0 does exist. The operation leading from the function g fulfilling the effectiveness condition to the function f will be called effective minimization operation (or, shortly, the effective minimum).

The above considerations justify (at least partially) the following inductive definition of the class of recursive functions.

Definition 1.2.1 (i) *The functions* I_i^n $(n \in \mathbb{N}, 1 \leqslant i \leqslant n)$, $+, \cdot$ *and* $K_<$ *are recursive.*

(ii) *If* $g, h_1, \ldots, h_k$ *are recursive then the function*

$$f(\overline{a}) = g(h_1(\overline{a}), \ldots, h_k(\overline{a}))$$

is also recursive.

(iii) *If a function* g *is recursive and*

$$\forall \overline{a} \exists x [g(\overline{a}, x) = 0]$$

then the function

$$f(\overline{a}) = \mu x [g(\overline{a}, x) = 0]$$

is also recursive.

(iv) *There are no other recursive functions except those given in* (i) *and those which can be obtained using the rules* (ii) *and* (iii). *The set of all recursive functions will be denoted by* $\mathcal{R}$.

Hence the set of recursive functions is the smallest set of functions containing the functions $I_i^n, +, \cdot, K_<$ and closed under the operations of composition and effective minimization.

Note that in (ii) of the above definition one demands that the functions $h_1, \ldots, h_k$ have the same arity as the function f being defined. To fulfill this demand we use the functions I_i^n. Let us explain this by giving a suitable example. Consider a function f given by $f(a, b, c) = (a+b)^2 + c$. It is a recursive function because all functions defined below are recursive (they are obtained from recursive functions by composition):

$$\begin{aligned} f_1(a,b,c) &= a+b = I_1^3(a,b,c) + I_2^3(a,b,c) \\ f_2(a,b,c) &= (a+b)^2 = f_1(a,b,c)\cdot f_1(a,b,c) \\ f(a,b,c) &= (a+b)^2 + c = f_2(a,b,c) + I_3^3(a,b,c) = \\ &= (I_1^3(a,b,c) + I_2^3(a,b,c))\cdot(I_1^3(a,b,c) + I_2^3(a,b,c)) + \\ &\quad + I_3^3(a,b,c). \end{aligned}$$

Definition 1.2.2 *A relation R is said to be recursive if and only if its characteristic function K_R is recursive.*

We now give several properties of recursive functions and relations. They will indicate new (secondary) methods of defining functions and relations belonging to the class $\mathcal{R}$.

Theorem 1.2.3 *If a relation Q and functions $h_1, \ldots, h_k$ are recursive then the relation R such that*

$$R(\overline{a}) \equiv Q(h_1(\overline{a}), \ldots, h_k(\overline{a}))$$

is recursive.

Proof. We have $K_R(\overline{a}) = K_Q(h_1(\overline{a}), \ldots, h_k(\overline{a}))$. Hence by Definitions 1.2.1(ii) and 1.2.2 the relation R is recursive. Q.E.D.

Theorem 1.2.4 *If a relation P is recursive and $\forall\overline{a}\exists x P(\overline{a}, x)$ then a function $f(\overline{a}) = \mu x P(\overline{a}, x)$ is recursive.*

Proof. Observe that $f(\overline{a}) = \mu x[K_P(\overline{a}, x) = 0]$. So by Definition 1.2.1(iii) the function f is recursive. Q.E.D.

Theorem 1.2.5 *Every constant function is recursive.*

Proof. Let f^k denote the n-ary function with the constant value k. We show by induction on k that f^k is recursive for every k. For $k = 0$ we have:

$$f^0(\overline{a}) = \mu x(I_{n+1}^{n+1}(\overline{a}, x) = 0).$$

Assume that the function f^k is recursive. We show that the function f^{k+1} is also recursive. Indeed, one has

$$f^{k+1}(\overline{a}) = \mu x(f^k(\overline{a}) < x).$$

By the induction hypothesis and Theorem 1.2.4 the function f^{k+1} is recursive. Q.E.D.

Recall the basic operations on relations analogous to connectives. If P and Q are relations then we define:

$$\begin{aligned}(\neg P)(\overline{a}) &\equiv \neg P(\overline{a}),\\ (P \vee Q)(\overline{a}) &\equiv P(\overline{a}) \vee Q(\overline{a})\end{aligned}$$

and similarly for $\rightarrow, \wedge, \equiv$.

Theorem 1.2.6 *If P is recursive then $\neg P$ is recursive. If P and Q are recursive then the relations $P \vee Q, P \rightarrow Q, P \wedge Q, P \equiv Q$ are recursive.*

Proof. The theorem follows by the following equations and equivalences:

$$\begin{aligned}K_{\neg P}(\overline{a}) &= K_{<}(0, K_P(\overline{a})),\\ K_{P\vee Q}(\overline{a}) &= K_P(\overline{a}) \cdot K_Q(\overline{a}),\\ (P \wedge Q)(\overline{a}) &\equiv \neg(\neg P(\overline{a}) \vee \neg Q(\overline{a})),\\ (P \rightarrow Q)(\overline{a}) &\equiv \neg P(\overline{a}) \vee Q(\overline{a}),\\ (P \equiv Q)(\overline{a}) &\equiv (P(\overline{a}) \rightarrow Q(\overline{a})) \wedge (Q(\overline{a}) \rightarrow P(\overline{a})).\end{aligned}$$

Theorem 1.2.7 *The relations $<$, $\leqslant$, $>$, $\geqslant$ and $=$ are recursive.*

Proof. The relation $<$ is recursive by Definition 1.2.1(i). Recursiveness of other indicated relations follows from equivalences given below by Theorem 1.2.6:

$$\begin{aligned} a \leqslant b &\equiv \neg(b < a), \\ a > b &\equiv b < a, \\ a \geqslant b &\equiv b \leqslant a, \\ a = b &\equiv a \leqslant b \wedge b \leqslant a. \end{aligned}$$

Q.E.D.

We now introduce a new operation on functions. Let a be a natural number and let $\varphi(x)$ be a formula in which x occurs free (there can be also other free variables in φ). We put

$$\mu x_{x<a}\varphi(x) = \mu x(\varphi(x) \vee x = a).$$

Observe that $\mu x_{x<a}\varphi(x)$ is always defined and equal either to the least number smaller than a and satisfying φ, if such a number exists, or to a. The operation leading from φ and a to $\mu x_{x<a}\varphi(x)$ will be called the bounded minimization and the operator $\mu x_{x<\ldots}$ will be called the bounded μ-operator.

Theorem 1.2.8 *If P is a recursive relation then a function*

$$f(a,\overline{a}) = \mu x_{x<a}P(\overline{a},x)$$

is recursive.

Proof. It follows from the definition of the bounded μ-operator and Theorems 1.2.4 and 1.2.6. Q.E.D.

We have stated above that the class of recursive relations is closed under operations similar to connectives. What about quantifiers? It will be shown in Section 1.5 that they do not have this property. But it turns out that there are certain "weaker" quantifiers — called bounded quantifiers — which do have the considered property.

Let a be a natural number and $\varphi(x)$ a formula with a free variable x (and maybe other free variables). We define two operations:

$$\exists x < a\varphi(x) \qquad \text{and} \qquad \forall x < a\varphi(x)$$

as follows:

$$\exists x < a\varphi(x) \equiv \mu x_{x<a}\varphi(x) < a,$$
$$\forall x < a\varphi(x) \equiv \neg\exists x < a\neg\varphi(x).$$

Observe that $\exists x < a\varphi(x)$ if and only if there exists an element $x < a$ such that $\varphi(x)$. Similarly $\forall x < a\varphi(x)$ if and only if for any x such that $x < a$ it holds $\varphi(x)$.

By the definition of the bounded quantifiers and Theorem 1.2.8 we obtain

Theorem 1.2.9 *If R is recursive then relations P and Q fulfilling the following conditions are also recursive:*

$$P(a, \overline{a}) \equiv \exists x < aR(\overline{a}, x),$$

$$Q(a, \overline{a}) \equiv \forall x < aR(\overline{a}, x).$$

It can easily be seen that the condition of the type $x < a$ can be replaced by a condition of the type $x \leqslant a$ (which is simply equivalent to the condition $x < a + 1$). Observe also that in Theorems 1.2.8 and 1.2.9 the conditions of the type $x < a$ can be replaced by conditions of the type $x < g(\overline{b})$ where g is any recursive function. Indeed, in the case of Theorem 1.2.8 we get

$$f(\overline{a}, \overline{b}) = \mu x_{x<g(\overline{b})} P(\overline{a}, x) = \mu x_{x<c}(c = g(\overline{b}) \wedge P(\overline{a}, x)),$$

which by the definition of the bounded minimum and Theorems 1.2.6 and 1.2.7 as well as Definitions 1.2.1 and 1.2.2 gives that f is recursive provided that P and g are recursive. In the case of Theorem 1.2.9 the argument is similar.

Definition 1.2.10

$$a \dot{-} b = \begin{cases} a - b, & \textit{if } a \geqslant b, \\ 0, & \textit{if } a < b. \end{cases}$$

Theorem 1.2.11 *The function $\dot{-}$ is recursive.*

Proof. $a \dot{-} b = \mu x(b + x = a \vee a < b)$. Q.E.D.

Theorem 1.2.12 *Let $g_1, \ldots, g_k$ be recursive functions and let $R_1, \ldots, R_k$ be recursive n-ary relations such that for any $\overline{a}$ exactly one of the relations $R_1(\overline{a}), \ldots, R_k(\overline{a})$ holds. If a function f satisfies the following condition*

$$f(\overline{a}) = \begin{cases} g_1(\overline{a}), & \textit{if } R_1(\overline{a}), \\ \vdots & \\ g_k(\overline{a}), & \textit{if } R_k(\overline{a}), \end{cases}$$

then f is recursive.

Proof. The following equation holds

$$f(\overline{a}) = g_1(\overline{a}) \cdot K_{\neg R_1}(\overline{a}) + \ldots + g_k(\overline{a}) \cdot K_{\neg R_k}(\overline{a}).$$

The function f is recursive by Theorem 1.2.6 and Definition 1.2.1. Q.E.D.

Theorem 1.2.13 *Let $Q_1, \ldots, Q_k$ and $R_1, \ldots, R_k$ be recursive relations such that for any arguments $\overline{a}$ exactly one of the relations $R_1(\overline{a}), \ldots, R_k(\overline{a})$ holds. Then the relation P defined by*

$$P(\overline{a}) = \begin{cases} Q_1(\overline{a}), & \textit{if } R_1(\overline{a}), \\ \vdots & \\ Q_k(\overline{a}), & \textit{if } R_k(\overline{a}) \end{cases}$$

is recursive.

Proof. The following equation holds

$$K_P(\overline{a}) = \begin{cases} K_{Q_1}(\overline{a}), & \text{if } R_1(\overline{a}), \\ \vdots & \\ K_{Q_k}(\overline{a}), & \text{if } R_k(\overline{a}). \end{cases}$$

Q.E.D.

Using the above theorems one can show that several functions known from number theory are recursive. Let us indicate some examples:

1. the predecessor function $P(\overline{a})$; indeed

$$P(\overline{a}) = \begin{cases} a\dot{-}1, & \text{if } a > 0, \\ 0, & \text{if } a = 0, \end{cases}$$

2. the function $|a - b|$; we have

$$|a-b| = \begin{cases} a\dot{-}b, & \text{if } a \geqslant b, \\ b\dot{-}a, & \text{if } a < b, \end{cases}$$

3. the function signum $\text{sg}(a)$:

$$\text{sg}(a) = \begin{cases} 1, & \text{if } a \neq 0, \\ 0, & \text{if } a = 0, \end{cases}$$

4. the function $\min(a, b) =$ the smaller of a, b:

$$\min(a, b) = a\dot{-}(a\dot{-}b),$$

5. for any natural number n the function $\min(a_1, \ldots, a_n) =$ the smallest of $a_1, \ldots, a_n$; we prove this by induction on n using the function defined in the example 4 and the following equation:

$$\min(a_1, \ldots, a_{n+1}) = \min(\min(a_1, \ldots, a_n), a_{n+1}),$$

6. the function $\max(a, b) =$ the bigger of a, b:

$$\max(a, b) = b + (a\dot{-}b),$$

7. for any n the function $\max(a_1, \ldots, a_n) =$ the greatest of $a_1, \ldots, a_n$,

8. the divisibility relation $\text{Div}(a, b) \equiv b$ divides a; indeed

$$\text{Div}(a, b) \equiv \exists z \leqslant a(a = b \cdot z),$$

9. the relation of being prime $\text{Pr}(a)$; indeed

$$\text{Pr}(a) \equiv a \neq 0 \wedge a \neq 1 \wedge a = \mu z_{z \leqslant a}(z > 1 \wedge \text{Div}(a, z)).$$

Corollary 1.2.14 (i) *The complement of a recursive set is recursive.*

(ii) *The union and intersection of recursive sets are recursive.*

(iii) *Every finite set is recursive.*

Proof. (i) and (ii) follow from Theorem 1.2.6 and the fact that sets can be considered as unary relations. Hence one needs to prove only (iii). If A is the empty set $\emptyset$ then we have $a \in A \equiv a < a$. By Theorem 1.2.7 the set A is recursive. Let now $A = \{k_1, \ldots, k_n\}$. Then

$$a \in A \equiv a = k_1 \vee \ldots \vee a = k_n$$

and A is recursive by Theorems 1.2.6 and 1.2.7. Q.E.D.

Notice that in the proof of Corollary 1.2.14 we used ... meaning "etc." It was simply an abbreviation for the disjunction of n equations of the form $a = k_i$. But in general it is not permitted to use ... in definitions of recursive functions and relations, in particular in cases when "dots" depend on the value of the argument. The reason is that it can lead to nonrecursive functions.

We now prove that the class $\mathcal{R}$ of recursive functions and relations has still another interesting property, namely that it is closed under the operation of primitive recursion. Before we do so we introduce a certain tool, namely the operation of coding finite sequences of natural numbers by single natural numbers in such a way that both coding and decoding are recursive. Coding will turn out to be a useful tool having many applications.

Theorem 1.2.15 (Gödel) *There exists a binary recursive function β such that*

$$\beta(a, i) \leqslant a \dot{-} 1$$

for all a and i and such that for any sequence $a_0, a_1, \ldots, a_{n-1}$ there exists a number a such that $\beta(a, i) = a_i$ for all $i < n$.

Proof. Introduce a relation RP defined as follows:

$$\mathrm{RP}(a,b) \equiv a \neq 0 \wedge b \neq 0 \wedge \forall x[\mathrm{Div}(ax,b) \rightarrow \mathrm{Div}(x,b)].$$

It can easily be seen that $\mathrm{RP}(a,b)$ if and only if a and b are relatively prime. We have

$$\mathrm{RP}(a,b) \rightarrow \mathrm{RP}(b,a). \tag{1.2}$$

Indeed, let $\mathrm{RP}(a,b)$ and let $\mathrm{Div}(bx,a)$. Then $bx = ay$ for some y. Hence $\mathrm{Div}(ay,b)$. By the assumption $\mathrm{RP}(a,b)$ we have $\mathrm{Div}(y,b)$, hence $y = bz$ for some z. So $bx = abz$, $x = az$ and consequently $\mathrm{Div}(x,a)$.

Now we show that

> for any numbers $a_1,\ldots,a_n,b_1,\ldots,b_m$ different from 0 and 1 and such that $\mathrm{RP}(a_i,b_j)$ for every i and j there exists a number c such that $\mathrm{Div}(c,a_i)$ for $i = 1,\ldots,n$ and $\neg\mathrm{Div}(c,b_j)$ for $j = 1,\ldots,m$.

Indeed, for $n = 1$ we take $c = a_1$. So let $n > 1$. Then by the induction hypothesis there exists a c' such that it is divisible by $a_1,\ldots,a_{n-1}$ and is not divisible by $b_1,\ldots,b_m$. So take $c = a_n c'$. It has the required property.

Observe that

$$k \neq 0 \wedge z \neq 0 \wedge \mathrm{Div}(z,k) \rightarrow \mathrm{RP}(1+(j+k)z, 1+jz). \tag{1.3}$$

Indeed, for any x, if $\mathrm{Div}(x+xjz,z)$ then $\mathrm{Div}(x,z)$. Hence

$$\mathrm{RP}(1+jz,z),$$

so, by (1.2), $\mathrm{RP}(z,1+jz)$. Assume now that

$$\mathrm{Div}(x+x(j+k)z, 1+jz).$$

Then $\mathrm{Div}(xkz,1+jz)$. But $\mathrm{RP}(z,1+jz)$, so $\mathrm{Div}(xk,1+jz)$. By the assumption $\mathrm{Div}(z,k)$, so $\mathrm{Div}(xz,1+jz)$. On the other hand $\mathrm{RP}(z,1+jz)$. Consequently $\mathrm{Div}(x,1+jz)$. So indeed $\mathrm{RP}(1+(j+k)z,1+jz)$.

Define now a function

$$\mathrm{OP}(a,b) = (a+b)(a+b) + a + 1.$$

It can easily be seen that OP is recursive and

$$\mathrm{OP}(a,b) = \mathrm{OP}(a_1,b_1) \rightarrow a = a_1 \wedge b = b_1. \tag{1.4}$$

Indeed, assume that $a + b < a_1 + b_1$. Then

$$\mathrm{OP}(a,b) \leqslant (a+b+1)^2 \leqslant (a_1+b_1)^2 < \mathrm{OP}(a_1,b_1).$$

But this contradicts our assumption, so it is not true that $a+b < a_1 + b_1$. In a similar way one can show that it is not true that $a_1+b_1 < a+b$. So consequently we get $a+b = a_1+b_1$. From this and the assumption that $\mathrm{OP}(a,b) = \mathrm{OP}(a_1,b_1)$ we get $a = a_1$. Consequently also $b = b_1$.

Now we can define the required Gödel function β. We put

$$\beta(a,i) = \mu x_{x<a\dot{-}1}[\exists y < a \exists z < a(a = \mathrm{OP}(y,z) \wedge \\ \wedge \mathrm{Div}(y, 1 + (\mathrm{OP}(x,i)+1)\cdot z))].$$

It can easily be seen that β is recursive and $\beta(a,i) \leqslant a\dot{-}1$. So it suffices to prove only the second property of β stated in the theorem.

Let $a_0, a_1, \ldots, a_{n-1}$ be given. We are looking for a number a such that $\beta(a,i) = a_i$ for $i < n$. Let c be the largest of the $\mathrm{OP}(a_i,i)+1$ for $i < n$ and let z be a non-zero number divisible by every x such that $x < c$. By (1.3) for $k = l - j$ we get that if $j < l < c$ then $\mathrm{RP}(1+jz, 1+lz)$. By the property proved above there exists a number y such that for $j < c$:

$$\mathrm{Div}(y, 1+jz)) \text{ if and only if } j \text{ is of the form } \mathrm{OP}(a_i,i)+1.$$

We put $a = \mathrm{OP}(y,z)$. This is the required number a. Indeed, by the definition of OP we have $a_i < y < a, z < a$. By (1.4) y and z are unique such numbers that $a = \mathrm{OP}(y,z)$. To prove that $\beta(a,i) = a_i$ for $i < n$ it is enough to show that

$$a_i = \mu x[\mathrm{Div}(y, 1 + (\mathrm{OP}(x,i)+1)\cdot z)].$$

To do this it suffices to show that if $x < a_i$ then $\mathrm{OP}(x,i) < c$ and $\mathrm{OP}(x,i)$ is not of the form $\mathrm{OP}(a_j,j)$ for $j < n$. But $\mathrm{OP}(x,i) \leqslant \mathrm{OP}(a_i,i) < c$ and $\mathrm{OP}(x,i) \neq \mathrm{OP}(a_j,j)$ by (1.4). Q.E.D.

Observe that by Theorem 1.2.15 it follows in particular that $\beta(0,i) = 0$ and $a \neq 0 \rightarrow \beta(a,i) < a$.

Let now $a_1,\ldots,a_n$ be any sequence of natural numbers. We assign to it the smallest number a such that $\beta(a,0) = n$ and $\beta(a,i) = a_i$ for $i = 1,\ldots,n$. Such a number exists by Theorem 1.2.15. We call this number the sequence number (the code) of $a_1,\ldots,a_n$ and denote it by $\langle a_1,\ldots,a_n\rangle$. For $n = 0$ we have $\langle\,\rangle = 0$.

Theorem 1.2.16 *For each fixed n the function $\langle a_1,\ldots,a_n\rangle$ is recursive.*

Proof. Indeed we have

$$\begin{aligned}\langle a_1,\ldots,a_n\rangle = &\\ &= \mu x[\beta(x,0) = n \wedge \beta(x,1) = a_1 \wedge \ldots\\ &\quad \ldots \wedge \beta(x,n) = a_n].\end{aligned}$$

Q.E.D.

We define now two decoding functions.

Definition 1.2.17

$$\begin{aligned} lh(a) &= \beta(a,0).\\ (a)_i &= \beta(a,i+1).\end{aligned}$$

The symbol lh is the abbreviation of the word "length". We read $lh(a)$ as "the length of the sequence (of the code) a". It can be easily seen that lh and $(.)_i$ are recursive functions. If $a = \langle a_0,\ldots,a_{n-1}\rangle$ then $lh(a) = n$ and $(a)_i = a_i$ for $i < n$. One also has that

$$a \neq \langle\,\rangle \rightarrow lh(a) < a \wedge (a)_i < a.$$

Denote by Seq the set of all sequence numbers. It is recursive, since it has the following definition

$$Seq(a) \equiv \forall x < a[lh(x) \neq lh(a) \vee \exists i < lh(a)((x)_i \neq (a)_i)].$$

We now introduce two auxiliary functions connected with the coding of sequences, namely the function In, giving for any sequence number a sequence number being a code of an initial segment of the given sequence, and the function $*$, which gives the sequence number of the sequence being the concatenation of two given sequences. Recall that the concatenation of the sequences $a_1, \ldots, a_n$ and $b_1, \ldots, b_m$ is the sequence $a_1, \ldots, a_n, b_1, \ldots, b_m$ of the length $n + m$.

Definition 1.2.18

$$In(a, i) = \mu x[lh(x) = i \wedge \forall j < i((x)_j = (a)_j)].$$

$$\begin{aligned} a * b \;\; = \;\; & \mu x[lh(x) = lh(a) + lh(b) \wedge \\ & \wedge \forall i < lh(a)((x)_i = (a)_i) \wedge \\ & \wedge \forall i < lh(b)((x)_{lh(a)+i} = (b)_i)]. \end{aligned}$$

The symbol In is derived from the words "initial segment". We read $In(a, i)$ as "the sequence number of the initial segment up to i of the sequence (of the code) a" and "$a * b$" as "concatenation of the sequence (of the code) a and the sequence (of the code) b".

It can easily be seen that the functions In and $*$ are recursive and have the following properties

$$In(\langle a_1, \ldots, a_n \rangle, i) = \langle a_1, \ldots, a_i \rangle \quad \text{for} \quad i \leqslant n,$$

$$\langle a_1, \ldots, a_n \rangle * \langle b_1, \ldots, b_m \rangle = \langle a_1, \ldots, a_n, b_1, \ldots, b_m \rangle.$$

Using sequence numbers one can replace n-ary functions and relations by unary functions and relations. We introduce the notion of contraction.

Definition 1.2.19 *Let $f : \mathbb{N}^n \to \mathbb{N}$. The contraction of the function f is the unary function $\langle f \rangle$ defined as follows*

$$\langle f \rangle(a) = f((a)_0, \ldots, (a)_{n-1}).$$

Let P be an n-ary relation in $\mathbb{N}$. The contraction of P is the unary relation $\langle P \rangle$ defined by

$$\langle P \rangle(a) \equiv P((a)_0, \ldots, (a)_{n-1}).$$

Notice that the following hold

$$\begin{aligned} f(a_1, \ldots, a_n) &= \langle f \rangle(\langle a_1, \ldots, a_n \rangle), \\ P(a_1, \ldots, a_n) &\equiv \langle P \rangle(\langle a_1, \ldots, a_n \rangle). \end{aligned}$$

Theorem 1.2.20 *The function f is recursive if and only if the function $\langle f \rangle$ is recursive. The relation P is recursive if and only if the relation $\langle P \rangle$ is recursive.*

Now we can consider the operation of primitive recursion announced above.

Definition 1.2.21 *Let f be an n-ary function ($n \neq 0$). We define a new n-ary function $\overline{f}$ by*

$$\overline{f}(\overline{a}, b) = \langle f(\overline{a}, 0), f(\overline{a}, 1), \ldots, f(\overline{a}, b-1) \rangle.$$

Hence the value of the function $\overline{f}$ on the arguments $\overline{a}, b$ is the code of the sequence (of length b) of values of f on arguments $(\overline{a}, 0), (\overline{a}, 1), \ldots, (\overline{a}, b-1)$. So $\overline{f}(\overline{a}, b)$ contains all the information on the values of $f(\overline{a}, x)$ for $x < b$. Therefore $\overline{f}$ is sometimes called the history-function.

Theorem 1.2.22 *The function f is recursive if and only if the function $\overline{f}$ is recursive.*

Proof. Let f be recursive. Observe that

$$\overline{f}(\overline{a},b) = \mu x[Seq(x) \wedge lh(x) = b \wedge \forall i < b((x)_i = f(\overline{a},i))].$$

So $\overline{f}$ is also recursive. Note that the equation from Definition 1.2.21 could not be used here since the dots ... occurring in it depend on argument b.

Conversely, assume that $\overline{f}$ is recursive. Then the recursiveness of f follows from the equation

$$f(\overline{a},b) = (\overline{f}(\overline{a},b+1))_b.$$

Q.E.D.

Theorem 1.2.23 *If g and h are recursive functions and f is a function satisfying the following condition*

$$f(\overline{a},b) = \begin{cases} g(\overline{a}), & \text{if } b = 0, \\ h(\overline{f}(\overline{a},b),\overline{a},b), & \text{if } b > 0, \end{cases} \tag{1.5}$$

then f is recursive.

Proof. Let

$$\begin{aligned} H(\overline{a},b) \quad = \quad & \mu x[Seq(x) \wedge lh(x) = b \wedge (x)_0 = g(\overline{a}) \wedge \\ & \wedge \forall i < b(i > 0 \rightarrow (x)_i = h(In(x,i),\overline{a},i))]. \end{aligned}$$

It can easily be seen that the function H is recursive and that it is equal to $\overline{f}$. Hence for $b \neq 0$ we have

$$f(\overline{a},b) = h(H(\overline{a},b),\overline{a},b).$$

So f is recursive. Q.E.D.

If a function f is defined by (1.5) then we say that it is defined by primitive recursion.

Corollary 1.2.24 *If functions g and g' are recursive then the function f defined by*

$$\begin{aligned} f(\overline{a},0) &= g(\overline{a}), \\ f(\overline{a},b+1) &= g'(f(\overline{a},b),\overline{a},b) \end{aligned}$$

is recursive.

Proof. It is enough to replace in Theorem 1.2.23 the function $h(c,\overline{a},b)$ by the function $g'((c)_{lh(c)},\overline{a},b)$. Q.E.D.

The operation of primitive recursion — very important and useful in proving the recursiveness of various particular number-theoretic functions — helps us also to explain the very name "recursive function". By Theorem 1.2.23 to calculate the value $f(\overline{a},b)$, where $b > 0$, one should know already the values of f for arguments $0, 1, \ldots, b-1$ (and the parameter $\overline{a}$), i.e., calculating the value of $f(\overline{a},b)$ one must return to the values of f for smaller arguments. It suggests that the source of the name "recursive function" can be seen in two Latin verbs: *recurrere* and *recursare* which mean run back, hasten back.

Using Theorem 1.2.23 and Corollary 1.2.24 it can be easily proved that the following functions are recursive:

1. the function $a!$ — indeed $0! = 1$, $(a+1)! = a! \cdot (a+1)$,

2. the function a^b — indeed $a^0 = 1, a^{b+1} = (a^b) \cdot a$,

3. the function $\mathrm{rm}(a,b)$ = the remainder of the division of b by a — indeed one has

$$\begin{aligned} \mathrm{rm}(a,0) &= 0, \\ \mathrm{rm}(a,b+1) &= (\mathrm{rm}(a,b)+1) \cdot \mathrm{sg}(|a-(\mathrm{rm}(a,b)+1)|), \end{aligned}$$

4. the function $\mathrm{qt}(a,b)$ = greatest integer not greater than the quotient of b by a — indeed one has

$$\begin{aligned} \mathrm{qt}(a,0) &= 0, \\ \mathrm{qt}(a,b+1) &= \mathrm{qt}(a,b) + \overline{\mathrm{sg}}(|a-(\mathrm{rm}(a,b)+1)|), \end{aligned}$$

where

$$\overline{\mathrm{sg}}(a) = \begin{cases} 1, & \text{if } a = 0, \\ 0, & \text{if } a \neq 0. \end{cases}$$

Observe that the function qt can be defined in a simpler way using the operation of effective minimum, i.e.,

$$\mathrm{qt}(a, b) = \mu x[a \cdot (x+1) > b \vee a = 0].$$

5. the function $p(a)$ = the ath prime — indeed one has

$$\begin{aligned} p(0) &= 2, \\ p(a+1) &= \mu x(x > p(a) \wedge \Pr(x)). \end{aligned}$$

Observe that the effective minimum can be replaced here by the bounded minimum and one can write:

$$p(a+1) = \mu x_{x \leqslant p(a)!+1}[x > p(a) \wedge \Pr(x)].$$

We end this section by mentioning two further (maybe more intuitive) ways of coding finite sequences of natural numbers. The first one is based on the theorem of the factorization of a natural number stating that any natural number can be represented in one and only one way as a product of primes. So let a sequence $a_1, \ldots, a_n$ of natural numbers be given. We assign to it a number $\langle a_1, \ldots, a_n \rangle$ called the code of it and defined as follows:

$$\langle a_1, \ldots, a_n \rangle = p_0^{a_1+1} \cdot \ldots \cdot p_{n-1}^{a_n+1},$$

where $p_i = p(i)$ = the ith prime — see example 5 above. The numbers a_i are increased here by 1 to distinguish the numbers belonging to the given sequence and maybe equal 0, and the zero exponents the number of which is infinite in every factorization of a natural number into primes. It is obvious that for any given n the function coding sequences of length n is recursive. The decoding functions as well as the function of length are also recursive. Indeed we have

$$\begin{aligned} lh(a) &= \mu x_{x<a}[\neg \mathrm{Div}(a, p_x)], \\ (a)_i &= \exp(a, p_{i \dot{-} 1}) \quad \text{for } i = 1, \ldots, n, \end{aligned}$$

where

$$\exp(a,b) = \mu x_{x<a}[b = 1 \vee b = 0 \vee \neg \mathrm{Div}(a, b^{x+2})].$$

Observe that if $a = \langle a_1, \ldots, a_n \rangle$ then $a_i = (a)_i$ for $i = 1, \ldots, n$.

The second way of coding sequences of natural numbers uses so-called pairing functions, i.e., bijections from $\mathbb{N}^2$ to $\mathbb{N}$. As examples of such functions can serve:

$$\begin{aligned} f_1(a,b) &= 2^a \cdot (2b+1) \dot{-} 1, \\ f_2(a,b) &= \frac{1}{2}[(a+b+1)(a+b)] + a. \end{aligned}$$

Denote by $\langle a, b \rangle$ the value of a given pairing function on a, b. The code of a sequence $a_1, \ldots, a_n$ can be defined now as follows

$$\langle a_1, \ldots, a_n \rangle = \langle \langle \ldots \langle \langle a_1, a_2 \rangle, a_3 \rangle, \ldots, a_{n-1} \rangle, a_n \rangle.$$

It can be easily seen that for a given n it is a recursive function provided that the given pairing function is recursive. The reader can prove (as an exercise) that the decoding functions as well as the function of length are also recursive.

Observe that if one uses a pairing function to code sequences of natural numbers then every natural number is a code of a sequence of natural numbers. This is not the case if one codes sequences using Gödel's function β.

1.3 Markov Algorithms and Turing Machines

In this section two further methods to specify the intuitive notion of computability will be described, namely Markov normal algorithms and Turing machines.

The idea of characterizing the notion of computability by using algorithms comes from the Soviet mathematician A.A. Markov (cf. Markov, 1951), its full description can be found in Markov (1954). The author of the second proposal using abstract calculating machines is an English mathematician A. Turing (his conception was given in Turing (1936)).

Let us begin by considering Markov algorithms. Our exposition will follow that of Markov (1954) and Mendelson (1964).

By an *alphabet* we mean any non-empty finite set of symbols called letters of this alphabet. A *word* in an alphabet A is any finite sequence of letters of A. The empty sequence of letters is called the *empty word* and is denoted by Λ. If P is a word $S_{j_1} \ldots S_{j_k}$ and Q is a word $S_{r_1} \ldots S_{r_n}$ then by PQ we denote the juxtaposition of P and Q, i.e., the word $S_{j_1} \ldots S_{j_k} S_{r_1} \ldots S_{r_n}$. So in particular, $P\Lambda = \Lambda P = P$ and $(P_1P_2)P_3 = P_1(P_2P_3)$. An alphabet A is said to be an *extension* of an alphabet B if and only if $B \subseteq A$. In such a case any word in the alphabet B is a word in the alphabet A.

By an *algorithm in* an alphabet A we mean any computable function $\mathfrak{A}$ whose domain is a subset of the set of all words in the alphabet A and whose values are also words in A. (Note that we are using here a notion of a computable function in a broader sense than in Section 1.1, where it was applied only to functions from the set of natural numbers to the set of natural numbers; this definition can be, of course, easily reformulated to meet the case of functions with arguments and values being words in a given alphabet). If a word P belongs to the domain of a given algorithm $\mathfrak{A}$ then it is said that the algorithm $\mathfrak{A}$ is applicable to P. If the alphabet A is an extension of the alphabet B then every algorithm in A is said to be an algorithm *over* B.

The general concept of an algorithm introduced above is imprecise because in its definition the imprecise notion of an effectively computable function was applied. We shall now try to define this concept accurately by fixing certain elementary operations to which any algorithm will be reduced. We start by a particularly simple operation, viz., by substitution of one word for another. If P and Q are two words in an alphabet A then the expressions $P \rightarrow Q$ and $P \rightarrow \cdot Q$ will be called *productions* in A. We assume that neither the arrow $\rightarrow$ nor the dot $\cdot$ are letters of the alphabet A. Words P and Q may be empty words. The rule $P \to Q$ will be called a *simple* production and the rule $P \to \cdot Q$ a *terminal* production. By a scheme of an algorithm we mean now a finite list of productions in A:

$$\begin{array}{lcl} P_1 & \to & (\cdot)Q_1 \\ P_2 & \to & (\cdot)Q_2 \\ & \vdots & \\ P_n & \to & (\cdot)Q_n \end{array}$$

where $P_i \to (\cdot)Q_i$ denotes either $P_i \to Q_i$ or $P_i \to \cdot Q_i$. This scheme determines the following algorithm in the alphabet A: We say that a word P occurs in a word Q if and only if there are (possibly empty) words U and V such that $Q = UPV$. Let a word P in A be given. Two cases are possible: (1) none of the words $P_1, \ldots, P_n$ occurs in P — in this case we write $\mathfrak{A} : P \sqsupset$, (2) at least one of the words $P_1, \ldots, P_n$ occurs in P: then let k be the smallest number l such that $1 \leqslant l \leqslant n$ and P_l occurs in P and let R be a word resulting from replacing the left-most occurrence of P_k in P by Q_k. If the rule $P_k \to (\cdot)Q_k$ is simple then we write

$$\mathfrak{A} : P \vdash R,$$

and if it is terminal then we write:

$$\mathfrak{A} : P \vdash \cdot R.$$

The expression $\mathfrak{A} : P \models R$ will mean that there exists a sequence $R_0, R_1, \ldots, R_m$ of words in A such that $R_0 = P$, $R_m = R$ and $\mathfrak{A} : R_j \vdash R_{j+1}$ for $j = 0, 1, \ldots, m-2$ and either $\mathfrak{A} : R_{m-1} \vdash R_m$ or $\mathfrak{A} : R_{m-1} \vdash \cdot R_m$ (in the second case we write $\mathfrak{A} : P \models \cdot R$). We write $\mathfrak{A}(P) = R$ if and only if either $\mathfrak{A} : P \models \cdot R$ or $\mathfrak{A} : P \models R$ and $\mathfrak{A} : R \sqsupset$. The algorithm thus defined is called a *normal* Markov algorithm in the alphabet A.

The action of a normal Markov algorithm $\mathfrak{A}$ can be described as follows: a word P in the alphabet A is given, we find the first production $P_i \to (\cdot)Q_i$ in the scheme of $\mathfrak{A}$ such that P_i occurs in the word P. We then substitute Q_i for the left-most occcurrence of P_i in P. Let R_1 denote the result of this substitution. If the rule $P_i \to (\cdot)Q_i$ is terminal then the process stops and the value of the algorithm $\mathfrak{A}$ on the word P is R_1. If it is a simple production then we apply the same procedure to the word R_1, etc. If we ever obtain a word R_k such that $\mathfrak{A} : R_k \sqsupset$, i.e., none of the words $P_1, \ldots, P_n$ occurs in R_k then the process stops and we put $\mathfrak{A}(P) = R_k$. It can happen that the process just described never stops — in that case we say that the algorithm $\mathfrak{A}$ is not applicable to the given word P.

Let us now consider some examples.

Example 1. Let A be the alphabet $\{b, c\}$. Consider the following scheme

$$\begin{cases} b \to \cdot\Lambda \\ c \to c. \end{cases}$$

It can be easily seen that the algorithm $\mathfrak{A}$ determined by this scheme transforms any word in A containing at least one occurrence of b into the word obtained from P by erasing the left-most occurrence of b and transforms the empty word Λ into itself. The algorithm $\mathfrak{A}$ is not applicable to non-empty words not containing the symbol b.

Example 2. Let $A = \{a_0, a_1, \ldots, a_n\}$. Consider the following scheme

$$\begin{cases} a_0 \to \Lambda \\ a_1 \to \Lambda \\ \quad\vdots \\ a_n \to \Lambda. \end{cases}$$

It can be simply written as

$$\{\xi \to \Lambda \quad (\xi \in A).$$

(The usage of such an abbreviation means also that the productions intended may be listed in any order). It can be seen that the algorithm $\mathfrak{A}$ determined by this scheme transforms any word into the empty word, hence in particular one has:

$$\mathfrak{A} : a_5a_2a_1a_0a_3 \vdash a_5a_2a_1a_3 \vdash a_5a_2a_3 \vdash a_5a_3 \vdash a_5 \vdash \Lambda$$

and $\mathfrak{A} : \Lambda \sqsupset$. So $\mathfrak{A}(a_5a_2a_1a_0a_3) = \Lambda$.

Example 3. We now describe an algorithm $\mathfrak{A}$ which transforms any word in an alphabet A into its inverse, i.e., it transforms the word $P = a_{i_0}a_{i_1} \dots a_{i_k}$ into the word

$$\breve{P} = a_{i_k}a_{i_{k-1}} \dots a_{i_1}a_{i_0}.$$

Consider the following scheme of an algorithm in the alphabet $B = A \cup \{\alpha, \beta\}$:

$$\alpha\alpha \to \beta \qquad (1.6)$$
$$\beta\xi \to \xi\beta \quad (\xi \in A) \qquad (1.7)$$
$$\beta\alpha \to \beta \qquad (1.8)$$
$$\beta \to \cdot\ \Lambda \qquad (1.9)$$
$$\alpha\eta\xi \to \xi\alpha\eta \quad (\eta, \xi \in A) \qquad (1.10)$$
$$\Lambda \to \alpha. \qquad (1.11)$$

Consider a word $P = a_{i_0}a_{i_1} \dots a_{i_k}$ in A. By (1.11) we have $\mathfrak{A} : P \vdash \alpha P$. Using (1.10) we get

$$\mathfrak{A} : \alpha P \vdash a_{i_1}a_{i_0}a_{i_2} \dots a_{i_k} \vdash a_{i_1}a_{i_2}\alpha a_{i_0}a_{i_3} \dots a_{i_k} \vdash \dots$$
$$\dots \vdash a_{i_1}a_{i_2} \dots a_{i_k}\alpha a_{i_0}.$$

Using again rule (1.11) we get

$$\mathfrak{A} : P \models \alpha a_{i_1}a_{i_2} \dots a_{i_k}\alpha a_{i_0}.$$

Iterating this process we finally obtain

$$\mathfrak{A} : P \models \alpha a_{i_k}\alpha a_{i_{k-1}} \dots \alpha a_{i_1}\alpha a_{i_0},$$

and

$$\mathfrak{A} : P \models \alpha\alpha a_{i_k}\alpha a_{i_{k-1}} \dots \alpha a_{i_1}\alpha a_{i_0}.$$

Using now (1.6) we get

$$\mathfrak{A} : P \models \beta a_{i_k} \alpha a_{i_{k-1}} \ldots \alpha a_{i_1} \alpha a_{i_0}.$$

By rules (1.7), (1.8) and (1.9) we obtain $\mathfrak{A} : P \models \cdot \check{P}$. Note that our algorithm $\mathfrak{A}$ was an algorithm **over** A and not an algorithm **in** A! Indeed, in the scheme determining $\mathfrak{A}$ we used additional letters α and β not belonging to A. Of course, every normal algorithm in an alphabet A is a normal algorithm over A but not *vice versa*, i.e., there exist algorithms in A which can be defined as normal algorithms over A but not as normal algorithms in A. An example of such an algorithm is the so-called doubling algorithm, i.e., the algorithm which transforms any word P in A into the word PP.

Example 4. If we want to consider functions with arguments and values in the set $\mathbb{N}$ of natural numbers then we need an appropriate alphabet in which the natural numbers could be written. So let A be an arbitrary alphabet containing the symbol S_1 which we shall abbreviate as 1. We define by induction a sequence of words $\overline{0}, \overline{1}, \ldots$ in the following way: $\overline{0} = 1$, $\overline{n+1} = \overline{n}1$. Hence $\overline{1} = 11, \overline{2} = 111, \overline{3} = 1111$, etc.

Consider the following scheme of an algorithm:

$$\begin{cases} \alpha 1 \to \cdot 11 \\ \Lambda \to \alpha. \end{cases}$$

It can easily be seen that it determines an algorithm $\mathfrak{A}$ over the alphabet $\{1\}$ and simultaneously in the alphabet $\{1, \alpha\}$. It can also be checked that $\mathfrak{A}(\overline{n}) = \overline{n+1}$ for any natural number n.

If $\mathfrak{A}$ and $\mathfrak{B}$ are two algorithms and P is a word then we write $\mathfrak{A}(P) \approx \mathfrak{B}(P)$ if and only if either $\mathfrak{A}$ and $\mathfrak{B}$ are both applicable to P and $\mathfrak{A}(P) = \mathfrak{B}(P)$ or neither $\mathfrak{A}$ nor $\mathfrak{B}$ is applicable to the word P. If $\mathfrak{A}$ and $\mathfrak{B}$ are two algorithms over an alphabet A, then it is said that they are *fully equivalent* with respect to A if and only if for any word P in A, $\mathfrak{A}(P) \approx \mathfrak{B}(P)$.

We now consider algorithms computing values of functions on the set $\mathbb{N}$ of natural numbers. So let $M = \{1, *\}$. As shown above in Example 4, every natural number n can be represented as the word $\overline{n}$ in M. A finite sequence of natural numbers of the length k, i.e. the sequence $n_1, \ldots, n_k$, will be represented as the word $\overline{n_1} * \overline{n_2} * \ldots * \overline{n_k}$. This word will be shortly denoted as $\overline{(n_1, \ldots, n_k)}$. For example the expression $\overline{(5, 1, 2)}$ denotes the word $111111 * 11 * 111$.

Let f be any partial computable k-ary function, i.e., a computable function whose domain $\mathrm{dom}(f)$ is a subset of the set $\mathbb{N}^k$. If $\mathrm{dom}(f) = \mathbb{N}^k$ then the function f is called total. Denote by $\mathfrak{B}_f$ an appropriate algorithm in the alphabet M, i.e., an algorithm such that $\mathfrak{B}_f(\overline{(n_1, \ldots, n_k)}) = \overline{f(n_1, \ldots, n_k)}$ whenever either of the two sides of the equation is defined. $\mathfrak{B}_f$ is assumed to be inapplicable to words not of the form $\overline{(n_1, \ldots, n_k)}$.

Definition 1.3.1 *A partial k-ary function f is said to be partially Markov-computable if and only if there exists a normal algorithm $\mathfrak{A}$ over M which is fully equivalent to $\mathfrak{B}_f$ with respect to M. If a partially Markov-computable function f is total then it is said to be Markov-computable.*

Let us generalize the notion of a recursive function introduced in Section 1.2. A partial function f is said to be *partial recursive* if and only if f belongs to the smallest set of functions containig the functions I_i^n ($n \in \mathbb{N}, 1 \leqslant i \leqslant n$), $+, \cdot, K_<$ and closed under composition and unbounded minimum. The operation of unbounded minimum is the operation leading from a function g to a function h defined as

$$h(\overline{a}) = \mu x \left[g(\overline{a}, x) = 0 \right].$$

Note that we do not asssume the effectiveness condition. Hence the function h may be undefined for certain $\overline{a}$, in particular for $\overline{a}$ such that there is no x with the property $g(\overline{a}, x) = 0$.

Any recursive function is of course partial recursive. It can also be shown that every total function which is partial recursive is recursive.

Having introduced two precise definitions of the intuitive notion of computability, i.e., the concept of (partial) recursive functions and the concept of (partially) Markov-computable functions, one should ask whether those definitions are equivalent, i.e., whether the appropriate classes of functions determined by those definitions are equal. The answer is positive. Indeed the following theorem holds.

Theorem 1.3.2 (i) *Every partial recursive function is partially Markov-computable and* vice versa, *i.e., every partial partially Markov-computable function is partial recursive.*

(ii) *Every recursive function is Markov-computable and* vice versa, *i.e., every Markov-computable function is recursive.*

Long and technically complicated proof of this theorem will be omitted here. It can be found in various books devoted to the theory of recursive functions, e.g., in Chapter 5 of Mendelson (1964).

Let us now describe the second attempt to specify the concept of effective computability, i.e., the definition of computability using Turing machines. Our exposition will follow that of Turing (1936), Davis (1958) and Mendelson (1964).

Before the precise definition of a Turing machine will be given let us describe its physical model which will help to better understand the abstract description in mathematical terms.

A Turing machine consists of a (potentially) infinite tape, and a read/write head connected to a control mechanism. The tape is divided into cells. A finite alphabet of symbols is given, denote them as $S_0, \ldots, S_n$. At any given time at most one symbol is written in a given cell, and the machine (or, more precisely, its control mechanism) is in one of a finite number of possible internal states $\{q_0, \ldots, q_m\}$. The cells are scanned, one at the time, by the read/write head. The head can move in both directions and it can read the symbol written in the given cell, erase it or write a new symbol. The behaviour of the head is governed by the present state of the machine and by the symbol in the cell under scan. The work of the machine can be described in the following way. Assume that at a given moment t the read/write head is scanning the cell in which the symbol S_i is written and that the machine is in the state q_j. Then the action of the machine can consists of one of the following four things: (1) it may erase the symbol S_i and write a new symbol S_k, (2) it may move left one cell, (3) it may move right one cell, (4) it may stop. In cases (1)–(3) the machine goes into a new internal state q_l and is ready to act at time $t+1$. Assume for simplicity that the symbol S_0 represents a blank, i.e., a cell in which no symbol is written. In this way one can assume that the read/write head is always scanning a symbol. The actions (1)–(3) can be represented by quadruples: (1) $q_jS_iS_kq_l$, (2) $q_jS_iLq_l$, (3) $q_jS_iRq_l$ (L and R denote the moves to left or right, resp.). The first two symbols in a quadruple stand for the present internal state and the scanned symbol, the third symbol represents the action of the machine (write S_k, move left, move right one cell), and the fourth symbol gives the internal state of the machine after the action has been performed.

With a Turing machine M described above certain algorithm $\mathfrak{A}$ in the alphabet A of M can be associated (it does not mean that the theory of Turing machines depends on the theory of normal Markov algorithms — only a general notion of an algorithm is used here). Let a word P in the alphabet A be given and write it on the tape from left to right (each symbol in a separate cell). Put the read/write head against the cell in which the first symbol of the word P is written and let the machine be in the internal state q_0. The machine begins to act according to its rules. If it ever stops, i.e., if it reaches the situation where no act can be performed, the word of the alphabet A appearing on the tape (and read from left to right) is the value of the algorithm $\mathfrak{A}$ on the word P.

So far for the intuitive descriptions. Now the precise definition of a Turing machine can be given.

Definition 1.3.3 *A Turing machine is a finite set M of quadruples satisfying the following conditions:* (1) *any quadruple of M*

is of one of the forms: $q_jS_iS_kq_l$, $q_jS_iLq_l$, $q_jS_iRq_l$, (2) *there are no two quadruples having identical first two symbols. The elements of* M *are called orders. The set of all symbols of the type* S_n *occurring in orders of* M *is called the alphabet of the machine* M, *and the set of all the symbols of the type* q_n *occurring in orders of* M *is called the set of internal states of* M. *It is assumed that a fixed state* q_0 (*called the initial state*) *is an internal state of every Turing machine.*

To describe the state of a machine and the sequence of symbols written on its tape at a given moment t let us introduce a notion of *configuration*. A configuration is any word of the form Pq_sQ where P is a (possibly empty) word in the alphabet of the machine M, q_s is an internal state of M, and Q is a non-empty word in the alphabet of M. We say that the machine M transforms a configuraton α into a configuration β (what is symbolically written as $\alpha \xrightarrow{M} \beta$ if and only if one of the following cases holds: (1) α is of the form Pq_jS_iQ, β is of the form Pq_lS_kQ and $q_jS_iS_kq_l \in M$, (2) α is of the form $PS_sq_jS_iQ$, β is of the form $Pq_lS_sS_iQ$ and $q_jS_iLq_l \in M$, (3) α is of the form q_jS_iQ, β is of the form $q_lS_0S_iQ$ and $q_jS_iLq_l \in M$, (4) α is of the form $Pq_jS_iS_kQ$, β is of the form $PS_iq_lS_kQ$ and $q_jS_iRq_l \in M$, (5) α is of the form Pq_jS_i, β is of the form $PS_iq_lS_0$ and $q_jS_iRq_l \in M$. We say that a machine M stops at a configuration α if and only if there exists no configuration β such that $\alpha \xrightarrow{M} \beta$. Observe that $\alpha \xrightarrow{M} \beta$ means that if the configuration α describes the state of the machine and of the tape at the moment t then the configuration β describes them at the moment $t + 1$.

A *computation* of a Turing machine M is a finite sequence of configurations $\alpha_0, \ldots, \alpha_n$ ($n \geqslant 0$) such that the internal state occurring in α_0 is q_0, $\alpha_i \xrightarrow{M} \alpha_{i+1}$ for $i < n$, and the machine M stops by the configuration α_n. We say that this computation begins with the configuration α_0 and ends with α_n. If C is any alphabet containing the alphabet A of M then one can define an algorithm $\mathfrak{A}_{M,C}$ in the alphabet C in the following way: for any two words P and Q in C the equality $\mathfrak{A}_{M,C}(P) = Q$ holds if and only if there exists a computation of the machine M beginning with the configuration q_0P and ending with the configuration $R_1q_jR_2$ where $R_1R_2 = Q$.

Consider now functions with arguments in the set $\mathbb{N}$ of natural numbers whose values can be calculated by Turing machines. To do this assume that S_1 is the symbol 1. Recall that the symbol $\overline{m}$ denotes the word $\underbrace{1\ldots1}_{m+1}$. Let M be a Turing machine whose alphabet contains symbols S_0 (blank, i.e., empty cell), S_1 and $*$. We say that the machine M *computes* a partial k-ary function f if and only if for any natural numbers $n_1, \ldots, n_k$ there exist words R_1, R_2 in the alphabet $\{S_0\}$ such that

$$\mathfrak{A}_{M,C}(\overline{n_1} * \ldots * \overline{n_k}) = R_1\overline{f(n_1, \ldots, n_k)}R_2.$$

The function f is said to be *Turing-computable* if and only if there exists a Turing machine M computing f.

Example 1. Consider a Turing machine defined by the following quadruples:

$$q_0 1 L q_1$$

$$q_1 S_0 1 q_2.$$

It can easily be checked that it computes the successor function. Indeed, for any natural number n:

$$q_0\overline{n} \xrightarrow{M} q_1 S_0 \overline{n} \xrightarrow{M} q_2\overline{n+1}.$$

Example 2. Consider a Turing machine defined by the following set of quadruples:

$$\begin{array}{c} q_0 1 S_0 q_0 \\ q_0 S_0 R q_1 \\ q_1 1 R q_1 \\ q_1 * 1 q_2 \\ q_2 1 R q_2 \\ q_2 S_0 L q_3 \\ q_3 1 S_0 q_3. \end{array}$$

One can check that this machine computes the addition function. Indeed, it acts as follows:

$$\begin{aligned}
&q_0\overline{m} * \overline{n} = q_0 1^{m+1} * 1^{n+1} \longrightarrow q_0 S_0 1^m * 1^{n+1} \longrightarrow \\
&\longrightarrow S_0 q_1 1^m * 1^{n+1} \longrightarrow S_0 1 q_1 1^{m \dot{-} 1} * 1^{n+1} \longrightarrow \\
&\longrightarrow \ldots \longrightarrow S_0 1^m q_1 * 1^{n+1} \longrightarrow S_0 1^m q_2 1 1^{n+1} \longrightarrow \\
&\longrightarrow S_0 1^{m+1} q_2 1^{n+1} \longrightarrow S_0 1^{m+1} 1 q_2 1^n \longrightarrow \\
&\longrightarrow \ldots \longrightarrow S_0 1^{m+1} 1^{n+1} q_2 S_0 \longrightarrow S_0 1^{m+1} 1^n q_3 1 S_0 \longrightarrow \\
&\longrightarrow S_0 1^{m+1} 1^n q_3 S_0 S_0 = S_0 1^{m+n+1} q_3 S_0 S_0 = S_0 \overline{m+n} q_3 S_0 S_0,
\end{aligned}$$

where 1^k denotes $\underbrace{1\ldots1}_{k}$.

Example 3. The identity function $I_1^1(x) = x$ is computed by a Turing machine given by the following quadruple:

$$q_0 1 L q_0.$$

Indeed, one has $q_0\overline{n} = q_0 1\overline{(n-1)} \xrightarrow{M} q_0 S_0 1 \overline{(n-1)} = q_0 S_0 \overline{n}$ and the machine stops.

Similarly as it was in the case of Markov-computable functions we should ask the question whether the class of Turing-computable functions is equal to the class of recursive functions. The answer is given by the following

Theorem 1.3.4 (i) *Every Turing-computable function f is partial recursive; if additionally f is total then f is recursive.*
(ii) *Every partial recursive function is Turing-computable.*

Proof of this theorem will not be given here. It can be found, e.g., in Chapter 3 of Davis (1958) or in Chapter 5 of Mendelson (1964).

Observe that Turing machines are better formalization of effective computability than the algebraical definition of recursive functions given in Section 1.2 (cf. Definition 1.2.1) in the sense that they provide a direct formalization of usual calculating practice. Still better formalization is given by equational systems of Kleene (which will be mentioned in next sections). However, we adopted the algebraical definition because it is the most convenient and useful in applications.

1.4 Primitive and Elementary Recursive Functions

The aim of this section is to define two subclasses of the class $\mathcal{R}$ of recursive functions (introduced in Section 1.2), namely the class $\mathcal{P}$ of primitive recursive functions and the class $\mathcal{E}$ of elementary recursive functions.

Definition 1.4.1 *The class $\mathcal{P}$ of primitive recursive functions is the smallest class of functions containing the projection functions I_i^n $(n \in \mathbb{N}, 1 \leqslant i \leqslant n)$, the constant function 0 and the successor function S and closed under composition and primitive recursion.*

Definition 1.4.2 *The class $\mathcal{E}$ of elementary recursive functions is the smallest class of functions containing the function $\dot{-}$, the exponentiation function a^b and the successor function S and closed under composition and bounded minimization.*

Using characteristic functions one can define now (in a similar way as in Definition 1.2.2) the notion of a primitive recursive relation and of an elementary recursive relation.

It can easily be seen that the functions $a + b$, $a \cdot b$ and a^b are primitive recursive. They can be defined as follows:

$$\begin{aligned} a + 0 &= a, \\ a + (b + 1) &= S(a + b), \end{aligned}$$

$$\begin{aligned} a \cdot 0 &= 0, \\ a \cdot (b + 1) &= a \cdot b + a, \end{aligned}$$

$$\begin{aligned} a^0 &= S(0), \\ a^{b+1} &= a^b \cdot a. \end{aligned}$$

It can also be shown that the functions $a + b$ and $a \cdot b$ are elementary recursive. We can write:

$$\begin{aligned} a \cdot b &= \mu z_{z \leqslant S(a)^{S(b)}}[(2^a)^b = 2^z], \\ a + b &= \mu z_{z \leqslant S(a) \cdot S(b)}[2^a \cdot 2^b = 2^z]. \end{aligned}$$

Before we start the investigation of the properties of primitive recursive functions and elementary recursive functions, we introduce three new classes of functions which will turn out to be equal to the class $\mathcal{E}$. In this way, we obtain new characterizations of elementary recursive functions. To do this we need three operations on functions.

The operation of finite sums is an operation which assigns with any function f a new function g given by the following equation

$$g(\overline{a}, x) = \sum_{i \leqslant x} f(\overline{a}, i).$$

and the operation of finite products leads from a function f to a new function h defined as follows

$$h(\overline{a}, x) = \prod_{i \leqslant x} f(\overline{a}, i).$$

The operation of bounded recursion assigns to any given functions g, h, j a function f defined in the following way

$$\begin{aligned} f(\overline{a}, 0) &= g(\overline{a}), \\ f(\overline{a}, x+1) &= h(\overline{a}, x, f(\overline{a}, x)), \\ f(\overline{a}, x) &\leqslant j(\overline{a}, x). \end{aligned}$$

Definition 1.4.3 (i) *Class $\mathcal{E}'$ is the smallest class of functions containing the functions S, $+$, $\dot{-}$ and closed under composition, finite sums and finite products.*

(ii) *Class $\mathcal{E}''$ is the smallest class of functions containing the functions S and a^b and closed under composition and bounded recursion.*

(iii) *Class $\mathcal{E}'''$ is the smallest class of functions containing the functions S, $\dot{-}$, $\cdot$ and a^b and closed under composition and finite sums.*

Theorem 1.4.4 *The following equations hold: $\mathcal{E} = \mathcal{E}' = \mathcal{E}'' = \mathcal{E}'''$.*

We will not prove this theorem here — the proof can be found in the paper by Grzegorczyk (1953).

Theorem 1.4.5 *The following inclusions hold: $\mathcal{E} \subseteq \mathcal{P} \subseteq \mathcal{R}$.*

Proof. Note that the inclusion $\mathcal{P} \subseteq \mathcal{R}$ follows from Theorem 1.2.22 and Corollary 1.2.23. Hence it suffices to prove the inclusion $\mathcal{E} \subseteq \mathcal{P}$. To do this we show that for any class $\mathcal{X}$ of functions, if $\mathcal{X}$ contains the functions S, $\dot{-}$, $+$ and $\cdot$ and is closed under composition and primitive recursion then it is closed under bounded minimization. So let $\mathcal{X}$ be such a class of functions and assume that $g \in \mathcal{X}$ and a function f is given by the equation

$$f(\overline{a}, x) = \mu z_{z<x}[g(\overline{a}, z) = 0].$$

The function f can be also defined in the following way

$$f(\overline{a}, 0) = 0, \tag{1.12}$$

$$f(\overline{a}, x+1) = \begin{cases} x+1, & \text{if } \forall t \leqslant x[g(\overline{a}, t) \neq 0 \wedge \\ & \quad \wedge\, g(\overline{a}, x+1) = 0] \vee \\ & \quad \vee\, \forall t \leqslant x+1[g(\overline{a}, t) \neq 0], \\ f(\overline{a}, x), & \text{if } \exists t \leqslant x[g(\overline{a}, t) = 0] \vee \\ & \quad \vee\, \exists t \leqslant x+1[g(\overline{a}, t) = 0 \wedge \\ & \quad \wedge\, g(\overline{a}, x+1) \neq 0]. \end{cases} \tag{1.13}$$

Analysing the proofs of Theorems 1.2.6 and 1.2.12 one can note that class $\mathcal{X}$ is closed under conditional definitions. Hence the conditions in (1.13) can be written as a single formula. Consequently the function f belongs to the class $\mathcal{X}$. Q.E.D.

It will be proved later (cf. Corollary 1.4.16 and Corollary 1.4.20) that the inclusions from the above Theorem 1.4.5 are proper, i.e., that $\mathcal{E} \neq \mathcal{P} \neq \mathcal{R}$.

Theorem 1.4.6 *If $\mathcal{X}$ is a class of functions containing the successor function S, subtraction $\dot{-}$ and closed under composition and bounded minimization, then the set of relations of the class $\mathcal{X}$ is closed under the operations $\neg, \wedge, \vee, \rightarrow, \equiv$ and the operations of bounded quantifiers.*

Proof. The operations $\wedge, \vee$ and $\equiv$ can be defined by $\neg$ and $\rightarrow$ and composition. Observe that the following equivalences hold:

$$(a = 0 \rightarrow b = 0) \equiv [(1 \dot{-} a) \dot{-} (1 \dot{-} b)] = 0,$$

$$\neg(a = 0) \equiv (1 \dot{-} a) = 0.$$

Combining those facts and the definition of a characteristic function of a relation we get that, if relations R and P belong to the class $\mathcal{X}$, then the relations $\neg R, R \wedge P, R \vee P, R \rightarrow P, R \equiv P$ also belong to $\mathcal{X}$.

To prove the second part of the theorem it suffices to note that the following equivalences hold:

$$\exists x < t \varphi(x) \equiv \varphi(\mu x_{x<t} \varphi(x)),$$

$$\forall x < t \varphi(x) \equiv \neg \exists x < t \neg \varphi(x).$$

Q.E.D.

Corollary 1.4.7 *The set of primitive recursive relations and the set of elementary recursive relations are closed under the operations $\neg, \wedge, \vee, \rightarrow, \equiv$ and the operations of bounded quantifiers.*

Observe that all three ways of coding finite sequences of natural numbers presented in Section 1.2, i.e., using Gödel's β function, by decomposition into primes and by a pairing function, are not only recursive but in fact elementary recursive and consequently also primitive recursive. It follows directly from their definitions. The only thing which needs to be proved is the elementary recursiveness of the function $p(a) = a$th prime. To show that let us introduce the following elementary recursive functions and relations:

$$
\begin{aligned}
x^{+} &= x^{x}, \\
x^{++} &= x^{x^{+}}, \\
xNy &\equiv \Pr(x) \wedge \Pr(y) \wedge x > y \wedge \\
&\qquad \wedge \forall z \leqslant x[y < z \wedge \Pr(z) \rightarrow z = x], \\
xPa &\equiv \exists v \leqslant (a+2)^{+}\{\exp(v,2) = 2 \wedge \Pr(x) \wedge \\
&\qquad\qquad \wedge \exp(v,x) = a + 2 \wedge \\
&\qquad \wedge \forall t \leqslant (a+2)^{++} \forall z < t[\Pr(z) \wedge \Pr(t) \wedge \\
&\qquad\qquad\qquad \wedge \operatorname{Div}(v,z) \wedge \operatorname{Div}(v,t) \rightarrow \\
&\qquad \rightarrow \exists w \leqslant t\{\operatorname{Div}(v,w) \wedge wNz \wedge \\
&\qquad\qquad\qquad \wedge \exp(v,w) = \exp(v,z) + 1\}]\}.
\end{aligned}
$$

We now have the following equation:

$$p(a) = \mu x_{x \leqslant (a+2)^{+}}[xPa]$$

which proves that the function $p(a)$ is elementary recursive.

The class $\mathcal{P}$ of primitive recursive functions is an important class of functions. It has been intensively studied — cf. for example Péter (1934, 1937, 1951) or Robinson R.M. (1947, 1948). The reason is that many definite functions known from the number theory are just primitive recursive. The main problem studied in connection with the class $\mathcal{P}$ is the problem of reducibility of various schemes of recursion (which turn out to be useful in applications) to the scheme of primitive recursion. Consider some of them.

1. Conditional recursive definitions, i.e., definitions by the following scheme

$$f(\overline{a},0) = g(\overline{a}),$$

$$f(\overline{a},x+1) = \begin{cases} g_1(\overline{a},x,f(\overline{a},x)), & \text{if } R_1(\overline{a},x), \\ \vdots & \\ g_k(\overline{a},x,f(\overline{a},x)), & \text{if } R_k(\overline{a},x), \end{cases}$$

where relations $R_1,\ldots,R_k$ are such that for any $\overline{a},x$ exactly one of $R_1(\overline{a},x)$, ..., $R_k(\overline{a},x)$ holds.

2. Recursion with k initial fixed values, i.e., definitions falling under the following scheme:

$$\begin{aligned} f(\overline{a},0) &= g_0(\overline{a}), \\ f(\overline{a},1) &= g_1(\overline{a}), \\ &\vdots \\ f(\overline{a},k) &= g_k(\overline{a}), \end{aligned}$$

$$\text{for } x \geqslant k, \quad f(\overline{a},x+1) = h(\overline{a},x,f(\overline{a},x)).$$

3. Recursion with respect to a course of values:

$$\begin{aligned} f(\overline{a},0) &= g(\overline{a}), \\ f(\overline{a},x+1) &= h(\overline{a},x,f(\overline{a},j_1(x)),\ldots,f(\overline{a},j_k(x))), \end{aligned}$$

where $j_i(x) < x$ for $i = 1,\ldots,k$.

4. Recursion progressing every $k+1$ numbers, i.e., definitions by the following scheme:

$$\begin{aligned} f(\overline{a},0) &= g_0(\overline{a}), \\ f(\overline{a},1) &= g_1(\overline{a}), \\ &\vdots \\ f(\overline{a},k) &= g_k(\overline{a}), \\ f(\overline{a},x+k+1) &= h(\overline{a},x,f(\overline{a},x),f(\overline{a},x+1),\ldots,f(\overline{a},x+k)). \end{aligned}$$

5. Simultaneous recursion, i.e., recursion falling under the following scheme:

$$\begin{aligned} f(\overline{a},0) &= g(\overline{a}), \\ F(\overline{a},0) &= G(\overline{a}), \\ f(\overline{a},x+1) &= h(\overline{a},x,f(\overline{a},x),F(\overline{a},x)), \\ F(\overline{a},x+1) &= H(\overline{a},x,f(\overline{a},x),F(\overline{a},x)). \end{aligned}$$

Theorem 1.4.8 *Schemes of recursion 1–5 listed above are reducible to the scheme of primitive recursion. Hence the class of primitive recursive functions is closed under defining functions by any of the above schemes 1–5.*

Proof. If a function f is defined by a conditional recursion (see Definition 1 above) then one has

$$\begin{aligned} f(\overline{a},0) &= g(\overline{a}), \\ f(\overline{a},x+1) &= g_1(\overline{a},x,f((\overline{a},x))\cdot K_{\neg R_1}(\overline{a},x)+\dots \\ &\quad \dots + g_k(\overline{a},x,f(\overline{a},x))\cdot K_{\neg R_k}(\overline{a},x). \end{aligned}$$

Hence scheme 1 can be reduced to the scheme of primitive recursion.

Scheme 2 of recursion with k initial fixed values can easily be reduced to scheme 1 of conditional recursion. So consider scheme 3 of recursion with respect to a course of values. Let

$$\begin{aligned} \overline{f}(\overline{a},0) &= 2^{g(\overline{a})+1}, \\ \overline{f}(\overline{a},x+1) &= \overline{f}(\overline{a},x)\cdot p_{x+1}^{h(\overline{a},x,\exp(\overline{f}(\overline{a},x),p_{j_1(x)}),\dots,\exp(\overline{f}(\overline{a},x),p_{j_k(x)})+1}. \end{aligned}$$

Then

$$f(\overline{a},x) = (\overline{f}(\overline{a},x))_{x+1}.$$

Scheme 4 of recursion progressing every $k+1$ numbers can be reduced to a schemes 2 and 3 — indeed, for $x \geqslant k$ one has:

$$f(\overline{a},x+1) = h(\overline{a},x,f(\overline{a},x\dot{-}k),f(\overline{a},(x\dot{-}k)+1),\dots,f(\overline{a},x))$$

and the functions $x \dot{-} k, (x \dot{-} k)+1, \ldots, (x \dot{-} k)+k$ are all bounded from above by the identity function I_1^1, i.e.,

$$(x \dot{-} k) + l \leqslant x$$

for $l = 0, \ldots, k$.

Consider now the scheme of simultaneous recursion. Let us define the function $j(\overline{a}, x)$ by the following scheme of primitive recursion:

$$\begin{aligned} j(\overline{a}, 0) &= 2^{g(\overline{a})+1} \cdot 3^{G(\overline{a})+1}, \\ j(\overline{a}, x+1) &= 2^{h(\overline{a},x,j(\overline{a},x))_1,(j(\overline{a},x))_2)+1} \cdot 3^{H(\overline{a},x,(j(\overline{a},x))_1,j(\overline{a},x))_2)+1}. \end{aligned}$$

Then

$$\begin{aligned} f(\overline{a}, x) &= (j(\overline{a}, x))_1, \\ F(\overline{a}, x) &= (j(\overline{a}, x))_2. \end{aligned}$$

Q.E.D.

One can ask if there exist schemes of recursion which are irreducible to the scheme of primitive recursion. The answer is positive — an example of such a recursion is the scheme of double recursion or of threefold recursion (Theorem 1.4.15 below provides an example of an application of the former). Even more, there exists a sequence of schemes of multiple recursion of increasing strength (cf. Péter, 1937, 1951). Each of them can be reduced to the effective minimization. On the other hand, schemes of recursion do not suffice to define all recursive functions — examples of recursive functions not definable by multiple recursion can be given.

We now present an example of an application of the scheme of double recursion: we shall prove the existence of a universal function for the class of primitive recursive functions of at most two arguments.

Definition 1.4.9 *Let $\mathcal{X}$ be a given class of functions. By $\mathcal{X}_n$ we shall denote a subclass of the class $\mathcal{X}$ consisting of functions of at most n arguments.*

Definition 1.4.10 *A function $f : \mathbb{N}^{n+1} \to \mathbb{N}$ is said to be universal for the class $\mathcal{X}_n$ if and only if for any function $g : \mathbb{N}^n \to \mathbb{N}$ the following equivalence holds:*

$$g \in \mathcal{X}_n \equiv \exists k \forall \overline{x}[g(\overline{x}) = f(k, \overline{x})].$$

Observe that if class $\mathcal{X}$ is closed under contraction then it suffices to consider only universal functions for class $\mathcal{X}_1$.

Theorem 1.4.11 *If class $\mathcal{X}$ contains the successor function S and is closed under composition then no universal function for class $\mathcal{X}_n$ belongs to class $\mathcal{X}$ (for any $n \in \mathbb{N}$).*

Proof. Let n be any given natural number and assume that a function h is universal for class $\mathcal{X}_n$ and $h \in \mathcal{X}$. Let

$$f(x_1, \ldots, x_n) = h(x_1, x_1, x_2, \ldots, x_n).$$

Of course $f \in \mathcal{X}_n$ and $f(\overline{x}) + 1 \in \mathcal{X}_n$. Since h is universal for $\mathcal{X}_n$, there exists a $k \in \mathbb{N}$ such that

$$\forall \overline{x}[f(\overline{x}) + 1 = h(k, \overline{x})],$$

i.e.,

$$\forall x_1 \ldots \forall x_n[h(x_1, x_1, x_2, \ldots, x_n) + 1 = h(k, x_1, x_2, \ldots, x_n)].$$

In particular for $x_1 = \ldots = x_n = k$:

$$h(k, k, k, \ldots, k) + 1 = h(k, k, k, \ldots, k),$$

which is impossible. Consequently $h \notin \mathcal{X}$. Q.E.D.

The above theorem provides a method of constructing a function not belonging to a given class of functions. Such a problem appears fairly often in recursion theory, in particular if one wants to show that a certain class of functions is a proper subclass of another one. We shall use this method in proving that there exists a recursive function being universal for the class of primitive recursive functions of at most two arguments — this function is not primitive recursive, hence the class $\mathcal{P}$ of primitive recursive functions is a proper subclass of the class $\mathcal{R}$ of recursive functions.

Another method of demonstrating that a class $\mathcal{X}$ of functions is a proper subclass of a class $\mathcal{Y}$ consists of showing that there exists a function in $\mathcal{Y}$ which increases faster than any function of $\mathcal{X}$. This method can be used to prove that the class $\mathcal{E}$ of elementary recursive functions is a proper subclass of the class $\mathcal{P}$ of primitive recursive functions — cf. the end of this section. It was also used by Ackermann (1928) where a recursive function majorizing all primitive recursive functions was constructed. Ackermann's function is defined by the following recursion scheme:[2]

$$\begin{aligned} f(0,y) &= y+1, \\ f(x+1,0) &= f(x,1), \\ f(x+1,y+1) &= f(x,f(x+1,y)). \end{aligned}$$

The function f is defined by double recursion. It can be shown that it is recursive and that for every primitive recursive func-

[2]This is the today's definition of Ackermann's function. The original definition given by Ackermann was as follows:

$$\begin{aligned} A(0,x,y) &= x+y, \\ A(1,x,0) &= 0, \\ A(2,x,0) &= 1, \\ A(n,x,0) &= x \text{ for } n>2, \\ A(n+1,x,y+1) &= A(n,x,A(n+1,x,y)). \end{aligned}$$

In 1935 Péter observed that Ackermann's 3-argument function can be simplified to a binary function, namely:

$$\begin{aligned} P(0,y) &= 2y+1, \\ P(n+1,0) &= P(n,1), \\ P(n+1,y+1) &= P(n,P(n+1,y)). \end{aligned}$$

In 1948 it was further simplified by R.M. Robinson who replaced in the above definition $P(0,y)$ by $P(0,y) = y+1$.

tion h there exist numbers m and k such that $f(m, y) > h(y)$ whenever $y \geqslant k$. Hence f is not primitive recursive.

Return to the proof of the existence of a universal function for the class $\mathcal{P}_2$. We need some lemmas.

Definition 1.4.12 *Let* $\mathbf{P}_2$ *be the smallest class of functions containing the constant function* 0 *and the functions* $I_1^1(x)$, $S(x) = x+1, p_y^x, (x)_y, x \cdot y, x+y$ *and closed under composition (restricted to cases which do not lead beyond the class of functions of at most two arguments) and the following scheme of recursion (which can be easily reduced to the primitive recursion):*

$$\begin{aligned} f(a, 0) &= g(a), \\ f(a, x+1) &= h(\langle a, x \rangle, f(a, x)). \end{aligned}$$

where $\langle \ldots \rangle$ *is a coding function based on the decomposition into primes.*

Lemma 1.4.13 *For any primitive recursive function*

$$f : \mathbb{N}^n \to \mathbb{N}$$

there exists a function $f' \in \mathbf{P}_2$ *such that*

$$f(x_1, \ldots, x_n) = f'(\langle x_1, \ldots, x_n \rangle).$$

Proof. It can easily be seen that the considered property holds for initial functions of class $\mathcal{P}$. Hence it suffices to show that it is hereditary with respect to composition and primitive recursion.

So let two functions f and h be given. By the assumption there are functions $f', h' \in \mathbf{P}_2$ such that

$$\begin{aligned} f(x_1, \ldots, x_n) &= f'(\langle x_1, \ldots, x_n \rangle), \\ h(y_1, \ldots, y_m) &= h'(\langle y_1, \ldots, y_m \rangle). \end{aligned}$$

Let

$$g(x_1, \ldots, x_{i-1}, y_1, \ldots, y_m, x_{i+1}, \ldots, x_n) = \\ = f(x_1, \ldots, x_{i-1}, h(y_1, \ldots, y_m), x_{i+1}, \ldots, x_n).$$

As the required function g' one can take:

$$g'(z) = f'(\langle (z)_1, \ldots, (z)_{i-1} \rangle \cdot p_{i-1}^{h'(\langle (z)_i, \ldots, (z)_{i+m} \rangle)+1} \cdot \\ \cdot p_i^{(z)_{i+m+1}+1} \cdot \ldots \cdot p_{n-1}^{(z)_{m+n}+1}).$$

Consider now the case that f is defined by primitive recursion using g and h. By the assumption there exist appropriate functions g' and h'. Let

$$\begin{aligned} H(x,y) &= h'(\langle (x)_1, (x)_2, y \rangle), \\ F(0,y) &= g'(\langle y \rangle), \\ F(x+1,y) &= H(\langle x, y \rangle, F(x,y)), \\ f'(x) &= F((x)_1, (x)_2). \end{aligned}$$

If $h'(\langle x,y,z \rangle) = h(x,y,z)$ and $g'(\langle y \rangle) = g(y)$ then it can be seen that $f'(\langle x,y \rangle) = f(x,y)$. Q.E.D.

Lemma 1.4.14 *The class $\mathcal{P}_2$ of primitive recursive functions of at most two arguments is equal to the class $\mathbf{P}_2$.*

Proof. The inclusion $\mathbf{P}_2 \subseteq \mathcal{P}_2$ is obvious. Hence it suffices to prove that $\mathcal{P}_2 \subseteq \mathbf{P}_2$. So let $f \in \mathcal{P}_2$. By Lemma 1.4.13 there exists a function $f' \in \mathbf{P}_2$ such that

$$f(x,y) = f'(\langle x,y \rangle).$$

But the function $\langle x,y \rangle = 2^{x+1} \cdot 3^{y+1}$ belongs to class $\mathbf{P}_2$. Hence $f \in \mathbf{P}_2$. Q.E.D.

Theorem 1.4.15 *There exists a 3-ary recursive function G universal for class $\mathbf{P}_2$ (and hence for class $\mathcal{P}_2$).*

Proof. Let us enumerate all the functions of class $\mathbf{P}_2$ by defining a 3-ary function $G(n,x,y)$ — the argument n will indicate the number of the appropriate function. We define $G(n,x,y)$ as follows:

1. Initial conditions:

$$\begin{aligned} G(0,x,y) &= 0, \\ G(1,x,y) &= x, \\ G(2,x,y) &= x+1, \\ G(3,x,y) &= p_y^x, \\ G(4,x,y) &= (x)_y, \\ G(5,x,y) &= x \cdot y, \\ G(6,x,y) &= x+y. \end{aligned}$$

2. Inductive conditions:

(a) if $n > 6$ is such that $(n)_2 < 3$ then we put

$$G(n+1,x,y) = \begin{cases} G((n)_0, y, x), & \text{if } (n)_2 = 0, \\ G((n)_0, x, x), & \text{if } (n)_2 = 1, \\ G((n)_0, x, G((n)_1, x, y)), & \text{if } (n)_2 = 2. \end{cases}$$

(b) if $n > 6$ is such that $(n)_2 \geqslant 3$ then we put

$$\begin{aligned} G(n+1,0,y)) &= G((n)_0, y, y), \\ G(n+1,x+1,y) &= G((n)_1, \langle x,y\rangle, G(n+1,x,y)). \end{aligned}$$

So the constant function 0 has the number 0, the identity function I_1^1 — number 1, the successor function S — number 2, etc. It can also be seen that case 2(a) corresponds to composition and case 2(b) to the scheme of primitive recursion. All the functions from $\mathbf{P}_2$ appear in this enumeration, even more, every such function appears infinitely many times.

Observe that for $n > 6$ the functions $(n)_0, (n)_1, (n)_2$ are defined and smaller or equal n. Hence the function $G(n,x,y)$ is well-defined. So it is universal for class $\mathbf{P}_2$. By Theorem 1.4.11, $G \notin \mathcal{P}$. On the other hand, the function G is a computable function defined by double recursion (with respect to n and x). We now show that this recursion can be reduced to the effective minimization. In this way we get that $G \in \mathcal{R}$. The double recursion cannot be reduced to the primitive recursion. Indeed, if it were so then the function G would be in $\mathcal{P}$, which is impossible.

Similarly as in the case of the function $\overline{f}$ (cf. Definition 1.2.21) we want to describe finite fragments of $G(n, x, y)$ by numbers. Since G is 3-ary we can present the initial values of it (for a fixed y) in a two-dimensional table:

$$\begin{array}{l} G(0,0,y),\ G(0,1,y),\ \ldots\ ,\ G(0,l_0,y),\ \ldots \\ G(1,0,y),\ G(1,1,y),\ \ldots\ ,\ G(1,l_1,y),\ \ldots \\ \qquad\qquad\qquad\qquad \vdots \\ G(n,0,y),\ G(n,1,y),\ \ldots\ ,\ G(n,l_n,y),\ \ldots \\ \qquad\qquad\qquad\qquad \vdots \end{array}$$

If y were not fixed we would get a three-dimensional table.

Our aim is to code finite fragments of the three-dimensional table of values of the function G by natural numbers. We say that the number m represents a table of the function G if and only if for any triple (n, x, y):

$$(m)_{\langle n,x,y\rangle} = G(n, x, y) + 1,$$

where $(m)_{\langle n,x,y\rangle} = \exp(n, p_{\langle n,x,y\rangle \dot{-} 1})$. The number 1 is added to the value of G in order to distinguish between those coordinates of the table for which the value of G occurs in the table and is equal 0, and those for which the value of G does not occur in the table represented by m.

In order to compute the value of $G(n, x, y)$ one has to know the values of G for a certain finite number of arguments. A table will be called closed if and only if for any triple (n, x, y) of coordinates from that table all those values of G which are necessary for computing the value $G(n, x, y)$ occur in that table. Let us introduce the following relation W:

$$W(m) \equiv m \text{ represents a closed table of the function } G.$$

The relation W is primitive recursive. Indeed, one has

$$\begin{aligned} W(m) \equiv \forall n \leqslant m \forall x \leqslant m \forall y \leqslant m\{(m)\langle n,x,y\rangle \neq 0 \to \\ \to [(n = 0 \to (m)\langle n,x,y\rangle = 1)\ \wedge \end{aligned}$$

$$\begin{aligned}
&\wedge (n = 1 \to (m)\langle n, x, y\rangle = x + 1) \wedge \\
&\wedge (n = 2 \to (m)\langle n, x, y\rangle = x + 2) \wedge \\
&\wedge (n = 3 \to (m)\langle n, x, y\rangle = p_y^x + 1) \wedge \\
&\wedge (n = 4 \to (m)\langle n, x, y\rangle = (x)_y + 1) \wedge \\
&\wedge (n = 5 \to (m)\langle n, x, y\rangle = x \cdot y + 1) \wedge \\
&\wedge (n = 6 \to (m)\langle n, x, y\rangle = x + y + 1)]\} \wedge
\end{aligned}$$

$$\begin{aligned}
&\wedge \forall n \leqslant m \forall x \leqslant m \forall y \leqslant m\{(m)\langle n + 1, x, y\rangle \neq 0 \to \\
&\qquad \to [(n \geqslant 6 \wedge (n)_2 < 3) \to \\
&\to [((n)_2 = 0 \to (m)\langle n + 1, x, y\rangle = (m)\langle (n)_0, y, x\rangle) \wedge \\
&\quad \wedge ((n)_2 = 1 \to (m)\langle n, x, y\rangle = (m)\langle (n)_0, x, x\rangle)) \wedge \\
&\quad \wedge ((n)_2 = 2 \to ((m)\langle (n)_1, x, y\rangle \neq 0 \wedge \\
&\quad \wedge (m)\langle n + 1, x, y\rangle = \\
&\qquad = (m)\langle (n)_0, x, (m)\langle (n)_1, x, y\rangle \dot{-} 1\rangle))]]\} \wedge
\end{aligned}$$

$$\begin{aligned}
&\wedge \forall n \leqslant m \forall x \leqslant m \forall y \leqslant m[n \geqslant 6 \wedge (n)_2 \geqslant 3 \to \\
&\qquad \to \{[(m)\langle n + 1, 0, y\rangle \neq 0 \to \\
&\quad \to (m)\langle n + 1, 0, y\rangle = (m)\langle (n)_0, y, y\rangle] \wedge \\
&\quad \wedge [(m)\langle n + 1, x + 1, y\rangle \neq 0 \to ((m)\langle n + 1, x, y\rangle \neq 0 \wedge \\
&\quad \wedge (m)\langle n + 1, x + 1, y\rangle = \\
&\qquad = (m)\langle (n)_1, \langle x, y\rangle, (m)\langle n + 1, x, y\rangle \dot{-} 1\rangle)]\}],
\end{aligned}$$

where $(m)\langle n, x, y\rangle$ denotes $(m)_{\langle n,x,y\rangle}$.

Observe that

$$[W(m) \wedge (m)\langle n, x, y\rangle \neq 0] \to \\ \to (m)\langle n, x, y\rangle = G(n, x, y) + 1. \tag{1.14}$$

Indeed, for $n \leqslant 6$ it is obvious by the following implication:

$$(m)\langle n, x, y\rangle \neq 0 \to n, x, y \leqslant m.$$

So assume that $n > 6$ and that the property (1.14) holds for all $k \leqslant n$ and any x, y. We show that (1.14) holds for $n + 1$. It follows easily from the facts that $(n)_0 \leqslant n$, $(n)_1 \leqslant n$ and the inductive assumption.

We prove now that

$$\forall n \forall x \forall y \exists m[W(m) \wedge (m)_{\langle n,x,y\rangle} \neq 0]. \tag{1.15}$$

Observe that by (1.14) the following property holds:

$$W(m_1 \wedge (m_1)\langle n,x,y\rangle \neq 0 \wedge W(m_2) \wedge \\ \wedge (m_2)\langle n,x,y\rangle \neq 0] \rightarrow (m_1)\langle n,x,y\rangle = (m_2)\langle n,x,y\rangle.$$

Hence two tables coincide in the common points. Consequently, if m_1 and m_2 represent two tables then the least common multiple $[m_1, m_2]$ represents a table being the least common extension of the given ones. So the following implication holds:

$$W(m_1) \wedge W(m_2) \rightarrow W([m_1, m_2]).$$

We can now prove by induction the property (1.15). For $n \leqslant 6$ it suffices to take, resp.:

$$\begin{aligned} &\text{for } n = 0, && m = p^1_{\langle 0,x,y\rangle}, \\ &\text{for } n = 1, && m = p^{x+1}_{\langle 1,x,y\rangle}, \\ &\text{for } n = 2, && m = p^{x+2}_{\langle 2,x,y\rangle}, \\ & && \vdots \\ &\text{for } n = 6, && m = p^{x+y+1}_{\langle 6,x,y\rangle}. \end{aligned}$$

Let now $n > 6$ and assume that for any $k \leqslant n$ and any x, y there exist tables m such that $W(m)$ and $(m)\langle k,x,y\rangle \neq 0$. Let $\mu(k,x,y)$ denote the smallest such m. The number m for $n+1$ and given x, y is constructed as follows. If $(n)_2 = 0$ then let $m_1 = \mu((n)_0, x, x)$ and

$$m = m_1 \cdot p^{(m_1)\langle (n)_0,y,x\rangle}_{\langle n+1,x,y\rangle}.$$

If $(n)_2 = 1$ then we take

$$m = m_1 \cdot p^{(m_1)\langle (n)_0,x,x\rangle}_{\langle n+1,x,y\rangle},$$

where $m_1 = \mu((n)_0, x, x)$. If $(n)_2 = 2$ then we take

$$m = m_1 \cdot p_{\langle n+1,x,y\rangle}^{(m_1)\langle (n)_0, x, (m_1)\langle (n)_1, x, y\rangle \dot{-} 1\rangle},$$

where

$$m_1 = [\mu((n)_1, x, y), \mu((n)_0, x, (\mu((n)_1, x, y))\langle (n)_1, x, y\rangle \dot{-} 1)].$$

If $(n)_2 \geqslant 3$ then for $x = 0$ we put

$$m = m_1 \cdot p_{\langle n+1,0,y\rangle}^{(m_1)\langle (n)_0, y, y\rangle}$$

where $m_1 = \mu((n)_0, y, y)$. For $x = z + 1$ we put

$$m = m_1 \cdot p_{\langle n+1,z+1,y\rangle}^{(m_1)\langle (n)_1, \langle z,y\rangle, (m_1)\langle n+1,z,y\rangle \dot{-} 1\rangle}$$

where

$$m_1 = [\mu(n+1, z, y), \mu((n)_1, \langle z, y\rangle, \mu(n+1, z, y))(\langle n+1, z, y\rangle \dot{-} 1)].$$

In each of the cases specified above it can be easily checked that the number m satisfies the condition:

$$W(m) \wedge (m)\langle n, x, y\rangle \neq 0.$$

The function G can now be defined as follows

$$G(n, x, y) = (\mu m[W(m) \wedge (m)\langle n, x, y\rangle \neq 0])_{\langle n,x,y\rangle} \dot{-} 1.$$

Hence G is recursive and universal for the class $\mathcal{P}_2$. Q.E.D.

Corollary 1.4.16 *The class $\mathcal{P}$ of primitive recursive functions is a proper subclass of the class $\mathcal{R}$ of recursive functions.*

Proof. The function G defined in the proof of Theorem 1.4.15 belongs to $\mathcal{R} - \mathcal{P}$. Q.E.D.

We have proved that there exist recursive functions being universal for the class of primitive recursive functions of at most two arguments. One can ask if there are universal functions for the class of all recursive functions. We shall show that they do exist — of course, they are not recursive. Before we do so we prove that every recursive function can be presented in a certain canonical form called the normal form. It reflects the property that every recursive function can be defined by a single application of the operation of effective minimization.

Theorem 1.4.17 (On normal form; Kleene, 1943) *For every recursive function f there exist elementary recursive relations R, R', R'' such that*

$$f(\overline{a}) = b \equiv \exists x R'(\overline{a}, x, b) \equiv \forall x R''(\overline{a}, x, b) \qquad (1.16)$$

and

$$f(\overline{a}) = (\mu y R(\overline{a}, y))_0,$$

and for G the effectiveness condition is satisfied, i.e.,

$$\forall \overline{a} \exists y R(\overline{a}, y).$$

Proof. First we prove that every recursive function f can be represented in the form (1.16). For initial functions of the class $\mathcal{R}$ (cf. Definition 1.2.1) it is obvious. We show that this property is preserved by the operation of composition. To simplify the notation let us consider only one particular case, i.e., assume that a function f is obtained by composition of g and h and that for g and h there exist elementary recursive relations P', P'', Q' and Q'' such that

$$g(a) = b \equiv \exists x P'(a, x, b) \equiv \forall x P''(a, x, b),$$

$$h(\overline{a}) = b \equiv \exists x Q'(\overline{a}, x, b) \equiv \forall x Q''(\overline{a}, x, b).$$

The following equivalences hold

$$f(\overline{a}) = b \equiv g(h(\overline{a})) = b \equiv \exists x(h(\overline{a}) = z \wedge g(z) = b),$$

$$f(\overline{a}) = b \equiv g(h(\overline{a})) = b \equiv \forall x(h(\overline{a}) = z \rightarrow g(z) = b).$$

Hence we have

$$\begin{aligned} f(\overline{a}) = b &\equiv \exists z\,[\exists x Q'(\overline{a}, x, z) \wedge \exists y P'(z, y, b)] \\ &\equiv \exists z \exists x \exists y [Q'(\overline{a}, x, z) \wedge P'(z, y, b)] \\ &\equiv \exists u [Q'(\overline{a}, (u)_1, (u)_0) \wedge P'((u)_0, (u)_2, b)]. \end{aligned}$$

Using the fact that the decoding functions are elementary recursive (cf. remarks after Corollary 1.4.7) and Corollary 1.4.7, we get that the relation $Q'(\overline{a}, (u)_1, (u)_0) \wedge P'(((u)_0, (u)_2, b)$ is elementary recursive which gives one of the required forms of $f(\overline{a}) = b$.

On the other hand we have

$$\begin{aligned} f(\overline{a}) = b &\equiv \forall z\,[\exists x Q'(\overline{a} x, z) \rightarrow \forall y P''(z, y, b)] \\ &\equiv \forall z \forall x \forall y [Q'(\overline{a}, x, z) \rightarrow P''(z, y, b)] \\ &\equiv \forall u [Q'(\overline{a}, (u)_1, (u)_0) \rightarrow P''((u)_0, (u)_2, b)]. \end{aligned}$$

This gives us the second required form for $f(\overline{a}) = b$.

In the case if $f(\overline{a}) = g(h_1(\overline{a}), \ldots, h_k(\overline{a}))$, $k > 1$, the proof is similar.

Assume now that $f(\overline{a}) = \mu x[g(\overline{a}, x) = 0]$ and that there exist elementary recursive relations P' and P'' such that

$$g(\overline{a}, x) = b \equiv \exists y P'(\overline{a}, x, y, b) \equiv \forall y P''(\overline{a}, x, y, b)$$

and the condition of effectiveness is satisfied, i.e.,

$$\forall \overline{a} \exists x g(\overline{a}, x) = 0.$$

We have

$$f(\overline{a}) = b \equiv [g(\overline{a}, b) = 0 \wedge \forall z < b\; g(\overline{a}, z) \neq 0].$$

Hence

$$\begin{aligned} f(\overline{a}) = b &\equiv [\forall y P''(\overline{a}, b, y, 0) \wedge \forall z < b \, \forall u \, \neg P'(\overline{a}, z, u, 0)] \\ &\equiv \forall y \forall u [P''(\overline{a}, b, y, 0) \wedge \forall z < b \, \neg P'(\overline{a}, z, u, 0)] \\ &\equiv \forall v [P''(\overline{a}, b, (v)_0, 0) \wedge \forall z < b \, \neg P'(\overline{a}, z, (v)_1, 0)]. \end{aligned}$$

Using Corollary 1.4.7 we obtain one of the required forms. To obtain the second one observe that

$$f(\overline{a}) = b \equiv [\exists y P'(\overline{a}, b, y, 0) \wedge \forall z < b \, \exists u \neg P''(\overline{a}, z, u, 0)] \, .$$

But the following equivalence holds

$$\forall z < b \, \exists x R(z, x) \equiv \exists y \forall z < b \, R(z, (y)_z). \tag{1.17}$$

Indeed, the implication from right to left is obvious. So assume that $\forall z < b \, \exists x R(z, x)$. To prove the right hand side of the equivalence (1.17) put $y = \langle x_0, \ldots, x_{b-1} \rangle$, where $x_z = \mu x R(z, x)$ for $z < b$.

Using (1.17) we have:

$$\begin{aligned} f(\overline{a}) = b &\equiv [\exists y P'(\overline{a}, b, y, 0) \wedge \exists v \forall z < b \, \neg P''(\overline{a}, z, (v)_z, 0)] \\ &\equiv \exists y \exists v \, [P'(\overline{a}, b, y, 0) \wedge \forall z < b \, \neg P''(\overline{a}, z, (v)_z, 0)] \\ &\equiv \exists w \, [P'(\overline{a}, b, (w)_0, 0) \wedge \forall z < b \, \neg P''(\overline{a}, z, ((w)_1)_z, 0)] \end{aligned}$$

which gives the second required form.

In this way the first part of the theorem was proved. To prove the second part observe that if

$$f(\overline{a}) = b \equiv \exists x R'(\overline{a}, x, b) \equiv \forall x R''(\overline{a}, x, b)$$

then

$$\forall \overline{a} \exists b \exists x R'(\overline{a}, x, b)$$

and for every $\overline{a}$ the number b is uniquely determined. Hence we get

$$\forall \overline{a} \exists y R'(\overline{a}, (y)_1, (y)_0)$$

which gives the required effectiveness condition. The element $(y)_0$ is uniquely determined because it is the value of the function $f(\overline{a})$. We have now

$$f(\overline{a}) = (\mu y R'(\overline{a}, (y)_1, (y)_0))_0,$$

which completes the proof. Q.E.D.

Now the existence of the universal function for the class $\mathcal{R}$ of recursive functions can be proved.

Theorem 1.4.18 *The function $F(n, x, y)$ defined as follows*

$$F(n, x) = (\mu y[G(n, x, y) = 0])_0,$$

where $G(n, x, y)$ is the recursive function universal for the class $\mathcal{P}_2$ of binary primitive recursive functions, is universal for the class $\mathcal{R}_1$, i.e., for the class of unary recursive functions.

Proof. By Theorem 1.4.17 for any function $f \in \mathcal{R}_1$ there exists an elementary recursive relation R such that

$$f(x) = (\mu y R(x, y)))_0.$$

But for any x, y:

$$R(x, y) \equiv K_R(x, y) = 0$$

where K_R denotes the characteristic function of the relation R. Hence $K_R \in \mathcal{E}$. In particular (cf. Theorem 1.4.5) $K_R \in \mathcal{P}_2$. Consequently by Theorem 1.4.15 there exists a number n_0 such that for any x, y

$$K_R(x, y) = G(n_0, x, y).$$

So finally we have

$$f(x) = (\mu y[G(n_0, x, y) = 0])_0.$$

From the above considerations it follows that the function F defined by

$$F(n,x) = (\mu y[G(n,x,y) = 0])_0$$

has the following property: for every recursive function $f \in \mathcal{R}_1$ there exists a number n_0 such that for any x

$$f(x) = F(n_0, x).$$

Hence the function F is universal for the class $\mathcal{R}_1$. Q.E.D.

Observe that by Theorem 1.4.11 the function F cannot be recursive. So the minimum operation used in its definition is not effective, and consequently F is not defined for all arguments n, x, y. Such functions are called 'partial functions' (to distinguish them from those defined for all n-tuples of natural numbers (for appropriate n) which are called 'total'). If for given arguments n, x, y the function F is defined then its value can be calculated in a mechanical way in a finite number of steps — hence F is recursive in its domain. Such functions are called 'partial recusive' (cf. Section 1.3). So one can finally state that there exists a partial recursive function universal for the class of binary recursive functions.

If there exists a universal function for, say, binary functions of a given class $\mathcal{X}$ then universal functions for the class of n-ary functions of $\mathcal{X}$ $(n \neq 2)$ can be easily obtained providing that a function coding finite sequences belongs to $\mathcal{X}$. Indeed, let $\mathcal{X}$ be such a class and let H be universal for the class $\mathcal{X}_2$. Then a function G universal for the class $\mathcal{X}_n$, $n \neq 2$, can be defined as follows:

$$G(n, x_1, \ldots, x_n) = H(n, \langle x_1, \ldots, x_k\rangle, \langle x_{k+1}, \ldots, x_n\rangle),$$

where k is any fixed number such that $1 \leqslant k < n$. In a similar way one proceeds when a universal function for the class of unary functions of a class $\mathcal{X}$ is given.

To end this section let us give some more information on various hierarchies of primitive recursive and elementary recursive functions, in particular on Grzegorczyk's hierarchy.

Grzegorczyk's hierarchy was introduced in (1953). Consider the following sequence of functions:

$$f_0(x, y) = y + 1,$$

$$f_1(x, y) = x + y,$$

$$f_2(x, y) = (x + 1) \cdot (y + 1),$$

for $n \geqslant 2$:

$$f_{n+1}(0, y) = f_n(y + 1, y + 1),$$

$$f_{n+1}(x + 1, y) = f_{n+1}(x, f_{n+1}(x, y)).$$

It can be proved that for $n > 1$ the following inequality holds: $f_n(x, y) > y$. Functions f_n (for $n > 0$) are strictly increasing with respect to every argument, i.e., $f_n(x + 1, y) > f_n(x, y)$ and $f_n(x, y+1) > f_n(x, y)$. The definition of Grzegorczyk's hierarchy is as follows.

Definition 1.4.19 *For any $n \in \mathbb{N}$, $\mathcal{E}^n$ is the smallest class of functions containing the successor function $x+1$, projection functions I_1^2 and I_2^2, the function f_n and closed under composition and the operation of bounded recursion.*

Classes $\mathcal{E}^n$ have several interesting properties. In particular one can prove that the class $\mathcal{E}^3$ is equal to the class $\mathcal{E}$ of elementary recursive functions. Every class $\mathcal{E}^n$ is closed under the operation of bounded minimum. The set of relations of the class $\mathcal{E}^n$ (i.e., relations whose characteristic functions belong to the class $\mathcal{E}^n$) is closed under connectives and bounded quantifiers. One can also prove that the classes $\mathcal{E}^n$ ($n \in \mathbb{N}$) form a hierarchy, i.e., for any $n \in \mathbb{N}$, $\mathcal{E}^n \subseteq \mathcal{E}^{n+1}$. What more, $\mathcal{E}^n \neq \mathcal{E}^{n+1}$ for any $n \in \mathbb{N}$ — indeed it can be shown that for any n the function $f_{n+1}(x, x)$ increases faster than all the functions from the class $\mathcal{E}^n$.

The meaning of Grzegorczyk's hierarchy follows among others from the fact that the sum of all the classes $\mathcal{E}^n$ for $n \in \mathbb{N}$ is equal to the class $\mathcal{P}$ of primitive recursive functions, i.e.,

$\bigcup_{n\in\mathbb{N}} \mathcal{E}^n = \mathcal{P}$. This together with the facts mentioned above that $\mathcal{E} = \mathcal{E}^3$ and $\mathcal{E}^n \neq \mathcal{E}^{n+1}$ (for any $n \in \mathbb{N}$) results in the following

Corollary 1.4.20 *The class $\mathcal{E}$ of elementary recursive functions is not equal to the class $\mathcal{P}$ of primitive recursive functions.*

It can also be proved that any recursive function f can be presented in the form

$$f(\overline{a}) = g(\iota x[h(\overline{a}, x) = 0])$$

for some g, $h \in \mathcal{E}^0$, where $\iota x\varphi(x)$ denotes the unique x such that $\varphi(x)$.

Let us mention at the end two other hierarchies of elementary recursive and primitive recursive functions. Ritchie (1963) considered classes of predictably computable functions. The definition of those classes is based on the idea of taking into account the length of the tape used by the Turing machine to calculate the value of a given function. Let us put

$\mathcal{F}^0$ = the class of functions computable by finite automata (i.e., by Turing machines in which the read/write head can move only in one direction and the head can only read the symbols on the tape but can neither change nor erase them),

$\mathcal{F}^{i+1}$ = $\{f : \mathbb{N}^k \to \mathbb{N} : k \in \mathbb{N}$, there exists a function $g \in \mathcal{F}^i$ such that $\forall \overline{x}$ [the length of the tape used to calculate the value of $f(\overline{x})$ is $\leqslant g(\overline{x})]\}$,

$\mathcal{F}$ = $\bigcup_{i\in\mathbb{N}} \mathcal{F}^i$.

It can be proved that the classes $\mathcal{F}^i$ form a hierarchy, i.e., $\mathcal{F}^i \subseteq \mathcal{F}^{i+1}$ and $\mathcal{F}^i \neq \mathcal{F}^{i+1}$ for $i \in \mathbb{N}$ and that $\mathcal{F} = \mathcal{E}$. Note that the classes $\mathcal{F}^i$ can be also characterized algebraically (i.e., not using Turing machines).

The second hierarchy we want to mention was introduced by Kleeve in 1963. We will not give all the complicated technical details of the appropriate definitions here and note only that to define the levels E^α (for $\alpha < \omega^2$) of the hierarchy Kleeve uses calculations on so-called 'unlimited register machines'. Kleeve's hierarchy has the following properties: $E^0 = \{+, \cdot, K_=\}$, $E^{\omega s} = \mathcal{E}^{s+2}$ for $s \geqslant 1$, $E^\omega = \mathcal{E}$, $E^{\omega^2} = \mathcal{P}$.

1.5 Arithmetical Hierarchy

It was shown in Section 1.2 that the class of recursive relations is closed under the operation of bounded quantifiers. One can ask whether this class is also closed under the operation of unbounded quantifiers. It will turn out that the answer is negative. The fact that quantifiers increase the degree of complexity of a given notion (a property, a relation) led to the idea of contructing a logical hierarchy of notions in which the measure of the complexity will be the number of quantifiers necessary to define the given notion. Such a classification consists of infinitely many levels each containing infinitely many steps. Most important are the first two levels which are called arithmetical hierarchy and analytical hierarchy, respectively. The topological counterpart of the arithmetical hierarchy is Borel hierarchy (cf. Kuratowski, 1948, p. 250), and of the analytical hierarchy — the projection hierarchy (cf. Kuratowski, 1948, p. 360).

In this section only the arithmetical hierarchy will be considered. It was introduced by Kleene (1943) and independently by Mostowski (1947). This hierarchy consists of two branches corresponding to the fact that the first quantifier can be either existential or universal. The levels of it are denoted by Σ^0_n, Π^0_n ($n \in \mathbb{N}$) and defined as follows:

Definition 1.5.1 (a) $\Sigma^0_0 = \Pi^0_0 =$ *set of recursive relations.*

(b) *A relation $R \subseteq \mathbb{N}^k$ is said to be of the class Σ^0_{n+1} if and only if there exists a relation $Q \subseteq \mathbb{N}^{k+1}$ of the class Π^0_n such that*

$$R(\overline{a}) \equiv \exists x Q(\overline{a}, x).$$

(c) *A relation $R \subseteq \mathbb{N}^k$ is said to be of the class Π^0_{n+1} if and only if there exists a relation $Q \subseteq \mathbb{N}^{k+1}$ of the class Σ^0_n such that*

$$R(\overline{a}) \equiv \forall x Q(\overline{a}, x).$$

Before general properties of the arithmetical hierarchy will be stated let us pay more attention to one particular class of it, namely to the class Σ^0_1. This class is very important in various applications and hence it was labeled. Relations of this class, i.e., relations R satisfying the condition $R(\overline{a}) \equiv \exists x Q(\overline{a}, x)$ for some recursive relation Q, are called 'recursively enumerable'. This notion is an exact mathematical counterpart of the intuitive notion of positive calculability (as the notion of recursiveness is a mathematical counterpart of the intuitive notion of computability). We say that a relation $R \subseteq \mathbb{N}^k$ is positively calculable if and only if there exists a mechanical method such that for any k-tuple $\overline{a} \in \mathbb{N}^k$, if $R(\overline{a})$ then this method will give in a finite number of steps an answer to the question: "Does $R(\overline{a})$ hold?" but if $\neg R(\overline{a})$ then this method will give no conclusion.

It can easily be seen that every recursive relation is recursively enumerable. But not *vice versa*. It can be shown that there are recursively enumerable relations being not recursive (cf. Theorem 1.5.10(ii) below where a more general fact will be proved). But the followng holds

Theorem 1.5.2 (Negation Theorem; Kleene, 1943) *A relation R is recursive if and only if the relations R and $\neg R$ are recursively enumerable.*

Proof. The implication to the right is obvious. Hence only the implication to the left needs to be proved. So let R and $\neg R$ be recursively enumerable. There exist relations Q and S such that for any $\overline{a}$

$$R(\overline{a}) \equiv \exists x Q(\overline{a}, x)$$

and

$$\neg R(\overline{a}) \equiv \exists x S(\overline{a}, x).$$

But for every $\overline{a}$: $R(\overline{a}) \vee \neg R(\overline{a})$. So

$$\forall \overline{a} \exists x \, [Q(\overline{a}, x) \;\vee\; S(\overline{a}, x)]$$

which means that the effectiveness condition for the relation $Q \vee S$ is satisfied. So a recursive function f can be defined in the following way:

$$f(\overline{a}) = \mu x[Q(\overline{a}, x) \vee S(\overline{a}, x)].$$

We claim that

$$R(\overline{a}) \equiv Q(\overline{a}, f(\overline{a})). \tag{1.18}$$

Indeed, if $Q(\overline{a}, f(\overline{a}))$ then $\exists x Q(\overline{a}, x)$, so $R(\overline{a})$. Conversely, if

$$\neg Q(\overline{a}, f(\overline{a}))$$

then $S(\overline{a}, f(\overline{a}))$, so $\exists x S(\overline{a}, x)$ and consequently $\neg R(\overline{a})$. The equivalence (1.18) implies that R is recursive. Q.E.D.

Return now to the general study of the arithmetical hierarchy.

Theorem 1.5.3 *If a relation R is of the class Σ^0_n (Π^0_n) and functions $f_1, \ldots, f_k$ are recursive then the following relation P*

$$P(\overline{a}) \equiv R(f_1(\overline{a}), \ldots, f_k(\overline{a}))$$

is also of the class Σ^0_n (Π^0_n).

Proof. This theorem follows from Definition 1.5.1 of the classes Σ^0_n and Π^0_n and from Theorem 1.2.3. Q.E.D.

Theorem 1.5.4 *Every class of the arithmetical hierarchy is closed under the operations $\wedge$ and $\vee$.*

Proof. We use induction on n. For the classes Σ^0_0 and Π^0_0 the theorem follows by Theorem 1.2.6. So assume that the theorem is true for Σ^0_k and Π^0_k and show that it is also true for Σ^0_{k+1} and Π^0_{k+1}. Let P, $R \in \Sigma^0_{k+1}$. Hence for some P_1, $R_1 \in \Pi^0_k$:

$$P(\overline{a}) \equiv \exists x P_1(\overline{a}, x),$$

$$R(\overline{a}) \equiv \exists x R_1(\overline{a}, x).$$

Consequently

$$\begin{aligned}(P \vee R)(\overline{a}) &\equiv \exists x P_1(\overline{a}, x) \vee \exists x R_1(\overline{a}, x)\\ &\equiv \exists x[P_1(\overline{a}, x) \vee R_1(\overline{a}, x)]\end{aligned}$$

and

$$\begin{aligned}(P \wedge R)(\overline{a}) &\equiv \exists x P_1(\overline{a}, x) \wedge \exists x R_1(\overline{a}, x)\\ &\equiv \exists x P_1(\overline{a}, x) \wedge \exists z R_1(\overline{a}, z)\\ &\equiv \exists x \exists z[P_1(\overline{a}, x) \wedge R_1(\overline{a}, z)]\\ &\equiv \exists y[P_1(\overline{a}, (y)_0) \wedge R_1(\overline{a}, (y)_1)].\end{aligned}$$

By the induction hypothesis, Theorem 1.5.3 and the fact that the decoding functions $(y)_0$ and $(y)_1$ are recursive (cf. remark after Definition 1.2.17), we get that $P \vee R$ and $P \wedge R$ are of the class Σ^0_{k+1}.

If P and R are of the class Π^0_{k+1} then the proof is similar. Indeed, there exist relations P_1 and $R_1 \in \Sigma^0_k$ such that

$$P(\overline{a}) \equiv \forall x P_1(\overline{a}, x),$$

$$R(\overline{a}) \equiv \forall x R_1(\overline{a}, x).$$

Hence

$$\begin{aligned}(P \vee R)(\overline{a}) &\equiv \forall x P_1(\overline{a}, x) \vee \forall x R_1(\overline{a}, x)\\ &\equiv \forall x P_1(\overline{a}, x) \vee \forall z R_1(\overline{a}, x)\\ &\equiv \forall x \forall z[P_1(\overline{a}, x) \vee R_1(\overline{a}, z)]\\ &\equiv \forall y[P_1(\overline{a}, (y)_0) \vee R_1(\overline{a}, (y)_1)]\\ (P \wedge R)(\overline{a}) &\equiv \forall x P_1(\overline{a}, x) \wedge \forall x R_1(\overline{a}, x)\\ &\equiv \forall x[P_1(\overline{a}, x) \wedge R_1(\overline{a}, x)].\end{aligned}$$

By induction hypothesis and Theorem 1.5.3 the relations $P \vee R$ and $P \wedge R$ are of the class Π^0_{k+1}. Q.E.D.

Definition 1.5.5 *Let $\mathcal{X}$ be any set of relations. A set $C\mathcal{X}$ of completions of relations of the class $\mathcal{X}$ is the set defined as follows*

$$R \in C\mathcal{X} \equiv \textit{there exists a relation } P \in \mathcal{X} \textit{ such that } \forall \overline{a}[R(\overline{a}) \equiv \neg P(\overline{a})].$$

Observe that $CC\mathcal{X} = \mathcal{X}$.

Theorem 1.5.6 *The class Σ_n^0 is equal to the class of completions of relations of the class Π_n^0 and vice versa.*

Proof. We use induction on n. For $n = 0$ the theorem follows from Theorem 1.2.6. So assume that it is true for k, i.e., $\Sigma_k^0 = C\Pi_k^0$ and $\Pi_k^0 = C\Sigma_k^0$. We show that $\Sigma_{k+1}^0 = C\Pi_{k+1}^0$. So let $R \in \Sigma_{k+1}^0$. It is true that

$$R \in \Sigma_{k+1}^0 \equiv \text{there exists } Q \in \Pi_k^0 \text{ such that } \forall \overline{a}[R(\overline{a}) \equiv \exists x Q(\overline{a}, x)].$$

Consequently

$$\forall \overline{a}[\neg R(\overline{a}) \equiv \forall x \neg Q(\overline{a}, x)].$$

By induction hypothesis $(\neg Q) \in \Sigma_k^0$. Hence $(\neg R) \in \Pi_{k+1}^0$. In this way we have shown that if $R \in \Sigma_{k+1}^0$ then $(\neg R) \in \Pi_{k+1}^0$. The converse implication can be proved in a similar way. Hence it was proved that $\Sigma_{k+1}^0 = C\Pi_{k+1}^0$. Using this and the fact that $CC\mathcal{X} = \mathcal{X}$ for any class $\mathcal{X}$ we get also the equality $\Pi_{k+1}^0 = C\Sigma_{k+1}^0$. Q.E.D.

Theorem 1.5.7 (i) *The class Π_n^0 is closed under universal quantifiers (for $n > 0$).*

(ii) *The class Σ_n^0 is closed under existential quantifiers (for $n > 0$).*

Proof. We shall prove only part (i). Proof of (ii) is similar. So let $R \in \Pi_n^0$. Hence there exists a relation $Q \in \Sigma_{n-1}^0$ such that

$$\forall \overline{a}\,[R(\overline{a}, x) \equiv \forall y Q(\overline{a}, x, y)]\,.$$

Consequently, for a relation P defined as $P(\overline{a}) \equiv \forall x R(\overline{a}, x)$ we have

$$P(\overline{a}) \equiv \forall x R(\overline{a}, x) \equiv \forall x \forall y Q(\overline{a}, x, y) \equiv \forall z Q(\overline{a}, (z)_0, (z)_1).$$

Using now Theorem 1.5.3 and the fact that the decoding functions are recursive we get that $P \in \Pi^0_n$. Q.E.D.

Theorem 1.5.8 *The following inclusions hold:*

$$\begin{array}{ll} \text{(i)} & \Pi^0_n \subseteq \Sigma^0_{n+1}, \quad \Sigma^0_n \subseteq \Pi^0_{n+1}, \\ \text{(ii)} & \Pi^0_n \subseteq \Pi^0_{n+1}, \quad \Sigma^0_n \subseteq \Sigma^0_{n+1}. \end{array}$$

Proof. Inclusion (i) follows from the fact that one can always add to a formula a quantifier bounding a variable not occurring in it. So if, e.g., $R \in \Pi^0_n$ then

$$R(\overline{a}) \equiv \exists x R(\overline{a}),$$

hence $R \in \Sigma^0_{n+1}$. Similarly for $\Sigma^0_n \subseteq \Pi^0_{n+1}$.

To prove (ii) we use induction on n. For $n = 0$ the inclusions follow from the part (i) and the fact that $\Sigma^0_0 = \Pi^0_0$. So assume that $\Pi^0_k \subseteq \Pi^0_{k+1}$ and $\Sigma^0_k \subseteq \Sigma^0_{k+1}$. Let $R \in \Pi^0_{k+1}$. Then there exists a relation $Q \in \Sigma^0_k$ such that

$$R(\overline{a}) \equiv \forall x Q(\overline{a}, x).$$

But by the induction hypothesis $Q \in \Sigma^0_{k+1}$. Consequently $R \in \Pi^0_{k+2}$. In a similar way one proves that $\Sigma^0_{k+1} \subseteq \Sigma^0_{k+2}$. Hence part (ii) was proved for any n. Q.E.D.

Theorem 1.5.8 can be illustrated by the following diagram in which arrows denote inclusion.

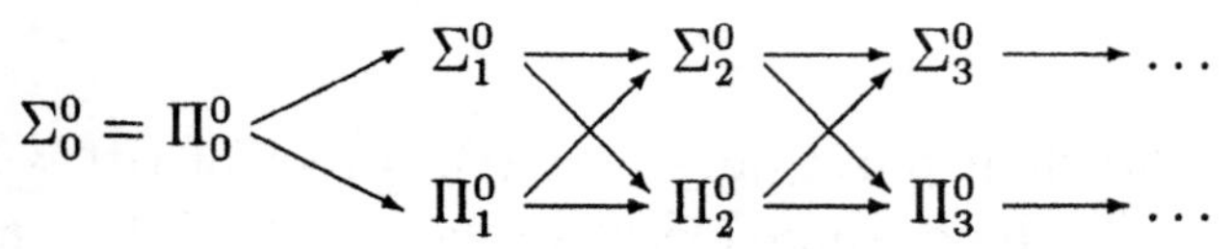

A natural question arises: are those inclusions proper? The answer is positive. It will follow from Theorem 1.5.10 which will be proved below. Before we do this, another auxiliary (though interesting in itself) theorem needs to be proved.

Theorem 1.5.9 *For any class* Σ_n^0 (Π_n^0) $(n > 0)$ *there exists in* Σ_n^0 (Π_n^0) *a universal relation for all relations of this class.*

Proof. At the beginning we show that for any n-ary relation $R \in \Sigma_1^0$ there exists an $n+1$-ary elementary recursive relation P such that for any $\overline{a}$

$$R(\overline{a}) \equiv \exists x P(\overline{a}, x). \tag{1.19}$$

Indeed, since $R \in \Sigma_1^0$, so there exists a relation $Q \in \mathcal{R}$ such that for every $\overline{a}$

$$R(\overline{a}) \equiv \exists x Q(\overline{a}, x).$$

Let K_Q be the characteristic function of the relation Q. By Theorem 1.4.17 on the normal form there exists an elementary recursive relation Q' such that for any $\overline{a}, x, b$

$$(K_Q(\overline{a}, x) = b) \equiv \exists y Q'(\overline{a}, x, y, b).$$

Hence

$$R(\overline{a}) \equiv \exists x \exists y Q'(\overline{a}, x, y, 0),$$

so

$$R(\overline{a}) \equiv \exists u Q'(\overline{a}, (u)_0, (u)_1, 0).$$

Using now the fact that the decoding functions are elementary recursive we get the required form of the relation $R \in \Sigma_1^0$.

By Theorem 1.4.15 on the existence of the universal function for the class $\mathcal{P}_2$ and remarks after Theorem 1.4.19 it follows that there exists an n-ary recursive function F universal for the class $\mathcal{P}_n$.

Let R be any given n-ary relation of the class Σ_1^0. There exists an elementary recursive relation P such that for every $\overline{a}$

$$R(\overline{a}) \equiv \exists x P(\overline{a}, x),$$

so

$$R(\overline{a}) \equiv \exists x\, [K_P(\overline{a}, x) = 0]\,.$$

Since there is a number k_0 such that $K_P(\overline{a}, x) = F(k_0, \overline{a}, x)$, so

$$R(\overline{a}) \equiv \exists x\, [F(k_0, \overline{a}, x) = 0]\,.$$

Define a relation U as follows

$$U(k, \overline{a}) \equiv \exists x\, [F(k, \overline{a}, x) = 0]\,.$$

This is a universal relation for the class Σ^0_1 and what more it belongs to the class Σ^0_1.

Universal relations for other classes Σ^0_n, Π^0_n can be now easily obtained. The relation $\neg U$ is of the class Π^0_1 and is universal for the class Π^0_1, the relation $\forall a_1 U$ is of the class Π^0_2 and is universal for Π^0_2, the relation $\exists a_1 \neg U$ is of the class Σ^0_2 and is universal for Σ^0_2, etc. Q.E.D.

Theorem 1.5.10 (i) *For every* $n > 0$*:* $\Pi^0_n \neq \Sigma^0_n$.

(ii) *For every* n*:* $\Sigma^0_n \neq \Sigma^0_{n+1}$ *and* $\Pi^0_n \neq \Pi^0_{n+1}$.

(iii) *For* $n > 0$ *the universal relation (defined in Theorem 1.5.9) for the class* Σ^0_n *belongs to the class* Σ^0_n *but neither to* Π^0_n *nor to* Σ^0_{n-1}*; the universal relation (defined in Theorem 1.5.9) for the class* Π^0_n *belongs to* Π^0_n *but neither to* Σ^0_n *nor to* Π^0_{n-1}.

Proof. (i) We prove a more general fact. We show that if a universal relation for relations of a certain class $\mathcal{X}$ belongs to class $\mathcal{X}$, then $C\mathcal{X} \neq \mathcal{X}$. Indeed, suppose that $\mathcal{X} = C\mathcal{X}$ and that $R(k, x, \overline{a})$ is universal for relations of the class $\mathcal{X}$ and $R \in \mathcal{X}$. Hence for any relation $P \in \mathcal{X}$ there exists a number k_0 such that for every $\overline{a}, x$

$$P(x, \overline{a}) \equiv R(k_0, x, \overline{a}).$$

Since $R \in \mathcal{X}$ and $C\mathcal{X} = \mathcal{X}$, so to $\mathcal{X}$ belongs also a relation R' defined as follows

$$R'(x, \overline{a}) \equiv \neg R(x, x, \overline{a}).$$

So it has its number, say k_1, i.e., for any $x, \overline{a}$

$$R'(x, \overline{a}) \equiv R(k_1, x, \overline{a}).$$

By the definition of R' we get for $x = k_1$ and any $\overline{a}$

$$R(k_1, k_1, \overline{a}) \equiv \neg R(k_1, k_1, \overline{a})$$

which is a contradiction. Consequently, the relation R universal for the class $\mathcal{X}$ does not belong to $\mathcal{X}$.

By Theorem 1.5.9 all classes Σ^0_n for $n > 0$ have universal relations of the class Σ^0_n, so $\Sigma^0_n \neq C\Sigma^0_n = \Pi^0_n$.

(ii) Assume that $\Pi^0_n = \Pi^0_{n+1}$. Then $C\Pi^0_n = C\Pi^0_{n+1}$, hence, by Theorem 1.5.6, $\Sigma^0_n = \Sigma^0_{n+1}$. So one can assume $\Pi^0_n = \Pi^0_{n+1}$ and $\Sigma^0_n = \Sigma^0_{n+1}$. But by Theorem 1.5.8, $\Pi^0_n \subseteq \Sigma^0_{n+1}$ and $\Sigma^0_n \subseteq \Pi^0_{n+1}$. So $\Pi^0_n \subseteq \Sigma^0_n$ and $\Sigma^0_n \subseteq \Pi^0_n$, i.e., $\Sigma^0_n = \Pi^0_n$ which contradicts part (i). Consequently $\Sigma^0_n \neq \Sigma^0_{n+1}$ and $\Pi^0_n \neq \Pi^0_{n+1}$.

(iii) It can be seen that the universal relation for class Σ^0_n does not belong to class Σ^0_{n-1}. If it were so, then the same property would hold for all relations of class Σ^0_n. This follows from the fact that any Σ^0_n relation can be obtained from the universal relation for class Σ^0_n by fixing the numbering argument, i.e., by substituting a constant function, hence a recursive function. So consequently it would hold that $\Sigma^0_n \subseteq \Sigma^0_{n-1}$ and $\Sigma^0_n = \Sigma^0_{n-1}$ which is impossible according to (ii).

To prove the rest of the part (iii) let P be a binary relation universal for unary relations of the class Σ^0_n. We show that $P \notin \Pi^0_n$. Indeed, let a relation Q be defined as follows

$$Q(a) \equiv \neg P(a, a).$$

The relation Q does not belong to class Σ^0_n. If it belonged to Σ^0_n then there would exist a number k_0 such that for any a: $Q(a) \equiv P(k_0, a)$. Hence for $a = k_0$ it would hold:

$$\neg P(k_0, k_0) \equiv Q(k_0) \equiv P(k_0, k_0)$$

which is impossible. So $Q \notin \Sigma^0_n$. On the other hand, the relation $Q'(a) \equiv P(a, a)$ is of the class Σ^0_n. So $\neg Q'$ belongs to Π^0_n and

$Q \in \Pi^0_n$. If the relation P belonged to the class Π^0_n then Q' would also belong to Π^0_n, so $\neg Q' \in \Sigma^0_n$ and $Q \in \Sigma^0_n$ which is impossible as we showed above. Conseqently $P \notin \Pi^0_n$.

The proof of (iii) for the universal relation for the class Π^0_n is similar. Q.E.D.

Having defined levels of the arithmetical hierarchy and having proved various properties of those levels one can ask now how particular concrete mathematical notions can be classified in this hierarchy. We will not study this problem in detail here. Say only that, for example, the concept of the convergence of a sequence of real numbers to a fixed limit belongs to class Π^0_3 and that it is neither of class Σ^0_3 nor Π^0_2 nor Σ^0_2. To define this property more precisely recall that the limit of a sequence (a_n) of real numbers is defined as follows:

$$a = \lim a_n \equiv \forall k \exists m \forall n \left(n > m \rightarrow |a_n - a| < \frac{1}{k+1} \right).$$

Instead of analysing arbitrary sequences we shall confine ourselves to analysing sequences of rational numbers (a_n) of the form $1/\alpha(n)$ convergent to zero, where $\alpha : \mathbb{N} \rightarrow \mathbb{N}$. A sequence $1/\alpha(n)$ is convergent to zero if and only if the function $\alpha(n)$ is convergent to the infinity, i.e., if

$$\forall k \exists m \forall n (n > m \rightarrow \alpha(n) > k).$$

It can be shown (cf., e.g., Mostowski, 1955) that the number of quantifiers occurring in this definition can neither be reduced nor can they be changed into dual ones. Hence the definition of the convergence of a sequence cannot be simplified.

In next chapters other examples of the estimation of the class of definite notions will be given (cf. remarks after Corollary 2.4.22 as well as Chapter 4).

1.6 Church's Thesis

Let us return to the beginning of our considerations. We started from the intuitive and imprecise notion of computability. Next we attempted to make this notion more precise. Therefore in Section 1.2 the notion of a recursive function and relation was introduced (in Section 1.3 two other notions equivalent to it, namely the notion of a function computable in the sense of Markov and of a function computable in the sense of Turing, were introduced). Now the following question should be asked: did we succeed? I.e., is the class $\mathcal{O}$ of computable functions equal to the class $\mathcal{R}$ of recursive functions? The positive answer to this question, i.e., the statement

$$\text{the class } \mathcal{O} = \text{the class } \mathcal{R},$$

is called Church's thesis. It was formulated for the first time by Church in 1936. Observe that the inclusion $\mathcal{R} \subseteq \mathcal{O}$ is obvious. Hence the essential part of Church's thesis is the inclusion $\mathcal{O} \subseteq \mathcal{R}$.

Church's thesis is a statement formulated rather in a colloquial language than in the precise language of mathematics — it is about the class of computable functions defined in the colloquial language using the imprecise notion of an effective mechanical method. Hence no precise mathematical proof of this thesis can be expected. The only thing one can do is to try to indicate evidences confirming it (or arguments against it). All such arguments must, at least in part, be of a philosophical character — they cannot be purely mathematical arguments.

Note that Church's thesis can also be treated simply as a nominal definition. Then its acceptance (or rejection) is a matter of taste, convenience, etc. If accepted, it is vacuously true, if rejected, there is no substantive issue. Add however that such an approach to Church's thesis cannot be found in the literature — it is a subject of philosophical and mathematical studies what proves that it is not treated as a nominal definition only.

Since one cannot give a precise mathematical proof of the thesis of Church, let us look for arguments in favour of it (arguments against it will be indicated in the sequel). They can be divided into three groups:

(A) heuristic arguments (no counterexamples arguments),

(B) direct arguments,

(C) arguments based on the existence of various specifications of the notion of computability.

Among arguments of group (A) one can find the following ones:

(A_1) all particular computable functions occurring in mathematics were shown to be recursive,

(A_1') no example of a computable function not being recursive was given,

(A_2) it was shown that all particular methods of obtaining computable functions from given computable ones lead also from recursive functions to recursive functions,

(A_2') no example of a method leading from computable functions to computable functions but not from recursive functions to recursive ones was given.

Arguments of group (B) consist of theoretical analyses of the process of computation and attempts to show in this way that only recursive functions can be computable. Such was the argumentation of Turing (1936) where a detailed analysis of the process of computing a value of a function was given and where the conclusion was formulated that any possible computation procedure has a faithful analogue in a Turing machine and that, therefore, every computable function is recursive.

Arguments of group (C) are based on the fact that in the second quarter of this century several somewhat different mathematical formulations of computability were given (more or less independently). All of them have been proven to be extensionally equivalent and equal to the class $\mathcal{R}$ of recursive functions. It suggests that their authors have the same intuitions connected with the notion of computability. The equality of all those classes of functions can serve as an argument in favour of the thesis that the class $\mathcal{R}$ comprises all computable functions.

All attempts to define the notion of computability in the language of mathematics can be classified in the following way:

(1) algebraical definitions — they consist of fixing certain initial functions and certain operations on functions. One considers then the smallest class of functions containing the initial functions and closed under the indicated operations. Such was, e.g., the definition of the class $\mathcal{R}$ of recursive functions given in Section 1.2;

(2) definitions using abstract mathematical machines — as an example of such a definition can serve the definition of the class of functions computable in the sense of Turing (cf. Section 1.3);

(3) definitions using certain formal systems — an example of a definition of this sort is the definition of functions computable in the sense of Markov (i.e., of functions computable by normal Markov's algorithms, cf. Section 1.3), Herbrand-Gödel-Kleene definition of computable functions (cf. Gödel, 1934 and Kleene, 1952), definition of computability by representability in a formal system, the theory of λ-definability by Church (cf. Church, 1941) or the theory of Post normal systems (cf. Post, 1943).

Observe that since there have been various attempts to define the notion of computability (not only through the notion of recursiveness), other theses analogous to Church's thesis can be also formulated. Indeed, already with Turing (1936) one finds implicitly a thesis stating the equality of the class of computable functions and the class called today 'the class of functions computable in the sense of Turing'. This thesis is of course equivalent to Church's thesis (it follows immediately from Theorem 1.3.4 stating that the class $\mathcal{R}$ of recursive functions is equal to the class of functions computable in the sense of Turing). Also Markov (1951) formulated a thesis equivalent to Church's thesis. It is the so-called 'normalization principle' which claims that every algorithm in the alphabet A is fully equivalent with respect to A to a normal algorithm over A.

Having considered arguments in favour of Church's thesis let us study now arguments against it. One can find various such arguments but, as Shapiro writes in 1981 (p. 354), "[they] seem to be supported only by their authors". We shall discuss here only arguments of Kalmár formulated by him in 1959. Note that Bowie (1973) also argued against Church's thesis.

Kalmár gives of course no example of a particular function which is not recursive but for which there exists a mechanical method of calculating its values. He shows only that Church's thesis implies certain peculiar consequences.

Let $F(k, x, y)$ be a ternary partially recursive function universal for binary recursive functions (cf. Theorem 1.4.18). Let $f(x, y)$ be its diagonalization, i.e., a function defined as:

$$f(x, y) = F(x, x, y).$$

The function f is recursive. Consider now the following example of a nonrecursive function (given by Kleene (1936), Theorem XIV).

$$g(x) = \mu y\,(f(x, y) = 0) = \begin{cases} \text{the smallest } y \text{ such that} \\ f(x, y) = 0, \text{ if such } y \text{ exists,} \\ \\ 0, \text{ otherwise.} \end{cases}$$

The function g is nonrecursive, hence by Church's thesis it is also noncomputable. On the other hand, for any natural number p for which there exists a number y such that $f(p, y) = 0$, there is a method of calculating the value of $g(p)$, i.e., of computing the smallest number y such that $f(p, y) = 0$. Indeed, it suffices to compute successively $f(p, 0)$, $f(p, 1)$, $f(p, 2)$, ... (each of them can be calculated in a finite number of steps since f is recursive) till we get the value 0. For any number p for which it can be proved (by any correct method) that there exists no y such that $f(p, y) = 0$, we also have a method of calculating the value of $g(p)$ in a finite number of steps — it suffices simply to prove that there exists no number y such that $f(p, y) = 0$ which requires a finite number of steps and leads to the result $g(p) = 0$. Hence assuming that the function g is not computable and using *tertium non datur* (note that it was already used in the definition of g) one comes to the conclusion that there are natural numbers p such that, on the one hand, there is no number y with the property $f(p, y) = 0$ and, on the other hand, this fact cannot be proved by any correct means. So Church's thesis implies the existence of absolutely undecidable propositions which can be decided! An example of such a proposition is any sentence of the form $\exists y[f(p, y) = 0]$ where p is a number for which there is no y such that $f(p, y) = 0$. Hence it is an absolutely undecidable proposition which we can decide, for we know that it is false!

In the above considerations we used a somewhat imprecise notion of provability by any correct means. This imprecision can be removed by introducing a particular formal system (more exactly: a system of equations) and showing that the considered sentence cannot be proved in any consistent extension of this system.

But the situation is even more peculiar. Church's thesis not only implies that the existence of false sentences of the form

$$\exists y[f(p, y) = 0]$$

is absolutely undecidable, but also that the absolute undecidability of those sentences cannot be proved by any correct means. Indeed, if P is a relation and the sentence $\exists y P$ is true then there exists a number q such that $P(q)$. So the question: "Does there exist a y such that $P(y)$?" is decidable and the answer is YES (because $P(q) \rightarrow \exists y P(y)$). Hence if the sentence $\exists y P(y)$ is undecidable then it is false. So if one proved the undecidability of the sentence $\exists y P(y)$ then one could also prove that $\neg \exists y P(y)$. Hence one would decide the undecidable sentence $\exists y P(y)$, which is a contradiction. Consequently, the undecidability of the sentence $\exists y P(y)$ cannot be proved by any correct means.

Kalmár comes to the following conclusion: "there are pre-mathematical notions which must remain pre-mathematical ones, for they cannot permit any restriction imposed by an exact mathematical definition" (cf. Kalmár, 1959, p. 79). Notions of effective computability, of solvability, of provability by any correct means can serve here as examples.

Observe that one can treat the argumentation of Kalmár given above not as an argumentation against Church's thesis but as an argumentation against the law of excluded middle (*tertium non datur*) — this law played a crucial role in Kalmár's argumentation. So did, for example, Markov.

Let us end this section by stating, following (Shapiro, 1981), that "one's view of Church's thesis and, in particular, one's view of the relevance of the various alleged evidence for Church's thesis depends, at least in part, on his philosophical view of the relationship between mathematics and non-mathematical reality" (p. 353). Shapiro (1981) presents two examples of the latter (more exactly: structuralistic interpretation and the view considering Church's thesis as a mathematical model) and considers Church's thesis and the relevance of the alleged evidence in light of each of them. For other opinions and interpretations of Church's thesis see Ross (1974), Kleene (1987), Gandy (1988), Mendelson (1990), Shapiro (1993), Sieg (1994 and 1997) and Folina (1998).

We would like to reassure the reader who might be uneasy about the not fully explained status of Church's thesis that in the sequel it will be never used to show that a given particular function is recursive. In any such case a precise mathematical proof of the recursiveness will be provided.

1.7 Historical Remarks

The idea of recursiveness can be traced back to the work by Richard Dedekind (1831–1916) *Was sind und was sollen die Zahlen?* (cf. Dedekind, 1888). He considered definitions of functions by the scheme of recursion treated as a generalization of definitions of addition and multiplication. In §9 (Satz 126) Dedekind justified definitions of this type. In this way the class of primitive recursive functions (without using the name) was introduced.

Giuseppe Peano (1858–1932) introduced the axiomatization of number theory and gave the recursive definitions of addition and multiplication without formal justification of the procedure (1889).

Thoralf Skolem (1887–1963) considered a theory called today 'primitive recursive arithmetic' — it was a system with bounded quantifiers only (1923, written in 1919) . Its aim was to develop arithmetic by means of recursive functions. He showed that the 'course-of-value' recursions do not lead outside the class of functions defined by ordinary recursion used by Dedekind and Peano, i.e., by recursion in which the definition of a function at a successor value $n+1$ makes reference only to its immediate predecessor n.

David Hilbert (1862–1943) allowed extended recursion schemas permitting functionals to appear in them (1926). He also introduced a hierarchy of functions and asked if there actually exist functions of level greater than 1 (functions of level 1 being primitive recursive functions). Wilhelm Ackermann (1896–1962) answered this question affirmatively in his paper of 1928 (it was written in 1927; Ackermann's result was mentioned already by Hilbert (1926, p. 185)). He exhibited an example of a function of level 2 increasing faster than any function of level 1. The proof was rather long. It was simplified by Rózsa Péter (Politzer) (1905–1977) in 1935. She also gave an alternative proof making use of the diagonal argument (Ackermann's construction makes use of this argument too). It should be added that Ackermann (1928) eliminated reference to functionals in favor of a simultaneous recursion.

It is worth noting here that at about the same time that Ackermann submitted his paper for publication, Gabriel Sudan independently gave an example of an effectively computable function that is not primitive recursive (cf. Sudan, 1927). It was an ordinal-valued function defined by nested recursion indexed by a transfinite ordinal. Ackermann's paper was cited by Sudan and *vice versa.* Unfortunately Sudan's result remained unknown. The reason for that can be the fact that his paper appeared in an obscure Romanian journal (cf. Calude *et al.*, 1979).

Ackermann's construction gave a stimulus to researches on the classification of recursion schemas (cf. Péter, 1951). It seems also to have guided Jacques Herbrand (1908–1931) in his studies towards a definition of the notion of effectively computable function (cf. Herbrand, 1931).

An explicit definition of recursiveness appeared for the first time[3] in a paper by Kurt Gödel (1906–1978) (1931). Studying incompleteness of a system of arithmetic of natural numbers Gödel defined explicitly a class of functions which he called 'recursive (*rekursiv*) functions'. This class coincides with the class of primitive recursive functions. The defined class was then applied to the study of syntax of the language of arithmetic. Using the theorem on representability of primitive recursive functions in the considered system of arithmetic Gödel could speak about the system within itself. This was one of the main ideas behind Gödel's proof of the incompleteness theorem (this problem will be considered in Chapter 2).

In the Spring of 1934, during one of his visits to Princeton, Gödel gave a lecture on effective computability (Stephen Cole Kleene (b. 1909) and Barkley J. Rosser (1907–1989), who were working then on their Ph.D. theses under the direction of Alonzo Church (1903–1995), prepared lecture notes which circulated in copies; the notes were published as the paper (Gödel, 1934)). In his lecture Gödel observed that to obtain the class of all recursive functions certain more general forms of recursion should be added to the definition of recursive functions given by him (1931). Hence he defined a class of functions which he called 'general recursive'. He used some ideas suggested to him in a letter by Jacques Herbrand in 1931 (cf. Dawson, 1993; Herbrand's letter and Gödel's reply will be published in the forthcoming vol. IV of Gödel's *Collected Works*). This class of functions

[3]For a more extensive remarks on the history of computability cf. Soare (1996).

coincides with the class $\mathcal{R}$ of recursive functions introduced in Section 1.2. It should be added that Gödel was not involved in the subsequent development of the theory of recursive functions (cf. Davis, 1982 and Feferman, 1988).

Church and his student Kleene are authors of a conception of λ-definable functions coming from the first half of the thirties. Among others they studied the problem of connections between λ-definablity and effective computability. In 1936 Kleene proved the equivalence of the notion of λ-definability and recursiveness (Kleene, 1936a).

The problem of effective computability was also studied by Alan Turing (1912–1954) who introduced the notion of the class of functions computable by certain abstract machines (today called Turing machines) (Turing, 1936). In an appendix to his paper the equivalence of his definition and of the notion of λ-definability was proved. Gödel set a high value on the idea of Turing writing:

> With this concept one has for the first time succeeded in giving an absolute definition of an interesting epistemological notion, i.e., one not depending on the formalism chosen. (1946, p. 1)

Also Church agreed early on that Turing's definition was conceptually superior. In his review of Turing's paper he wrote:

> As a matter of fact, there is involved here the equivalence of three different notions: computability by a Turing machine, general recursiveness in the sense of Herbrand-Gödel-Kleene, and λ-definability in the sense of Kleene and the present reviewer. Of these, the first has the advantage of making the identification with effectiveness in the ordinary (not explicitly defined) sense evident immediately — i.e., without the necessity of proving preliminary theorems. The second and third have the advantage of suitability for embodiment in a system of symbolic logic. (1937, pp. 42–43)

This quotation indicates also that Turing's definition is very close to the intuition and the practice of calculating while the algebraical definition is convenient from the point of view of applications in mathematical logic.

Another approach to the problem of effective computability was proposed by Andrei Andreevich Markov (1903–1979) who introduced functions computable by normal algorithms (cf. Markov, 1951 and 1954). He founded the school of Soviet constructivists in which the theory of algorithms was applied to the analysis of notions and constructions of the classical mathematics.

The class of elementary recursive functions was introduced by László Kalmár (1905–1976) in 1943.

The Negation Theorem (cf. Theorem 1.5.2) proved by Kleene (1943) was independently proved by Post (1944) and by Janiczak (1955). Mostowski (1947, footnote 6) writes: "From letters I understand that A. Tarski has also found the same theorem".

Andrzej Mostowski (1913–1975), when introducing the arithmetical hierarchy, was inspired by researches of a group of set-theoretically minded mathematicians around Wacław Sierpiński (1882–1969) in Warsaw. Their work contributed to the development of the descriptive set theory. In their studies considerable space was devoted to the estimation of the Borel class or projection class of given sets. Using an analogy between the operation of projection (applied in the descriptive set theory) and the strictly logical operation of the existential quantifier Mostowski constructed a hierarchy of arithmetical notions. Mostowski's work was done during the war. His results could be published only in 1947. Earlier Kleene's paper (1943) appeared — it contained results similar to those of Mostowski. This paper was of course unknown in Poland at that time — it became available only when Mostowski's paper was under press (cf. Mostowski, 1947, p. 112). Mostowski (1947) also admitted that Tarski informed him that he had found in 1942 results very similar to his own. It should be added here that the inspiration of Kleene was quite different from that of Mostowski. Kleene was motivated by some recursion-theoretic considerations rooted in the incompleteness theorem of Gödel.

Chapter 2

Gödel's Incompleteness Theorems

2.1 Arithmetic of Natural Numbers

The first system of axioms for the arithmetic of natural numbers was proposed in 1889 by the Italian mathematician Giuseppe Peano. Later it has been modified and improved. Its today's form is called Peano arithmetic. A main difference between Peano arithmetic and the original system of Peano is the fact that in the former no set-theoretical notions appear (such as the notion of a set and the predicate $\in$ of being an element of a set).

Peano arithmetic PA is a formalized first-order theory based on the classical predicate calculus. The following symbols are signs of the language of PA:

(a) logical constants: $\neg, \wedge, \vee, \rightarrow, \forall, \exists$,

(b) individual variables: $x_1, x_2, x_3, \ldots$,

(c) binary predicate: $=$ (identity),

(d) individual constant: 0 (zero),

(e) unary function symbol: S (successor function),

(f) binary function symbols: $+$ (addition), $\cdot$ (multiplication),

(g) technical signs: (,) (parentheses).

Expressions of the language of Peano arithmetic are defined in the following way.

Definition 2.1.1 (i) *Every individual variable* x_i *and the individual constant 0 are terms of the language of Peano arithmetic.* (ii) *If* α *is a term then* $S(\alpha)$ *is also a term.* (iii) *If* α *and* β *are terms then the expressions* $(\alpha) + (\beta)$ *and* $(\alpha) \cdot (\beta)$ *are also terms.* (iv) *There are no other terms of the language of Peano arithmetic except those listed in* (i) *and those which can be consructed according to the rules* (ii) *and* (iii).

Definition 2.1.2 (i) *If* α *and* β *are any terms of the language of Peano arithmetic then the expression* $\alpha = \beta$ *is a formula of the language of Peano arithmetic.* (ii) *If* φ *is a formula of the language of Peano arithmetic then the expressions* $\neg(\varphi)$, $\forall x_i(\varphi)$ *and* $\exists x_i(\varphi)$ *are also formulas of the language of Peano arithmetic.* (iii) *If* φ *and* ψ *are formulas of the language of Peano arithmetic then the expressions* $(\varphi) \wedge (\psi)$, $(\varphi) \vee (\psi)$, $(\varphi) \rightarrow (\psi)$ *and* $(\varphi) \equiv (\psi)$ *are also formulas of the language of Peano arithmetic.* (iv) *There are no other formulas of the language of Peano arithmetic except those which can be constructed according to the rules* (i)–(iii).

To simplify notation some parentheses will be omitted according to the following rules: (1) neither single formulas nor negated formulas will be put in brackets, (2) the signs $\wedge$ and $\vee$ connect stronger than the signs $\rightarrow$ and $\equiv$, (3) the sign $\cdot$ connects stronger than the sign $+$. The parentheses (and) will be sometimes replaced by square brackets [ ] or by braces { }. Instead of $x_1, x_2, x_3, \ldots$ we shall sometimes write simply $x, y, z, \ldots$ The language of Peano arithmetic will be denoted by L(PA).

Axioms of Peano arithmetic can be divided into three groups: logical axioms, identity axioms and nonlogical axioms. Logical axioms are all those formulas of the language of Peano arithmetic which can be obtained from the following formulas of the propositonal calculus by replacing propositional variables by formulas of the language of PA:

(1) $p \to (q \to p)$,
(2) $[p \to (q \to r)] \to [(p \to q) \to (p \to r)]$,
(3) $(p \to q) \to (\neg q \to \neg p)$,
(4) $\neg\neg p \to p$,
(5) $p \to \neg\neg p$,
(6) $p \wedge q \to p$,
(7) $p \wedge q \to q$,
(8) $(p \to q) \to [(p \to r) \to (p \to q \wedge r)]$,
(9) $p \to p \vee q$,
(10) $q \to p \vee q$,
(11) $(p \to r) \to [(q \to r) \to (p \vee q \to r)]$,
(12) $(p \equiv q) \to (p \to q)$,
(13) $(p \equiv q) \to (q \to p)$,
(14) $(p \to q) \to [(q \to p) \to (p \equiv q)]$.

Those formulas form a complete set of axioms for the propositional calculus, i.e., every tautology of the propositional calculus can be derived from those axioms by substitution rule and *modus ponens.* In the sequel substitutions of arbitrary tautologies of the propositional calculus, and not only of those indicated above as (1)–(14), will be used in formal proofs in PA — one can do it because any such tautology can be replaced by its formal proof based on axioms (1)–(14).

Identity axioms of PA are the following formulas:

(i) $x = x$,
(ii) $x = y \to y = x$,
(iii) $x = y \wedge y = z \to x = z$,
(iv) $x = y \to S(x) = S(y)$,
(v) $x = y \to x + z = y + z$,
(vi) $x = y \to z + x = z + y$,
(vii) $x = y \to x \cdot z = y \cdot z$,
(viii) $x = y \to z \cdot x = z \cdot y$.

Nonlogical axioms of PA, hence the proper axioms of arithmetic, are the following formulas:

(A1) $S(x) = S(y) \to x = y$,

(A2) $\neg(0 = S(x))$,

(A3) $x + 0 = x$,

(A4) $x + S(y) = S(x + y)$,

(A5) $x \cdot 0 = 0$,

(A6) $x \cdot S(y) = x \cdot y + x$,

(A7) $\varphi(0) \wedge \forall x[\varphi(x) \to \varphi(S(x))] \longrightarrow \forall x \varphi(x)$.

Axiom (A1) states that the successor function is injective, axiom (A2) says that 0 is not a successor of any number, axioms (A3)–(A6) characterize addition and multiplication (they are sometimes treated as recursive definitions of those number-theoretic functions). Axiom (A7) is the axiom scheme of induction — it is a scheme of infinitely many axioms: for any particular formula $\varphi(x)$ of the language of Peano arithmetic one gets an axiom stating that if 0 has the property φ and if for any x the assumption that x has the property φ implies that also the successor $S(x)$ has the property φ then any x has the property φ. Note that the formula φ may contain free variables other than x.

As inference rules we adopt the following ones:

- substitution rule: $\dfrac{\varphi}{\varphi(x/\alpha)}$, if term α is substitutable for the variable x_i in the formula φ,

- *modus ponens*: $\dfrac{\varphi \to \psi, \varphi}{\psi}$

- elimination of the universal quantifier: $\dfrac{\varphi \to \forall x_i \psi}{\varphi \to \psi}$

- introduction of the universal quantifier: $\dfrac{\varphi \to \psi}{\varphi \to \forall x_i \psi}$, if x_i does not occur free in φ,

- elimination of the existential quantifier: $\dfrac{\exists x_i \varphi \to \psi}{\varphi \to \psi}$

- introduction of the existential quantifier: $\dfrac{\varphi \to \psi}{\exists x_i \varphi \to \psi}$, if x_i does not occur free in ψ,

- generalization: $\dfrac{\varphi}{\forall x_i \varphi}$.

Observe that the generalization rule is not necessary because a formula $\forall x_i \varphi$ can be obtained from φ using appropriate substitutions of tautologies and *modus ponens*. Since it is useful in constructing formal proofs, we adopt it in our system. Recall that a term α is substitutable for a variable x_i in a formula φ if and only if for each variable y occurring in α, no part of φ of the form $\forall y \psi$ or $\exists y \psi$ contains an occurrence of x which is free in φ.

Using the described axioms and rules of inference one can prove in Peano arithmetic PA all fundamental theorems of the elementary theory of natural numbers. If a formula φ is a theorem of PA then we write simply PA $\vdash \varphi$.

Peano arithmetic, as shown above, is based on an infinite set of axioms. It was shown that this set cannot be replaced by a finite set of axioms. Indeed, Ryll-Nardzewski (1953) proved that the axiom scheme of induction cannot be replaced by any finite number of axioms. It was also shown that this axiom cannot be replaced by any (even infinite) set of particular cases obtained by taking the axiom of induction for formulas of a fixed bounded number of quantifiers.

Observe that new predicates and function symbols can be added to the language of Peano arithmetic by definitions. In particular one can add a binary predicate $<$ (smaller than) which will be used later. One does it by the following definition:

$$x < y \equiv \exists z (z \neq 0 \wedge x + z = y).$$

An expression $x \leqslant y$ will be an abbreviation for $x < y \vee x = y$. In PA various properties of the predicate $<$ can be proved. In particular the following facts, which will be used in the sequel, are true.

Lemma 2.1.3 $\mathrm{PA} \vdash \neg(x < 0)$.

Lemma 2.1.4 $\mathrm{PA} \vdash x < S(y) \equiv x < y \vee x = y$.

Lemma 2.1.5 $\mathrm{PA} \vdash x < y \vee x = y \vee y < x$.

In the sequel the following equivalence theorem and equality theorem will be applied. We shall formulate them here for the case of Peano arithmetic though they hold for any first-order theory.

Theorem 2.1.6 (Equivalence Theorem) *Let a formula* φ' *be obtained from a formula* φ *by replacing some occurrences (not necessary all) of subformulas* $\psi_1, \ldots, \psi_n$ *by* $\psi_1', \ldots, \psi_n'$ *respectively. If*

$$\mathrm{PA} \vdash \psi_1 \equiv \psi_1', \ldots, \mathrm{PA} \vdash \psi_n \equiv \psi_n',$$

then

$$\mathrm{PA} \vdash \varphi \equiv \varphi'.$$

Theorem 2.1.7 (Equality Theorem) *Let term* α' *be obtained from a term* α *by replacing some occurrences (not necessary all) of subterms* $\beta_1, \ldots, \beta_n$ *not within quantifiers by terms* $\beta_1', \ldots, \beta_n'$, *respectively, and let a formula* φ' *be obtained from a formula* φ *by the same type of replacement. If*

$$\mathrm{PA} \vdash \beta_1 = \beta_1', \ \ldots, \ \mathrm{PA} \vdash \beta_n = \beta_n',$$

then

$$\mathrm{PA} \vdash \alpha = \alpha' \text{ and } \mathrm{PA} \vdash \varphi \equiv \varphi' .$$

The language of Peano arithmetic can be interpreted in various ways. We shall fix one natural interpretation denoted by $\mathfrak{N}_0$ and called standard. Let $\mathfrak{N}_0 = \langle \mathbb{N}, 0, S, +, \cdot \rangle$ where $\mathbb{N}$ is the set of natural numbers $\{0, 1, 2, 3, \ldots\}$, 0 is the number zero, S is the successor function and + and · are addition and multiplication of natural numbers, respectively. All axioms of Peano arithmetic PA are satisfied in the interpretation $\mathfrak{N}_0$ and therefore $\mathfrak{N}_0$ is called the standard model of PA. The theory PA was constructed just to describe properties of the structure $\mathfrak{N}_0$ of natural numbers. Is this description adequate? The answer is negative. It will be provided by Gödel's theorems (cf. Sections 2.4 and 2.5 as well as Chapter 4). There exists a variety of models of PA — among them are also models nonisomorphic to $\mathfrak{N}_0$. The latter are called nonstandard. Their existence can be proved by Compactness Theorem (we return to this problem in Chapter 4).

2.2 Representability in Peano Arithmetic

In this section certain relations between recursive functions and relations on the one hand, and the formal system of Peano arithmetic on the other will be exhibited — it will be shown that all recursive functions and and relations are representable in PA, i.e., certain statements about values of recursive functions and about recursive relations can be proved in PA. (Observe that the Herbrand-Gödel's definition of the class of recursive functions mentioned in Section 1.6 is based on this property). Hence the system PA is, in a certain sense, adequate with respect to the class $\mathcal{R}$ of recursive functions and relations.

Definition 2.2.1 *Term 0 is a numeral. If a term α is a numeral then the term $S(\alpha)$ is a numeral. Only terms obtained in this way are numerals.*

Hence the following terms are numerals:

$$0,\ S(0),\ S(S(0)),\ S(S(S(0))),\ldots$$

The numeral $\underbrace{S(S(\ldots S}_{n}(0)\ldots))$, written simply as $\underbrace{SS\ldots S}_{n}(0)\ldots$, will be denoted by $\overline{n}$. Hence in particular $\overline{1}$ denotes the term $S0$,

$\overline{2}$ denotes the term $SS0$, $\overline{3}$ the term $SSS0$, etc. Observe that $\overline{n}$ is simply the name of the natural number n in the language of Peano arithmetic PA.

Definition 2.2.2 (i) *It is said that a formula φ of the language* L(PA) *with n free variables weakly represents a relation $R \subseteq \mathbb{N}^n$ in Peano arithmetic* PA (*or: a relation R is weakly represented in* PA *by a formula φ*) *if and only if for any natural numbers $k_1, \ldots, k_n$ the following equivalence holds*

$$R(k_1, \ldots, k_n) \textit{ if and only if } \mathrm{PA} \vdash \varphi(\overline{k_1}, \ldots, \overline{k_n}).$$

(ii) *A relation $R \subseteq \mathbb{N}^n$ is said to be weakly representable in Peano arithmetic* PA *if and only if there exists a formula φ of* L(PA) *with n free variables such that φ weakly represents R in* PA.

(iii) *It is said that a formula φ of the language* L(PA) *with n free variables strongly represents a relation $R \subseteq \mathbb{N}^n$ in Peano arithmetic* PA (*or: a relation R is strongly represented in* PA *by a formula φ*) *if and only if for any natural numbers $k_1, \ldots, k_n$ the following implications hold*

$$\textit{if } R(k_1, \ldots, k_n) \textit{ then } \mathrm{PA} \vdash \varphi(\overline{k_1}, \ldots, \overline{k_n}), \tag{2.1}$$

$$\textit{if } \neg R(k_1, \ldots, k_n) \textit{ then } \mathrm{PA} \vdash \neg\varphi(\overline{k_1}, \ldots, \overline{k_n}). \tag{2.2}$$

(iv) *A relation $R \subseteq \mathbb{N}^n$ is said to be strongly representable in Peano arithmetic* PA *if and only if there exists a formula φ of* L(PA) *with n free variables such that φ strongly represents R in* PA.

Observe that he following theorem holds.

Theorem 2.2.3 *If Peano arithmetic* PA *is consistent and a relation R is strongly represented in* PA *by a formula φ then the implications (2.1) and (2.2) can be extended to equivalences, i.e., the following equivalences hold:*

$$R(k_1, \ldots, k_n) \textit{ if and only if } \mathrm{PA} \vdash \varphi(\overline{k_1}, \ldots, \overline{k_n}), \tag{2.3}$$

$$\neg R(k_1, \ldots, k_n) \textit{ if and only if } \mathrm{PA} \vdash \neg\varphi(\overline{k_1}, \ldots, \overline{k_n}). \tag{2.4}$$

Hence a relation R is strongly represented in PA if and only if there exists a formula φ of the language L(PA) such that R is weakly represented by φ and its complement $\neg R$ is weakly represented by the negation $\neg\varphi$ of the formula φ.

Definition 2.2.4 (i) *It is said that a formula φ of the language* L(PA) *with $n+1$ free variables represents in Peano arithmetic* PA *a function $f\colon \mathbb{N}^n \to \mathbb{N}$ (or: a function f is represented in* PA *by a formula φ) if and only if for any natural numbers $k_1, \ldots, k_n$*

$$\mathrm{PA} \vdash \forall y[\varphi(\overline{k_1}, \ldots, \overline{k_n}, y) \equiv (y = \overline{f(k_1, \ldots, k_n)})].$$

(ii) *A function $f : \mathbb{N}^n \to \mathbb{N}$ is said to be representable in Peano arithmetic* PA *if and only if there exists a formula φ of the language* L(PA) *with $n+1$ free variables such that φ represents f in* PA.

We shall now give some simple examples of functions and relations representable in Peano arithmetic PA.

Lemma 2.2.5 *The identity relation is strongly represented in Peano arithmetic* PA *by the formula $x_1 = x_2$.*

Proof. By Definition 2.2.2 one should prove that for any natural numbers m, n the following implications hold:

$$\textit{if } m = n \textit{ then } \mathrm{PA} \vdash \overline{m} = \overline{n}, \tag{2.5}$$

$$\textit{if } m \neq n \textit{ then } \mathrm{PA} \vdash \neg(\overline{m} = \overline{n}). \tag{2.6}$$

If $m = n$ then of course the numerals $\overline{m}$ and $\overline{n}$ are identical. Hence by the identity axiom $x = x$ one gets $\mathrm{PA} \vdash \overline{m} = \overline{n}$. To show that (2.6) holds for any m and n we prove by induction on n the formula $\forall m(m \neq n \to \mathrm{PA} \vdash \neg(\overline{m} = \overline{n}))$. So let $n = 0$. Axiom (A2) of PA states that $\neg(0 = S(x))$. Hence in particular $\mathrm{PA} \vdash \neg(\overline{m} = 0)$ for $m \neq 0$. Assume now that $n > 0$. By induction hypothesis

$$\forall m(m \neq n - 1 \to \mathrm{PA} \vdash \neg(\overline{m} = \overline{n-1})).$$

Let m be any given natural number and assume that $m \neq n$. Then $m-1 \neq n-1$. By induction hypothesis

$$m-1 \neq n-1 \to \mathrm{PA} \vdash \neg(\overline{m-1} = \overline{n-1}).$$

Consequently $\mathrm{PA} \vdash \neg(\overline{m-1} = \overline{n-1})$. But

$$\mathrm{PA} \vdash S(x) = S(y) \to x = y$$

(axiom (A1)), so in particular

$$\mathrm{PA} \vdash \overline{m} = \overline{n} \to \overline{m-1} = \overline{n-1}.$$

By contraposition

$$\mathrm{PA} \vdash \neg(\overline{m-1} = \overline{n-1}) \to \neg(\overline{m} = \overline{n}).$$

Consequently $\mathrm{PA} \vdash \neg(\overline{m} = \overline{n})$. Since m was arbitrary we proved in fact $\forall m(m \neq n \to \mathrm{PA} \vdash \neg(\overline{m} = \overline{n}))$. Q.E.D.

Observe that the induction used in the above proof was the induction in the metasystem and not the axiom of induction (A7) of Peano arithmetic PA.

Lemma 2.2.6 *Addition of natural numbers is represented in Peano arithmetic* PA *by the formula* $x_1 + x_2 = x_3$.

Proof. First we show that for any natural numbers m and n

$$\mathrm{PA} \vdash \overline{m} + \overline{n} = \overline{m+n}. \tag{2.7}$$

We prove it by induction on n. If $n = 0$ then from the axiom (A3) by substitution rule one gets

$$\mathrm{PA} \vdash \overline{m} + 0 = \overline{m}.$$

Assume now that (2.7) holds for n and show that it also holds for $n+1$. Using induction hypothesis, the definition of numerals, Equality Theorem and the identity axiom $x = x$ one obtains

$$\mathrm{PA} \vdash S(\overline{m} + \overline{n}) = \overline{m+n+1}.$$

By the axiom (A4) one has

$$\mathrm{PA} \vdash \overline{m} + (\overline{n+1}) = S(\overline{m} + \overline{n}).$$

Using now the identity axiom (iii) and *modus ponens* one gets

$$\mathrm{PA} \vdash \overline{m} + (\overline{n+1}) = \overline{m+n+1}.$$

Hence (using again induction in the metasystem) we proved that for any numbers m and n the property (2.7) holds.

Let now $m + n = l$. We show that

$$\mathrm{PA} \vdash (\overline{m} + \overline{n} = y) \equiv (y = \overline{l}). \tag{2.8}$$

Applying the generalization rule to (2.8) we obtain the required property from Definition 2.2.4.

Using identity axioms (ii) and (iii), substitution rule and the appropriate tautology (more exactly: exportation rule) one gets

$$\mathrm{PA} \vdash (x = z) \rightarrow (x = y \rightarrow y = z).$$

Substitute in this formula the term $\overline{m} + \overline{n}$ for x and the term $\overline{l}$ for z. Then

$$\mathrm{PA} \vdash (\overline{m} + \overline{n} = \overline{l}) \rightarrow (\overline{m} + \overline{n} = y \rightarrow y = \overline{l}).$$

But we assumed that $m + n = l$. Hence by Lemma 2.2.5

$$\mathrm{PA} \vdash \overline{m+n} = \overline{l}.$$

We have shown above that

$$\mathrm{PA} \vdash \overline{m} + \overline{n} = \overline{m+n}.$$

Hence by the identity axioms (ii) and (iii)

$$\mathrm{PA} \vdash \overline{m} + \overline{n} = \overline{l}.$$

Using *modeus ponens* one finally gets

$$\mathrm{PA} \vdash (\overline{m} + \overline{n} = y) \rightarrow (y = \overline{l}).$$

In much the same way one proves that

$$\mathrm{PA} \vdash (y = \bar{l}) \rightarrow (\overline{m} + \overline{n} = y).$$

So finally

$$\mathrm{PA} \vdash (\overline{m} + \overline{n} = y) \rightarrow (y = \bar{l}).$$

Q.E.D.

Lemma 2.2.7 *Multiplication of natural numbers is represented in Peano arithmetic* PA *by the formula* $x_1 \cdot x_2 = x_3$.

Proof. First we prove that for any natural numbers m and n

$$\mathrm{PA} \vdash \overline{m} \cdot \overline{n} = \overline{m \cdot n}. \tag{2.9}$$

As in the proof of the previous lemma we do it using induction in the metasystem. So let $n = 0$. By axiom (A5) and substitution rule we get

$$\mathrm{PA} \vdash \overline{m} \cdot 0 = 0.$$

Assume that (2.9) holds for a number n, i.e.

$$\mathrm{PA} \vdash \overline{m} \cdot \overline{n} = \overline{m \cdot n}.$$

By axiom (A6) using the substitution rule we obtain

$$\mathrm{PA} \vdash \overline{m} \cdot (\overline{n} + 1) = \overline{m} \cdot \overline{n} + \overline{m}.$$

By Definition 2.2.1 and the axiom $x = x$ we have

$$\mathrm{PA} \vdash \overline{m \cdot (n + 1)} = \overline{m \cdot n + m}.$$

By (2.7) (see the proof of Lemma 2.2.6) we get

$$\mathrm{PA} \vdash \overline{m \cdot n + m} = \overline{m \cdot n} + \overline{m}.$$

Using now identity axioms (ii) and (iii) and *modus ponens* we obtain

$$\mathrm{PA} \vdash \overline{m} \cdot (\overline{n + 1}) = \overline{m \cdot (n + 1)}.$$

To finish the proof we should show that

$$\mathrm{PA} \vdash \forall y[(\overline{m} \cdot \overline{n} = y) \equiv (y = \bar{l})]$$

where $l = m \cdot n$. We do this just as in the proof of Lemma 2.2.6 now using (2.9). Q.E.D.

Lemma 2.2.8 *The relation "smaller than" is strongly represented in Peano arithmetic* PA *by the formula* $x_1 < x_2$.

Proof. According to the Definition 2.2.2 of strong representability one must prove that for any natural numbers m and n the following implications hold

$$\textit{if } m < n \textit{ then } \mathrm{PA} \vdash \overline{m} < \overline{n}, \tag{2.10}$$

$$\textit{if } \neg(m < n) \textit{ then } \mathrm{PA} \vdash \neg(\overline{m} < \overline{n}). \tag{2.11}$$

We prove them by induction (in the metasystem) on n. So let $n = 0$ and let $m < n$, i.e., $m < 0$ which is impossible. Hence the implication (2.10) is true for $n = 0$. Let now $\neg(m < 0)$, i.e., $m \geqslant 0$. Then by Lemma 2.1.3 using substitution rule one gets

$$\mathrm{PA} \vdash \neg(\overline{m} < 0),$$

hence (2.11) holds for $n = 0$.

Assume now that implications (2.10) and (2.11) hold for n (induction hypothesis). We show that they also hold for $n + 1$. By Lemma 2.1.4 and Definition 2.2.1

$$\mathrm{PA} \vdash \overline{m} < \overline{n+1} \equiv (\overline{m} < \overline{n} \vee \overline{m} = \overline{n}). \tag{2.12}$$

To prove (2.10) assume that $m < n + 1$. Then either $m < n$ or $m = n$. If $m < n$ then by the induction hypothesis $\mathrm{PA} \vdash \overline{m} < \overline{n}$. By the tautology $p \to p \vee q$ we obtain $\mathrm{PA} \vdash \overline{m} < \overline{n} \vee \overline{m} = \overline{n}$. Using (2.12) one gets now

$$\mathrm{PA} \vdash \overline{m} < \overline{n+1}.$$

If $m = n$ then by Lemma 2.2.5 one has $\mathrm{PA} \vdash \overline{m} = \overline{n}$. As before $\mathrm{PA} \vdash \overline{m} < \overline{n} \vee \overline{m} = \overline{n}$, hence $\mathrm{PA} \vdash \overline{m} < \overline{n+1}$. In this way (2.10) has been proved for $n + 1$ and any given m.

To prove (2.11) assume that $\neg(m < n + 1)$. So $m \geqslant n + 1$, i.e., $m > n$. By the induction hypothesis $\mathrm{PA} \vdash \neg(\overline{m} < \overline{n})$. By Lemma 2.2.5 we have $\mathrm{PA} \vdash \neg(\overline{m} = \overline{n})$. Using now the conjuction law $p \to (q \to p \wedge q)$ one obtains

$$\mathrm{PA} \vdash \neg(\overline{m} < \overline{n}) \wedge \neg(\overline{m} = \overline{n}).$$

Applying (2.12) and De Morgan's law one finally gets

$$\mathrm{PA} \vdash \neg(\overline{m} < \overline{n+1}).$$

Hence (2.11) has been proved for $n+1$ and any given m.

Consequently the implications (2.10) and (2.11) have been proved for all m and n. Q.E.D.

We can prove now the main theorem of this section, i.e., the Representability Theorem. To simplify its proof consider the following lemmas.

Lemma 2.2.9 *A relation $R \subseteq \mathbb{N}^n$ is strongly representable in Peano arithmetic* PA *if and only if its characteristic function K_R is representable in* PA.

Proof. Assume that a formula $\varphi(x_1, \ldots, x_n)$ strongly represents R in PA. We show that a formula $\psi(x_1, \ldots, x_n, y)$ of the following form

$$(\varphi(x_1, \ldots, x_n) \wedge y = 0) \vee (\neg\varphi(x_1, \ldots, x_n) \wedge y = \overline{1})$$

represents the characteristic function K_R in PA.

Indeed, let $k_1, \ldots, k_n$ be any given natural numbers and assume that $K_R(k_1, \ldots, k_n) = 0$. Hence $R(k_1, \ldots, k_n)$. So

$$\mathrm{PA} \vdash \varphi(\overline{k_1}, \ldots, \overline{k_n}).$$

We show that

$$\mathrm{PA} \vdash \psi(\overline{k_1}, \ldots, \overline{k_n}, y) \equiv (y = 0). \tag{2.13}$$

Applying the conjunction law one gets

$$\mathrm{PA} \vdash \varphi(\overline{k_1}, \ldots, \overline{k_n}) \to [y = 0 \to (\varphi(\overline{k_1}, \ldots, \overline{k_n}) \wedge y = 0)].$$

So using the fact that $\mathrm{PA} \vdash \varphi(\overline{k_1}, \ldots, \overline{k_n})$ one obtains

$$\mathrm{PA} \vdash y = 0 \to (\varphi(\overline{k_1}, \ldots, \overline{k_n}) \wedge y = 0).$$

Using now an appropriate tautology one has

$$\mathrm{PA} \vdash y = 0 \rightarrow [(\varphi(\overline{k_1}, \ldots, \overline{k_n}) \wedge y = 0) \vee (\neg\varphi(\overline{k_1}, \ldots, \overline{k_n}) \wedge y = \overline{1})],$$

i.e.,

$$\mathrm{PA} \vdash y = 0 \rightarrow \psi(\overline{k_1}, \ldots, \overline{k_n}, y).$$

To show that the converse implication can also be proved in PA observe that the following formula is a tautology:

$$p \rightarrow [(p \wedge r) \vee (\neg p \wedge q) \rightarrow r].$$

Hence substituting $\varphi(\overline{k_1}, \ldots, \overline{k_n})$ for p, $y = \overline{1}$ for q and $y = 0$ for r one gets

$$\mathrm{PA} \vdash \varphi(\overline{k_1}, \ldots, \overline{k_n}) \rightarrow [\psi(\overline{k_1}, \ldots, \overline{k_n}, y) \rightarrow y = 0].$$

By *modus ponens* one finally obtains

$$\mathrm{PA} \vdash \psi(\overline{k_1}, \ldots, \overline{k_n}, y) \rightarrow (y = 0).$$

In this way (2.13) was proved. In much the same way one proves that if $K_R(k_1, \ldots, k_n) = 1$ then

$$\mathrm{PA} \vdash \psi(\overline{k_1}, \ldots, \overline{k_n}, y) \equiv (y = \overline{1}).$$

Consequently it has been proved that if the relation R is strongly representable in Peano arithmetic PA then its characteristic function is representable in PA. To prove now the converse implication assume that the function K_R is represented in PA by a formula $\chi(x_1, \ldots, x_n, y)$. We show that the formula $\chi(x_1, \ldots, x_n, 0)$ strongly represents R in PA.

Indeed, let $R(k_1, \ldots, k_n)$. Hence $K_R(k_1, \ldots, k_n)$. So

$$\mathrm{PA} \vdash \chi(\overline{k_1}, \ldots, \overline{k_n}, y) \equiv (y = 0).$$

By the substitution rule

$$\mathrm{PA} \vdash \chi(\overline{k_1}, \ldots, \overline{k_n}, 0) \equiv (0 = 0).$$

Using the fact that PA $\vdash 0 = 0$ and applying *modus ponens* one gets

$$\text{PA} \vdash \chi(\overline{k_1}, \ldots, \overline{k_n}, 0).$$

Now assume that $\neg R(k_1, \ldots, k_n)$. Then $K_R(k_1, \ldots, k_n) = 1$. Hence

$$\begin{aligned} \text{PA} &\vdash \chi(\overline{k_1}, \ldots, \overline{k_n}, y) \equiv (y = \overline{1}), \\ \text{PA} &\vdash \chi(\overline{k_1}, \ldots, \overline{k_n}, 0) \equiv (0 = \overline{1}), \\ \text{PA} &\vdash \neg(0 = \overline{1}) \rightarrow \neg\chi(\overline{k_1}, \ldots, \overline{k_n}, 0). \end{aligned}$$

By Lemma 2.2.5: PA $\vdash \neg(0 = \overline{1})$. Hence PA $\vdash \neg\chi(\overline{k_1}, \ldots, \overline{k_n}, 0)$.

Q.E.D.

Lemma 2.2.10 *For any formula φ of the language* L(PA) *of Peano arithmetic and any natural number n*

$$\text{PA} \vdash \varphi(0) \wedge \varphi(\overline{1}) \wedge \ldots \wedge \varphi(\overline{n-1}) \wedge x < \overline{n} \longrightarrow \varphi(x).$$

Proof will use induction on n in the metasystem. If $n = 0$ then one should prove that PA $\vdash x < 0 \rightarrow \varphi(x)$. It follows from Duns Scotus' law, i.e., from the tautology $p \rightarrow (\neg p \rightarrow q)$, and Lemma 2.1.3. So assume that the considered lemma holds for n. By Lemma 2.1.4 and the substitution rule one gets

$$\text{PA} \vdash x < \overline{n+1} \equiv x < \overline{n} \vee x = \overline{n}. \tag{2.14}$$

By the Equality Theorem

$$\text{PA} \vdash x = \overline{n} \rightarrow [\varphi(x) \equiv \varphi(\overline{n})]. \tag{2.15}$$

Using now the induction hypothesis, (2.14), (2.15) and appropriate tautologies one obtains

$$\text{PA} \vdash \varphi(0) \wedge \varphi(\overline{1}) \wedge \ldots \wedge \varphi(\overline{n-1}) \wedge \varphi(\overline{n}) \wedge x < \overline{n+1} \longrightarrow \varphi(x).$$

Q.E.D.

Lemma 2.2.11 *For any formula* φ *of the language* L(PA) *of Peano arithmetic and any natural number* n*, if for every* $i < n$, $\mathrm{PA} \vdash \neg\varphi(\bar{i})$ *and* $\mathrm{PA} \vdash \varphi(\bar{n})$ *then*

$$\mathrm{PA} \vdash \varphi(x) \wedge \forall y[y < x \rightarrow \neg\varphi(y)] \equiv (x = \bar{n}). \tag{2.16}$$

Proof. Let $\psi(x)$ denote the following formula

$$\varphi(x) \wedge \forall y[y < x \rightarrow \neg\varphi(y)].$$

By the Equality Theorem one has

$$\mathrm{PA} \vdash x = \bar{n} \rightarrow [\psi(x) \equiv \varphi(\bar{n}) \wedge \forall y(y < \bar{n} \rightarrow \neg\varphi(y))].$$

By Lemma 2.2.10 and the assumption that $\mathrm{PA} \vdash \neg\varphi(\bar{i})$ for every $i < n$, one gets, by using the generalization rule:

$$\mathrm{PA} \vdash \forall y[y < \bar{n} \rightarrow \neg\varphi(y)].$$

Consequently using the assumption that $\mathrm{PA} \vdash \varphi(\bar{n})$ and an appropriate tautology one obtains

$$\mathrm{PA} \vdash x = \bar{n} \rightarrow \psi(x). \tag{2.17}$$

In this way one half of the equivalence (2.16) has been proved. To prove the other half observe that by the tautology *dictum de omni* one has

$$\mathrm{PA} \vdash \forall y[y < x \rightarrow \neg\varphi(y)] \rightarrow [\bar{n} < x \rightarrow \neg\varphi(\bar{n})].$$

This, together with the assumption that $\mathrm{PA} \vdash \varphi(\bar{n})$, implies

$$\mathrm{PA} \vdash \psi(x) \rightarrow \neg(\bar{n} < x). \tag{2.18}$$

But by Lemma 2.2.10 we have that $\mathrm{PA} \vdash x < \bar{n} \rightarrow \neg\varphi(x)$. Using that and an appropriate tautology one obtains

$$\mathrm{PA} \vdash \psi(x) \rightarrow \neg(x < \bar{n}). \tag{2.19}$$

But it was shown in Lemma 2.1.5 that

$$\mathrm{PA} \vdash x < \bar{n} \vee x = \bar{n} \vee \bar{n} < x.$$

Using this, (2.18), (2.19) and an appropriate tautology one gets

$$\mathrm{PA} \vdash \psi(x) \rightarrow x = \overline{n}. \tag{2.20}$$

So finally by (2.17) and (2.20) we have

$$\mathrm{PA} \vdash \psi(x) \equiv (x = \overline{n}).$$

Q.E.D.

Theorem 2.2.12 (Representability Theorem) *Every recursive function is representable in Peano arithmetic* PA. *Every recursive relation is strongly representable in Peano arithmetic* PA.

Proof. By Lemma 2.2.9 it suffices to prove that every recursive function is representable in PA. We show this using the algebraical definition of the class $\mathcal{R}$ of recursive functions (cf. Definition 1.2.1). It has been proved in Lemmas 2.2.6 and 2.2.7 that addition and multiplication are representable in PA and in Lemma 2.2.8 the strong representability of the relation $<$ (hence the representability of the function $K_<$) has been shown. Observe that the projection functions I_i^n are represented in PA by formulas: $x_1 = x_1 \wedge \ldots \wedge x_n = x_n \wedge x_{n+1} = x_i$ $(1 \leqslant i \leqslant n)$.

Assume now that

$$f(k_1, \ldots, k_n) = g(h_1(k_1, \ldots, k_n), \ldots, h_k(k_1, \ldots, k_n))$$

where functions $g, h_1, \ldots, h_k$ are represented in PA by

$$\psi(y_1, \ldots, y_k, z), \varphi_1(x_1, \ldots, x_n, y_1), \ldots, \varphi_k(x_1, \ldots, x_n, y_k),$$

resp. Consider a formula $\chi(x_1, \ldots, x_n, z)$ of the following form

$$\exists y_1 \ldots \exists y_k[\varphi_1(x_1, \ldots, x_n, y_1) \wedge \ldots$$
$$\ldots \wedge \varphi_k(x_1, \ldots, x_n, y_k) \wedge \psi(y_1, \ldots, y_k, z)].$$

We claim that χ represents the function f. Indeed, assume that $f(k_1, \ldots, k_n) = l$ and let $h_i(k_1, \ldots, k_n) = n_i$ $(i = 1, \ldots, k)$. Then

$$g(n_1, \ldots, n_k) = l.$$

Hence for $i = 1, \ldots, k$

$$PA \vdash \varphi_i(\overline{k_1}, \ldots, \overline{k_n}, y_i) \equiv (y_i = \overline{n_i}), \tag{2.21}$$

$$PA \vdash \psi(\overline{n_1}, \ldots, \overline{n_k}, z) \equiv (z = \overline{l}). \tag{2.22}$$

By the Equivalence Theorem one gets

$$PA \vdash \chi(\overline{k_1}, \ldots, \overline{k_n}, z) \equiv \\ \equiv \exists y_1 \ldots \exists y_k [y_1 = \overline{n_1} \wedge \ldots \wedge y_k = \overline{n_k} \wedge \psi(y_1, \ldots, y_k, z)].$$

But

$$PA \vdash \exists y_i [y_i = \overline{n_i} \wedge \psi(y_1, \ldots, y_k, z)] \equiv \\ \equiv \psi(y_1, \ldots, y_{i-1}, \overline{n_i}, y_{i+1}, \ldots, y_k, z).$$

Hence

$$PA \vdash \chi(\overline{k_1}, \ldots, \overline{k_n}, z) \equiv \psi(\overline{n_1}, \ldots, \overline{n_k}, z).$$

Using (2.22) one gets now

$$PA \vdash \chi(\overline{k_1}, \ldots, \overline{k_n}, z) \equiv (z = \overline{l}).$$

In this way it was shown that the operation of composition leads from functions representable in PA to functions representable in PA. Now we show that the operation of effective minimum has the same property. So assume that

$$f(k_1, \ldots, k_n) = \mu x [g(k_1, \ldots, k_n, x) = 0]$$

where g is represented in PA by a formula $\varphi(x_1, \ldots, x_n, y, z)$ and the appropriate effectiveness condition for g is satisfied. Consider a formula $\psi(x_1, \ldots, x_n, y)$ of the following form

$$\varphi(x_1, \ldots, x_n, y, 0) \wedge \forall x_{n+1}[x_{n+1} < y \rightarrow \neg\varphi(x_1, \ldots, x_n, x_{n+1}, 0)].$$

We claim that the formula ψ represents the function f i PA. Indeed, let $f(k_1 \ldots, k_n) = l$ and let $g(k_1, \ldots, k_n, i) = n_i$. Then

$$PA \vdash \varphi(\overline{k_1}, \ldots, \overline{k_n}, \overline{i}, z) \equiv (z = \overline{n_i}).$$

Hence

$$\mathrm{PA} \vdash \varphi(\overline{k_1}, \ldots, \overline{k_n}, \overline{i}, 0) \equiv (0 = \overline{n_i}).$$

Observe that $n_i \neq 0$ for $i < l$. Hence for $i < l$

$$\mathrm{PA} \vdash \neg\varphi(\overline{k_1}, \ldots, \overline{k_n}, \overline{i}, 0).$$

Since $n_l = 0$, we conclude that $\mathrm{PA} \vdash \varphi(\overline{k_1}, \ldots, \overline{k_n}, \overline{l}, 0)$. Using now Lemma 2.2.11 we get

$$\mathrm{PA} \vdash \psi(\overline{k_1}, \ldots, \overline{k_n}, y) \equiv (y = \overline{l}).$$

Q.E.D.

2.3 Arithmetization of Syntax

In this section one of the main ideas of Gödel (1931) will be described. It was the idea of a one-to-one correspondence between expressions of the language of arithmetic and natural numbers. Using this correspondence one can translate metamathematical statements about arithmetic as a formal system into statements about natural numbers, i.e., one can speak about arithmetic as a theory speaking about natural numbers. What more, fundamental metamathematical relations can be translated in this way into certain recursive relations, hence into relations representable in Peano arithmetic. Consequently, one can speak about a formal system of arithmetic of natural numbers and about its properties as a theory in the system itself! This is the essence of Gödel's idea (called the 'arithmetization').

The principles of an arithmetization of syntax will be described here in a general setting, i.e., in the case of any first-order theory T formulated in a language with finitely many non-logical symbols (this assumption can be weakened, i.e., one can consider arithmetization of syntax of theories formalized in a language with recursive, hence possibly infinite, set of nonlogical notions).

We start by assigning natural numbers in a one-to-one way to logical symbols (i.e., connectives and quantifiers) and non-logical symbols (i.e., predicates, function symbols, individual constants and technical signs) of a language of a given first-order theory T. A natural number corresponding to a symbol v will be denoted by $SN(v)$. To different symbols different numbers are assigned. In particular we assign the number $2i$ to the variable x_i. To assign numbers to complex expressions of the language L(T) of T assume (without loss of generality) that terms and formulas of L(T) are of the form $vv_1 \ldots v_n$, where v is either a function symbol or a predicate, a connective or a quantifier. It is an inessential change of the usually accepted definition of terms and formulas (cf. Definitions 2.2.1 and 2.2.2). This notation does not need parentheses. If a term or a formula u is of the form $vv_1 \ldots v_n$ then the natural number assigned to it (denoted by $\ulcorner u \urcorner$ and called the Gödel number of u) is defined as follows

$$\ulcorner u \urcorner = \langle SN(v), \ulcorner v_1 \urcorner, \ldots, \ulcorner v_n \urcorner \rangle$$

where $\ulcorner v_i \urcorner$ $(i = 1, \ldots, n)$ are previously defined Gödel numbers of terms or formulas v_i and $\langle \ldots \rangle$ is a recursive function coding finite sequences of natural numbers defined in Section 1.2 — cf. Gödel's Theorem 1.2.15 (any other recursive function coding finite sequences, not necessarily the function based on Gödel's function β, can be used here, of course). The assumption that any term or formula are of the form $vv_1 \ldots v_n$ where v is either a function symbol, a predicate, a connective or a quantifier, enables us to "reflect" in a Gödel number $\ulcorner u \urcorner$ the inductive structure of u by separating on the first place the main functor and then putting behind it subformulas of u. It will simplify the arithmetization of syntax which will be described in the sequel.

We now give an example which will illustrate how to calculate the Gödel number of a given formula. Consider the language L(PA) of Peano arithmetic and calculate the Gödel number of a formula $\varphi(x)$ of the following form

$$\neg \exists y (S(x) + y = 0 \vee x = y).$$

We used a usual notation with parentheses here. To calculate

$\ulcorner\varphi(x)\urcorner$ one should treat φ as written in the following way

$$\neg\exists y \vee = +Sxy0 = xy.$$

Start by defining numbers $SN(v)$ for symbols v of L(PA). Let $SN(\neg) = 3$, $SN(\vee) = 5$, $SN(\wedge) = 7$, $SN(\rightarrow) = 9$, $SN(\equiv) = 11$, $SN(\exists) = 13$, $SN(\forall) = 15$, $SN(S) = 17$, $SN(+) = 19$, $SN(\cdot) = 21$, $SN(0) = 23$, $SN(=) = 25$, $SN(x_i) = 2i$. Then

$$\ulcorner x = y\urcorner = \langle 25, 2, 4\rangle,$$

$$\ulcorner Sx\urcorner = \langle 17, 2\rangle,$$

$$\ulcorner Sx + y\urcorner = \langle 19, \langle 17, 2\rangle, 4\rangle,$$

$$\ulcorner Sx + y = 0\urcorner = \langle 25, \langle 19, \langle 17, 2\rangle, 4\rangle, 23\rangle,$$

$$\ulcorner Sx + y = 0 \vee x = y\urcorner = \langle 5, \ulcorner Sx + y = 0\urcorner, \ulcorner x = y\urcorner\rangle,$$

$$\ulcorner\varphi(x)\urcorner = \langle 3, \langle 13, 4, \ulcorner Sx + y = 0 \vee x = y\urcorner\rangle\rangle.$$

Note that if a natural number a is given, one can effectively find a formula φ such that $a = \ulcorner\varphi\urcorner$ provided that a is a Gödel number of a formula. Indeed, one should proceed as follows: (1) check if $Seq(a)$ where Seq is the recursive relation defined in Section 1.2, (2) if $Seq(a)$ then calculate $lh(a)$ and $(a)_i$ for $i = 0, 1, \ldots, lh(a)$ (cf. Definition 1.2.17), (3) we get $a = \langle a_0, \ldots, a_{lh(a)}\rangle$; check if $a_0 = SN(v)$ where v is a logical or non-logical symbol of the language L(T) of the considered theory T, (4) if the answer is affirmative then we analyse a_i $(i = 1, \ldots, lh(a))$ in a similar way as we did with a. Since $a_i < a$ (see the properties of the function β — cf. Theorem 1.2.15), we conclude that in a finite number of steps one can find a formula φ such that $a = \ulcorner\varphi\urcorner$ (provided that a is a Gödel number of a formula of the language L(T) and this can be effectively checked). A similar procedure can be applied to terms.

Consequently, we showed above that one can effectively assign natural numbers to terms and formulas of the language L(T) of a first-order theory T (with finitely many non-logical symbols) in such a way that the reverse procedure, i.e., the reconstruction of a term or a formula from a given number is also effective. Observe that we are talking here about effective computability and not about recursiveness. The reason is that we are considering relations between terms and formulas (hence expressions) on the one hand and natural numbers on the other, and the mathematical notion of recursiveness was defined only for number-theoretical functions and relations.

Using the arithmetization of language described above we shall now define the arithmetization of metamathematics and show that certain functions and relations corresponding to certain metamathematical functions and relations are recursive. To restrict our attention we shall consider Peano arithmetic. It can be easily seen that the procedure we describe can be applied to any first-order theory formalized in a language with a recursive set (of Gödel numbers) of non-logical symbols.

We now define functions and relations which correspond to metamathematical functions and relations indicated below. All those functions and relations are recursive, what follows from their definitions and from properties of functions and relations of the class $\mathcal{R}$ proved in Section 1.2.

$$Vble(a) \equiv a = \langle (a)_0 \rangle \wedge \exists y \leqslant a((a)_0 = 2y).$$

$Vble(a)$ means that $a = \ulcorner x_i \urcorner$ for some variable x_i.

$$\begin{aligned} Term(a) &\equiv 0 = 0, \text{ if } a = \langle SN(0) \rangle, \\ &\equiv Term((a)_1), \text{ if } a = \langle SN(S), (a)_1 \rangle, \\ &\equiv Term((a)_1) \wedge Term((a)_2), \\ &\qquad \text{if } a = \langle SN(+), (a)_1, (a)_2 \rangle \vee \\ &\qquad\quad \vee a = \langle SN(\cdot), (a)_1, (a)_2 \rangle, \\ &\equiv Vble(a), \text{ otherwise.} \end{aligned}$$

$Term(a)$ means that $a = \ulcorner \alpha \urcorner$ for a term α. Observe that the recursiveness of the relation $Term$ follows from Theorem 1.2.13 and 1.2.23 and Corollary 1.2.24 as well as from the fact that

$(a)_i < a$ (cf. Theorem 1.2.15). Those properties will be frequently used in the sequel.

$$\begin{aligned} AForm(a) \ \equiv\ & a = \langle (a)_0, (a)_1, (a)_2 \rangle \wedge (a)_0 = SN(=) \wedge \\ & \wedge Term((a)_1) \wedge Term((a)_2). \end{aligned}$$

$AForm(a)$ means that $a = \ulcorner \varphi \urcorner$, where φ is an atomic formula.

$$\begin{aligned} Form(a) \ \equiv\ & Form((a)_1), \text{ if } a = \langle SN(\neg), (a)_1 \rangle, \\ \equiv\ & Form((a)_1) \wedge Form((a)_2), \\ & \text{if } a = \langle SN(\vee), (a)_1, (a)_2 \rangle \vee \\ & \vee a = \langle SN(\wedge), (a)_1, (a)_2 \rangle \vee \\ & \vee a = \langle SN(\rightarrow), (a)_1, (a)_2 \rangle \vee \\ & \vee a = \langle SN(\equiv), (a)_1, (a)_2 \rangle, \\ \equiv\ & Vble((a)_1) \wedge Form((a)_2), \\ & \text{if } a = \langle SN(\exists), (a)_1, (a)_2 \rangle \vee \\ & \vee a = \langle SN(\forall), (a)_1, (a)_2 \rangle, \\ \equiv\ & AForm(a), \text{ otherwise.} \end{aligned}$$

$Form(a)$ means that $a = \ulcorner \varphi \urcorner$ for a formula φ.

$$\begin{aligned} Sub(a,b,c) \ =\ & c, \text{ if } Vble(a) \wedge a = b, \\ =\ & \langle (a)_0, Sub((a)_1, b, c) \rangle, \text{ if } a = \langle (a)_0, (a)_1 \rangle, \\ =\ & \langle (a)_0, Sub((a)_1, b, c), Sub((a)_2, b, c) \rangle, \\ & \text{if } a = \langle (a)_0, (a)_1, (a)_2 \rangle \wedge \\ & \wedge (a)_0 \neq SN(\exists) \wedge (a)_0 \neq SN(\forall), \\ =\ & \langle (a)_0, (a)_1, Sub((a)_2, b, c) \rangle, \\ & \text{if } [a = \langle SN(\exists), (a)_1, (a)_2 \rangle \vee \\ & \vee a = \langle SN(\forall), (a)_1, (a)_2 \rangle] \wedge (a)_1 \neq b, \\ =\ & a, \text{ otherwise.} \end{aligned}$$

Sub corresponds to the operation of substitution and has the following properties:

$$Sub(\ulcorner\alpha\urcorner, \ulcorner x\urcorner, \ulcorner\beta\urcorner) = \ulcorner\alpha(x/\beta)\urcorner,$$

$$Sub(\ulcorner\varphi\urcorner, \ulcorner x\urcorner, \ulcorner\alpha\urcorner) = \ulcorner\varphi(x/\alpha)\urcorner.$$

$$\begin{aligned} Fr(a,b) \;&\equiv\; a = b, \text{ if } Vble(a),\\ &\equiv\; F((a)_1, b), \text{ if } a = \langle (a)_0, (a)_1\rangle,\\ &\equiv\; Fr((a)_1, b) \vee Fr((a)_2, b),\\ &\qquad \text{if } a = \langle (a)_0, (a)_1, (a)_2\rangle \wedge\\ &\qquad\quad \wedge (a)_0 \neq SN(\exists) \wedge (a)_0 \neq SN(\forall),\\ &\equiv\; Fr((a)_2, b) \wedge (a)_1 \neq b, \text{ otherwise.} \end{aligned}$$

Fr is a relation such that if φ is a formula then $Fr(\ulcorner\varphi\urcorner, \ulcorner x\urcorner)$ if and only if the variable x occurs free in the formula φ.

$$\begin{aligned} Subtl(a,b,c) \;&\equiv\; Subtl((a)_1, b, c) \text{ if } a = \langle (a)_0, (a)_1\rangle,\\ &\equiv\; Subtl((a)_1, b, c) \wedge Subtl((a)_2, b, c),\\ &\qquad \text{if } a = \langle (a)_0, (a)_1, (a)_2\rangle \wedge\\ &\qquad\quad \wedge (a)_0 \neq SN(\exists) \wedge (a)_0 \neq SN(\forall),\\ &\equiv\; Subtl((a)_2, b, c) \wedge [\neg Fr((a)_2, b) \vee \neg Fr(c, (a)_1))],\\ &\qquad \text{if } [a = \langle SN(\exists), (a)_1, (a)_2\rangle \vee\\ &\qquad\quad \vee a = \langle SN(\forall), (a)_1, (a)_2\rangle] \wedge (a)_1 \neq b,\\ &\equiv\; 0 = 0, \text{ otherwise.} \end{aligned}$$

The relation $Subtl$ is such that $Subtl(\ulcorner\varphi\urcorner, \ulcorner x\urcorner, \ulcorner\alpha\urcorner)$ if and only if term α is substitutable in the formula φ for the variable x.

$$\begin{aligned} LAx(x) \;&\equiv\; \exists x < a \exists y < a[Form(x) \wedge Form(y) \wedge\\ &\qquad \wedge a = \langle 9, x, \langle 9, y, x\rangle\rangle] \vee\\ &\quad \vee \exists x < a \exists y < a \exists z < a[Form(x) \wedge Form(y) \wedge \end{aligned}$$

$$
\begin{aligned}
&\wedge Form(z) \wedge a = \langle 9, \langle 9, x, \langle 9, y, z \rangle \rangle, \\
&\langle 9, \langle 9, x, y \rangle, \langle 9, x, z \rangle \rangle \rangle] \vee \\
&\vee \exists x < a \exists y < a [Form(x) \wedge Form(y) \wedge \\
&\wedge a = \langle 9, \langle 9, x, y \rangle, \langle 9, \langle 3, y \rangle, \langle 3, x \rangle \rangle \rangle] \vee \\
&\vee \exists x < a [Form(x) \wedge a = \langle 9, \langle 3, \langle 3, x \rangle \rangle, x \rangle] \vee \\
&\vee \exists x < a [Form(x) \wedge a = \langle 9, x, \langle 3, \langle 3, x \rangle \rangle \rangle] \vee \\
&\vee \exists x < a \exists y < a [Form(x) \wedge Form(y) \wedge \\
&\wedge a = \langle 9, \langle 7, x, y \rangle, x \rangle] \vee \\
&\vee \exists x < a \exists y < a [Form(x) \wedge Form(y) \wedge \\
&\wedge a = \langle 9, \langle 7, x, y \rangle, y \rangle] \vee \\
&\vee \exists x < a \exists y < a \exists z < a [Form(x) \wedge Form(y) \wedge \\
&\wedge Form(z) \wedge \\
&\wedge a = \langle 9, \langle 9, x, y \rangle, \langle 9, \langle 9, x, z \rangle, \\
&\langle 9, x, \langle 7, y, z \rangle \rangle \rangle \rangle] \vee \\
&\vee \exists x < a \exists y < a [Form(x) \wedge Form(y) \wedge \\
&\wedge a = \langle 9, x, \langle 5, x, y \rangle \rangle] \vee \\
&\vee \exists x < a \exists y < a [Form(x) \wedge Form(y) \wedge \\
&\wedge a = \langle 9, y, \langle 5, x, y \rangle \rangle] \vee \\
&\vee \exists x < a \exists y < a \exists z < a [Form(x) \wedge Form(y) \wedge \\
&\wedge Form(z) \wedge \\
&\wedge a = \langle 9, \langle 9, x, z \rangle, \langle 9, \langle 9, y, z \rangle, \\
&\langle 9, \langle 5, x, y \rangle, z \rangle \rangle \rangle] \vee \\
&\vee \exists x < a \exists y < a [Form(x) \wedge Form(y) \wedge \\
&\wedge a = \langle 9, \langle 11, x, y \rangle, \langle 9, x, y \rangle \rangle] \vee \\
&\vee \exists x < a \exists y < a [Form(x) \wedge Form(y) \wedge \\
&\wedge a = \langle 9, \langle 11, x, y \rangle, \langle 9, y, x \rangle \rangle] \vee \\
&\vee \exists x < a \exists y < a [Form(x) \wedge Form(y) \wedge \\
&\wedge a = \langle 9, \langle 9, x, y \rangle, \langle 9, \langle 9, y, x \rangle, \langle 11, x, y \rangle \rangle \rangle].
\end{aligned}
$$

$LAx(a)$ holds if and only if a is a Gödel number of a logical axiom of Peano arithmetic PA.

$$
\begin{aligned}
EAx(a) \equiv\ & \exists x < a[a = \langle 25, 2x, 2x\rangle] \vee \\
& \vee \exists x < a \exists y < a[a = \langle 9, \langle 25, 2x, 2y\rangle, \\
& \quad \langle 25, 2y, 2x\rangle\rangle] \vee \\
& \vee \exists x < a \exists y < a \exists z < a[a = \langle 9, \langle 7, \langle 25, 2x, 2y\rangle, \\
& \quad \langle 25, 2y, 2z\rangle\rangle, \langle 25, 2x, 2z\rangle\rangle] \vee \\
& \vee \exists x < a \exists y < a[a = \langle 9, \langle 25, 2x, 2y\rangle, \\
& \quad \langle 25, \langle 17, 2x\rangle, \langle 17, 2y\rangle\rangle\rangle] \vee \\
& \vee \exists x < a \exists y < a \exists z < a[a = \langle 9, \langle 25, 2x, 2y\rangle, \\
& \quad \langle 25, \langle 19, 2x, 2z\rangle, \langle 19, 2y, 2z\rangle\rangle\rangle] \vee \\
& \vee \exists x < a \exists y < a \exists z < a[a = \langle 9, \langle 25, 2x, 2y\rangle, \\
& \quad \langle 25, \langle 19, 2z, 2x\rangle, \langle 19, 2z, 2y\rangle\rangle\rangle] \vee \\
& \vee \exists x < a \exists y < a \exists z < a[a = \langle 9, \langle 25, 2x, 2y\rangle, \\
& \quad \langle 25, \langle 21, 2x, 2z\rangle, \langle 21, 2y, 2z\rangle\rangle\rangle] \vee \\
& \vee \exists x < a \exists y < a \exists z < a[a = \langle 9, \langle 25, \langle 25, 2x, 2y\rangle, \\
& \quad \langle 25, \langle 21, 2z, 2x\rangle, \langle 21, 2z, 2y\rangle\rangle\rangle\rangle].
\end{aligned}
$$

$EAx(a)$ holds if and only if a is a Gödel number of an identity axiom of Peano arithmetic PA.

$$
\begin{aligned}
S(a, b) \equiv\ & \exists x < a \exists y < a[Term(x) \wedge Vble(y) \wedge Form(a) \wedge \\
& \wedge Form(b) \wedge Subtl(a, y, x) \wedge b = Sub(a, y, x)].
\end{aligned}
$$

S is a relation such that $S(\ulcorner\varphi\urcorner, \ulcorner\psi\urcorner)$ if and only if the formula ψ is obtained from the formula φ by substituting a term for an individual variable.

$$
\begin{aligned}
MP(a, b, c) \equiv\ & Form(a) \wedge Form(b) \wedge Form(c) \wedge \\
& \wedge a = \langle 9, (a)_1, (a)_2\rangle \wedge b = (a)_1 \wedge c = (a)_2.
\end{aligned}
$$

The relation $MP(a, b, c)$ holds if and only if a formula with Gödel number c is obtained from formulas with Gödel numbers a and b by *modus ponens*.

In a similar way one defines relations corresponding to other rules of inference of the arithmetic PA, i.e., the following relations:

$I\exists(a,b)$	corresponding to the introduction of the existential quantifier,
$E\exists(a,b)$	corresponding to the elimination of the existential quantifier,
$I\forall(a,b)$	corresponding to the introduction of the universal quantifier,
$E\forall(a,b)$	corresponding to the elimination of the universal quantifier,
$GR(a)$	corresponding to the generalization.

We write only the precise and exact definition of the relation $I\exists$ — definitions of other relations can be easily written as an exercise. We have:

$$\begin{aligned} I\exists(a,b) \equiv\ & \exists x < a \exists y < a \exists z < a[Form(x) \wedge Form(y) \wedge \\ & \wedge Vble(z) \wedge a = \langle 9, x, y\rangle \wedge \\ & \wedge b = \langle 9, \langle 13, z, x\rangle, y\rangle \wedge \neg Fr(y,z)]. \end{aligned}$$

One can also easily write the definition of the following relation:

$NLAx(a) \equiv a$ is a Gödel number of a nonlogical axiom of Peano arithmetic.

At last we put

$$Ax(a) \equiv LAx(a) \vee EAx(a) \vee NLAx(a).$$

Hence $Ax(a)$ holds if and only if a is a Gödel number of an axiom of Peano arithmetic.

Recall that a formal proof in Peano arithmetic PA is a finite sequence of formulas such that the first formula is an axiom (i.e., either a logical axiom or an identity axiom or a nonlogical axiom) and every other formula in this sequence is either an axiom or can be obtained from proceeding formulas by one of the rules of inference of PA. If one assigns a Gödel number to every formula of a formal proof, one will obtain a finite sequence of natural numbers which can be coded by a single natural number — this code will be called the Gödel number of the proof. Hence one can define a recursive relation Prf such that $Prf(a, b)$ holds if and only if a is a Gödel number of a proof of the formula with a Gödel number b. The relation Prf can be defined in the following way

$$\begin{aligned} Prf(a,b) \equiv\ & Seq(a) \wedge lh(a) \neq 0 \wedge \\ & \wedge \forall i < lh(a)\{Form((a)_i) \wedge [Ax((a)_i) \vee \\ & \quad \vee \exists j < i \exists k < i[S((a)_j, (a)_i) \vee \\ & \quad \vee MP((a)_j, (a)_k, (a)_i)) \vee \\ & \quad \vee I\exists((a)_j, (a)_i) \vee E\exists((a)_j, (a)_i) \vee \\ & \quad \vee I\forall((a)_j, (a)_i) \vee E\forall((a)_j, (a)_i) \vee \\ & \quad \vee GR((a)_j, (a)_i)]] \wedge \\ & \wedge (a)_{lh(a)-1} = b. \end{aligned}$$

All relations and functions defined above are recursive. So we can formulate the following

Theorem 2.3.1 *The relations $Vble$, $Term$, $AForm$, $Form$, Fr, $Subtl$, LAx, EAx, S, MP, $I\exists$, $E\exists$, $I\forall$, $E\forall$, GR, $NLAx$, Ax, Prf and the function Sub are recursive.*

We need one more relation connected with metamathematics of PA, namely the relation

$$Pr(a) \equiv \exists x Prf(x, a). \tag{2.23}$$

$Pr(a)$ means that a formula with a Gödel number a is provable in Peano arithmetic PA, hence it is a theorem of PA. Since the length of possible proofs in PA cannot be bounded, the quantifier

$\exists x$ in the formula (2.23) cannot be bounded. Consequently, the relation Pr need not be recursive.

Recall the notion of a recursively enumerable relation introduced in Section 1.5. Definition (2.23) implies that the set of Gödel numbers of theorems of Peano arithmetic PA is just a recursively enumerable set. It will be shown in the sequel that it is not recursive.

At the beginning of this section we remarked that the method of arithmetization described here is a general one, i.e., it can be applied to any first-order theory T formalized in a language with finitely many nonlogical symbols. Arithmetization of syntax of such a theory can be done in much the same way as in the case of Peano arithmetic. A difference can appear in the case of the relation $NLAx_{\mathrm{T}}$ — in general the set $NLAx_{\mathrm{T}}$ of Gödel numbers of nonlogical axioms of T need not be recursive and, consequently, the set Ax_{T} of Gödel numbers of axioms of T and the relation Prf_{T} of being a proof in T does not have to be recursive and the set Pr_{T} of Gödel numbers of theorems of the theory T need not be recursively enumerable. This justifies the following definition.

Definition 2.3.2 *If the set $NLAx_{\mathrm{T}}$ of Gödel numbers of nonlogical axioms of the theory T is recursive then the theory T is said to be (recursively) axiomatizable.*

Hence Peano arithmetic PA is (recursively) axiomatizable. It can be easily seen that any first-order theory (based on the classical first-order logic) formalized in a language with finitely many nonlogical symbols and having a finite number of nonlogical axioms and a finite number of axiom schemes is always (recursively) axiomatizable, hence the set of Gödel numbers of its theorems is recursively enumerable. Observe that the assumption that the language of the considered theory contains only finitely many nonlogical symbols can be weakened — it suffices to assume that the set of (Gödel numbers of) nonlogical symbols is recursive.

2.4 Gödel's Theorems

In this section we formulate and prove Gödel's incompleteness theorems. We also discuss various strengthenings and generalizations of them as well as general theorems on the form of undecidable sentences.

Let us start by introducing a necessary new notion.

Definition 2.4.1 *Peano arithmetic* PA *is said to be* ω*-consistent if and only if for any formula* $\varphi(x)$*:*

$$\text{if PA} \vdash \varphi(0), \text{PA} \vdash \varphi(\overline{1}), \text{PA} \vdash \varphi(\overline{2}), \ldots \text{ then PA} \nvdash \exists x \neg \varphi(x).$$

Observe that this definition can be generalized to any first-order theory which contains numerals or, more generally, names for all natural numbers in its language.

Theorem 2.4.2 *If Peano arithmetic* PA *is* ω*-consistent then it is consistent.*

Proof. It suffices to prove that there exists at least one formula which is not a theorem of PA. So assume that PA is ω-consistent and consider a formula $x = x$. Since it is one of the identity axioms of PA, PA $\vdash x = x$. Hence PA $\vdash \overline{n} = \overline{n}$ for any natural number n. By ω-consistency PA $\nvdash \exists x \neg (x = x)$. Q.E.D.

We now construct Gödel's sentence undecidable in PA. Let Num be a unary recursive function defined by the following scheme of primitive recursion:

$$\begin{aligned} Num(0) &= \langle SN(0) \rangle, \\ Num(a+1) &= \langle SN(S), Num(a) \rangle. \end{aligned}$$

Hence $Num(k)$ is simply a Gödel number of the numeral $\overline{k}$. Consider now a binary recursive relation R defined as follows:

$$R(a, b) \equiv Form(a) \wedge Fr(a, 2) \wedge Prf(b, Sub(a, 2, Num(a))).$$

Recursiveness of the relation R follows from the properties of recursive relations considered in Section 1.2 and from the fact that the relations $Form$, Fr, Prf and the function Sub are recursive (cf. Theorem 2.3.1 in the previous section). Note that the relation $R(a, b)$ holds if and only if a is the Gödel number of a formula ϑ of the language L(PA) in which the variable x occurs free and b is the Gödel number of a proof in PA of the formula obtained from ϑ by substituting the numeral corresponding to the number a (i.e., to the Gödel number of the formula ϑ itself) for the variable x.

Since R is recursive, it is strongly representable in Peano arithmetic PA (cf. Theorem 2.2.12). So let $\varphi(x, y)$ be a formula strongly representing R in PA. Consider a formula ψ of the form: $\forall y \neg\varphi(x, y)$. Let $m = \ulcorner \forall y \neg\varphi(x, y) \urcorner$. Consider a sentence χ of the following form

$$\forall y \neg\varphi(\overline{m}, y). \tag{2.24}$$

Observe that formula (2.24) states that the formula with Gödel number m, hence this formula itself, is not a theorem of Peano arithmetic PA. So formula (2.24) says about itself: "I am not a theorem of PA". The sentence χ is called the Gödel's sentence.

Theorem 2.4.3 (First Incompleteness Theorem) *If Peano arithmetic* PA *is* ω*-consistent then the sentence* χ*, i.e., the sentence* $\forall y \neg\varphi(\overline{m}, y)$ *is undecidable in* PA*, i.e.,* PA $\nvdash \chi$ *and* PA $\nvdash \neg\chi$*. Hence Peano arithmetic* PA *is incomplete.*

Proof. Suppose that PA $\vdash \chi$. Then χ has a proof in PA. Let k be a Gödel number of a proof of χ. Hence $R(m, k)$. We assumed that the formula φ strongly represents the relation R. So in particular PA $\vdash \varphi(\overline{m}, \overline{k})$. But from the assumption PA $\vdash \chi$, i.e. from the assumption PA $\vdash \forall y \neg\varphi(\overline{m}, y)$, it follows that

$$\text{PA} \vdash \neg\varphi(\overline{m}, \overline{k}).$$

As a consequence, Peano arithmetic PA is inconsistent which is impossible since we assumed that PA is ω-consistent. In this way we proved that PA $\nvdash \chi$.

Now we prove that PA $\nvdash \neg\chi$. It was shown above that

$$\text{PA} \nvdash \chi.$$

Hence no natural number is a Gödel number of a proof of χ in PA. Consequently for any number n the relation $\neg R(m,n)$ holds. By strong representability of R we get PA $\vdash \neg\varphi(\overline{m},\overline{n})$ for any n. By ω-consistency of PA we obtain PA $\nvdash \exists y \neg\neg\varphi(\overline{m},y)$, hence PA $\nvdash \neg\forall y \neg\varphi(\overline{m},y)$, i.e., PA $\nvdash \neg\chi$. Q.E.D.

Note that for the proof that PA $\nvdash \chi$ the assumption of consistency of PA suffices. The stronger assumption of the ω-consistency of PA was needed only in the proof that PA $\nvdash \neg\chi$. Observe also that since PA $\nvdash \chi$, we can conclude that sentence χ (stating that χ is not provable in PA) is true (in the standard model $\mathfrak{N}_0$).

Gödel's sentence χ is of the form $\forall y \varphi_1(y)$ where $\varphi_1(y)$ is $\neg\varphi(\overline{m},y)$. From the fact that PA $\nvdash \chi$ and the fact that formula φ strongly represents the relation R in PA, it follows that for any given natural number n, PA $\vdash \neg\varphi(\overline{m},\overline{n})$, i.e., PA $\vdash \varphi_1(\overline{n})$. Hence every particular instance of the undecidable universal sentence χ of the form $\forall y \varphi_1(y)$ is provable in PA, hence decidable in PA.

Gödel's First Incompleteness Theorem shows that Peano arithmetic PA is incomplete provided it is ω-consistent. Rosser (1936) extended this theorem. He constructed an appropriate sentence undecidable in Peano arithmetic PA provided PA is consistent. So Rosser proved the incompleteness of PA under the weaker assumption of its consistency.

Let P be the following binary relation

$$P(a,b) \equiv Form(a) \wedge Fr(a,2) \wedge Prf(b, Sub(\langle 3,a\rangle, 2, Num(a))).$$

Now $P(a,b)$ holds if and only if a is the Gödel number of a formula ϑ with one free variable x and b is the Gödel number of a proof of a formula obtained by substituting for x in $\neg\vartheta$ the numeral corresponding to the number a, i.e., to the Gödel number of the formula ϑ.

Let $\varphi(x, y)$ be a formula of the language L(PA) of Peano arithmetic strongly representing the relation R in PA (see the construction of Gödel's sentence χ above) and let $\psi(x, y)$ be a formula of L(PA) strongly representing the relation P in PA. Consider the following formula $\chi(x)$:

$$\forall y[\varphi(x, y) \to \exists z(z \leqslant y \wedge \psi(x, z))]. \tag{2.25}$$

Let $n = \ulcorner \chi(x) \urcorner$ and let λ be the following sentence

$$\forall y[\varphi(\overline{n}, y) \to \exists z(z \leqslant y \wedge \psi(\overline{n}, z))]. \tag{2.26}$$

This is the Rosser's sentence. Observe that for any natural number y:

$R(n, y)$ if and only if y is the Gödel number of a proof in PA of the formula (2.26),

$P(n, y)$ if and only if y is the Gödel number of a proof in PA of the formula $\neg$(2.26).

Hence Rosser's sentence λ states that if there exists a proof (in PA) of the formula λ then there exists also a proof (with even a smaller or equal Gödel number) of the formula $\neg\lambda$, hence it says about itself that if it is a theorem of the arithmetic PA then its negation can also be proved in PA.

Theorem 2.4.4 (Rosser, 1936) *If Peano arithmetic* PA *is consistent then the sentence λ is undecidable in* PA, *i.e.* PA $\nvdash \lambda$ *and* PA $\nvdash \neg\lambda$. *Consequently* PA *is incomplete.*

Proof. First we prove that PA $\nvdash \lambda$. So suppose that PA $\vdash \lambda$. It means that

$$\text{PA} \vdash \forall y[\varphi(\overline{n}, y) \to \exists z(z \leqslant y \wedge \psi(\overline{n}, z))]. \tag{2.27}$$

Let k be a Gödel number of a proof of λ in PA. Hence $R(n, k)$. So PA $\vdash \varphi(\overline{n}, \overline{k})$. But by (2.27)

$$\text{PA} \vdash \varphi(\overline{n}, \overline{k}) \to \exists z(z \leqslant \overline{k} \wedge \psi(\overline{n}, z)).$$

Hence

$$\mathrm{PA} \vdash \exists z(z \leqslant \overline{k} \wedge \psi(\overline{n}, z)). \tag{2.28}$$

By the assumption Peano arithmetic PA is consistent. So the formula $\neg\lambda$ is not provable in PA. Hence $\neg P(n, y)$ for any natural number y. But the formula ψ strongly represents P in PA, so for any natural number j, $\mathrm{PA} \vdash \neg\psi(\overline{n}, \overline{j})$. Hence in particular

$$\mathrm{PA} \vdash \neg\psi(\overline{n}, 0) \wedge \neg\psi(\overline{n}, \overline{1}) \wedge \ldots \wedge \neg\psi(\overline{n}, \overline{k}).$$

But by Lemma 2.2.10 we have

$$\mathrm{PA} \vdash \neg\psi(\overline{n}, 0) \wedge \neg\psi(\overline{n}, \overline{1}) \wedge \ldots \wedge \neg\psi(\overline{n}, \overline{k}) \longrightarrow \\ \longrightarrow \forall z(z \leqslant \overline{k} \rightarrow \neg\psi(\overline{n}, z)).$$

Consequently

$$\mathrm{PA} \vdash \forall z(z \leqslant \overline{k} \rightarrow \neg\psi(\overline{n}, z)).$$

This together with (2.28) implies that PA is inconsistent which is impossible. So finally the assumption that $\mathrm{PA} \vdash \lambda$ is false, hence $\mathrm{PA} \nvdash \lambda$.

Now we prove that $\mathrm{PA} \nvdash \neg\lambda$. So suppose that $\mathrm{PA} \vdash \neg\lambda$, i.e.,

$$\mathrm{PA} \vdash \neg\forall y[\varphi(\overline{n}, y) \rightarrow \exists z(z \leqslant y \wedge \psi(\overline{n}, z))].$$

Let r be a Gödel number of a proof of the formula $\neg\lambda$ in PA. Hence $P(n, r)$. So $\mathrm{PA} \vdash \psi(\overline{n}, \overline{r})$. By the assumption that PA is consistent we get $\mathrm{PA} \nvdash \lambda$. Hence no natural number is a Gödel number of a proof of the sentence λ, i.e., $\neg R(n, y)$ for any y. So $\mathrm{PA} \vdash \neg\varphi(\overline{n}, \overline{j})$ for any j. Consequently

$$\mathrm{PA} \vdash \neg\varphi(\overline{n}, 0) \wedge \neg\varphi(\overline{n}, \overline{1}) \wedge \ldots \wedge \neg\varphi(\overline{n}, \overline{r}).$$

Using now Lemma 2.2.10 we obtain

$$\mathrm{PA} \vdash y \leqslant \overline{r} \rightarrow \neg\varphi(\overline{n}, y). \tag{2.29}$$

Since $\mathrm{PA} \vdash \psi(\overline{n}, \overline{r})$, we get, using laws of logic, that

$$\mathrm{PA} \vdash \overline{r} \leqslant y \rightarrow \exists z(z \leqslant y \wedge \psi(\overline{n}, z)). \tag{2.30}$$

Observe that by Lemma 2.1.5

$$\mathrm{PA} \vdash y \leqslant \overline{r} \vee \overline{r} \leqslant y. \tag{2.31}$$

By (2.29), (2.30) and (2.31) we immediately obtain

$$\mathrm{PA} \vdash \neg\varphi(\overline{n}, y) \vee \exists z(z \leqslant y \wedge \psi(\overline{n}, z)).$$

Using that, using the laws of logic and the generalization rule we get

$$\mathrm{PA} \vdash \forall y[\varphi(\overline{n}, y) \to \exists z(z \leqslant y \wedge \psi(\overline{n}, z))],$$

i.e.,

$$\mathrm{PA} \vdash \lambda.$$

But this contradicts the assumption that PA is consistent. So finally PA $\nvdash \neg\lambda$ and Rosser's sentence is undecidable in Peano arithmetic PA. Q.E.D.

Observe that, just as it was in the case of Gödel's sentence, Rosser's sentence is true (in the standard model $\mathfrak{N}_0$).

Note that both Gödel's sentence and Rosser's sentence are self-referential sentences. They were constructed by diagonalization. We shall now describe this procedure in detail and consider other self-referential sentences.

To simplify the statement of theorems let us make the following **general assumption**:

Peano arithmetic PA is consistent.

Recall that in Section 2.3 a ternary recursive function *Sub* was defined. Let now **Sub**(x, y, z, w) be a formula of the language L(PA) of Peano arithmetic strongly representing *Sub* in PA.

Lemma 2.4.5 (Diagonalization Lemma) *Let $\varphi(x)$ be a formula of the language* L(PA) *with one free variable. Then there exists a sentence ψ of* L(PA) *such that*

$$\mathrm{PA} \vdash \psi \equiv \varphi(\overline{\ulcorner\psi\urcorner}).$$

Proof. Let $Sub'(x, y, z) = Sub(x, y, Num(z))$. So Sub' is recursive. Let $\mathbf{Sub'}$ be a formula of the language L(PA) representing the function Sub' in PA and let θ be the following formula

$$\forall y[\mathbf{Sub'}(x, \overline{2}, x, y) \rightarrow \varphi(y)]$$

(recall that 2 is the Gödel number of the variable x). Let now $m = \ulcorner \theta(x) \urcorner$ and let ψ be the sentence $\theta(\overline{m})$. Then the following equivalences can be proved in PA:

$$\begin{aligned} \psi &\equiv \theta(\overline{m}) \\ &\equiv \forall y[\mathbf{Sub'}(\overline{m}, \overline{2}, \overline{m}, y) \rightarrow \varphi(y)] \\ &\equiv \forall y[\mathbf{Sub'}(\overline{\ulcorner \theta(x) \urcorner}, \overline{2}, \overline{m}, y) \rightarrow \varphi(y)] \\ &\equiv \varphi(\overline{\ulcorner \theta(\overline{m}) \urcorner}) \\ &\equiv \varphi(\overline{\ulcorner \psi \urcorner}). \end{aligned}$$

Q.E.D.

Observe that the sentence ψ whose existence is proved by Diagonalization Lemma says about itself that it has the property φ. Hence it is a self-referential sentence.

We can formulate now Gödel's First Incompleteness Theorem in full generality. Let $\mathbf{Prf}$ be a formula of the language L(PA) of Peano arithmetic with two free variables strongly representing in PA the recursive relation Prf (cf. Section 2.3 for its definition) and let $\mathbf{Pr}$ be the formula $\exists x \mathbf{Prf}(x, y)$. It can easily be seen that for any formula φ of the language L(PA), if PA $\vdash \varphi$ then PA $\vdash \mathbf{Pr}(\overline{\ulcorner \varphi \urcorner})$. The converse implication is not always true.

Theorem 2.4.6 (First Incompleteness Theorem) *Let φ_G be a sentence of the language* L(PA) *of Peano arithmetic such that* PA $\vdash \varphi_G \equiv \neg\mathbf{Pr}(\overline{\ulcorner \varphi_G \urcorner})$. *Then*

(i) PA $\nvdash \varphi_G$,

(ii) *if the formula* $\mathbf{Pr}$ *has the property that for any sentence* ψ *the following implication holds*

$$\text{if PA} \vdash \mathbf{Pr}(\overline{\ulcorner \psi \urcorner}) \text{ then PA} \vdash \psi, \tag{2.32}$$

then PA $\nvdash \neg\varphi_G$.

Proof. Suppose that PA $\vdash \varphi_G$. Then PA $\vdash \mathbf{Pr}(\overline{\ulcorner\varphi_G\urcorner})$. Hence PA $\vdash \neg\varphi_G$. But this means that PA is inconsistent which contradicts our general assumption. Consequently PA $\nvdash \varphi_G$.

To prove (ii) suppose that PA $\vdash \neg\varphi_G$. Then

$$\text{PA} \vdash \neg\neg\mathbf{Pr}(\overline{\ulcorner\varphi_G\urcorner}),$$

so PA $\vdash \mathbf{Pr}(\overline{\ulcorner\varphi_G\urcorner})$. By assumption (2.32) we get that PA $\vdash \varphi_G$ which contradicts (i). So finally PA $\nvdash \varphi_G$. Q.E.D.

Note that in the above version of Gödel's First Incompleteness Theorem we did not assume that PA is ω-consistent (cf. Theorem 2.4.3), instead the (simple) consistency of PA and implication (2.32) were assumed. Observe also that in fact we used implication (2.32) for a single particular sentence φ_G defined in the theorem. One can prove that ω-consistency is a consequence of implication (2.32).

The sentence φ_G is of the form $\neg\exists x\mathbf{Prf}(x, \overline{\ulcorner\varphi_G\urcorner})$, hence it is equivalent to a universal sentence of the form $\forall x\psi(x)$. By the construction of φ_G and the fact that PA $\nvdash \varphi_G$ it follows that for any natural number n, PA $\vdash \psi(\overline{n})$. Hence sentence φ_G is an undecidable universal sentence of the form $\forall x\psi(x)$ such that all its particular instances $\psi(\overline{n})$ are decidable in PA.

To formulate and prove Gödel's Second Incompleteness Theorem further assumptions must be made, more exactly, certain conditions must be put on the formula **Pr**, hence in fact on the formula **Prf** strongly representing the relation Prf in PA (recall that $Prf(a, b)$ if and only if a is the Gödel number of a proof of a formula with the Gödel number b). Those additional conditions are called derivability conditions and are as follows:

(D1) if PA $\vdash \varphi$ then PA $\vdash \mathbf{Pr}(\overline{\ulcorner\varphi\urcorner})$,

(D2) PA $\vdash \mathbf{Pr}(\overline{\ulcorner\varphi\urcorner}) \rightarrow \mathbf{Pr}(\overline{\ulcorner\mathbf{Pr}(\overline{\ulcorner\varphi\urcorner})\urcorner})$,

(D3) PA $\vdash \mathbf{Pr}(\overline{\ulcorner\varphi\urcorner}) \wedge \mathbf{Pr}(\overline{\ulcorner\varphi \rightarrow \psi\urcorner}) \rightarrow \mathbf{Pr}(\overline{\ulcorner\psi\urcorner})$.

Observe that condition (D1) is satisfied if the formula **Prf** strongly represents the relation Prf in PA. Note also that (D1) was used in the proof of Theorem 2.4.6 and that condition (2.32) used in the proof of (ii) of this Theorem is simply an implication converse to implication (D1). One can easily see that (D2) is a formalized version of (D1).

Let $\mathrm{Con}_{\mathrm{PA}}$ denote the formula: $\neg\mathbf{Pr}(\overline{\ulcorner 0 = \overline{1} \urcorner})$.

Theorem 2.4.7 (Second Incompleteness Theorem)

$$\mathrm{PA} \nvdash \mathrm{Con}_{\mathrm{PA}}.$$

Before we prove this theorem note that since all inconsistent sentences are equivalent in PA, it follows that the choice of the sentence '$0 = \overline{1}$' to express in the language L(PA) of Peano arithmetic the consistency of Peano arithmetic is in fact inessential.

Proof. Let φ_G be a sentence such that $\mathrm{PA} \vdash \varphi_G \equiv \neg\mathbf{Pr}(\overline{\ulcorner \varphi_G \urcorner})$. We show that $\mathrm{PA} \vdash \varphi_G \equiv \mathrm{Con}_{\mathrm{PA}}$. By Theorem 2.4.6 it will imply that $\mathrm{PA} \nvdash \mathrm{Con}_{\mathrm{PA}}$.

Observe that

$$\mathrm{PA} \vdash (0 = \overline{1}) \to \varphi_G.$$

So by (D1)

$$\mathrm{PA} \vdash \mathbf{Pr}(\overline{\ulcorner (0 = \overline{1}) \to \varphi_G \urcorner}),$$

and consequently by (D3)

$$\mathrm{PA} \vdash \mathbf{Pr}(\overline{\ulcorner 0 = \overline{1} \urcorner}) \to \mathbf{Pr}(\overline{\ulcorner \varphi_G \urcorner}).$$

Hence

$$\mathrm{PA} \vdash \neg\mathbf{Pr}(\overline{\ulcorner \varphi_G \urcorner}) \to \neg\mathbf{Pr}(\overline{\ulcorner 0 = \overline{1} \urcorner}).$$

By the definition of the sentence φ_G we have

$$\mathrm{PA} \vdash \varphi_G \to \neg\mathbf{Pr}(\overline{\ulcorner \varphi_G \urcorner}).$$

So

$$\mathrm{PA} \vdash \varphi_G \to \neg\mathbf{Pr}(\overline{\ulcorner 0 = \overline{1} \urcorner}),$$

i.e.,

$$\mathrm{PA} \vdash \varphi_G \to \mathrm{Con}_{\mathrm{PA}}.$$

To prove that $\mathrm{PA} \vdash \mathrm{Con}_{\mathrm{PA}} \to \varphi_G$ observe that by (D2)

$$\mathrm{PA} \vdash \mathbf{Pr}(\overline{\ulcorner \varphi_G \urcorner}) \to \mathbf{Pr}(\overline{\ulcorner \mathbf{Pr}(\overline{\ulcorner \varphi_G \urcorner}) \urcorner}) \qquad (2.33)$$

By the definition of φ_G we have

$$\mathrm{PA} \vdash \varphi_G \to \neg \mathbf{Pr}(\overline{\ulcorner \varphi_G \urcorner}),$$

so

$$\mathrm{PA} \vdash \mathbf{Pr}(\overline{\ulcorner \varphi_G \urcorner}) \to \neg \varphi_G.$$

Hence by (D1) and (D2) we have

$$\mathrm{PA} \vdash \mathbf{Pr}(\overline{\ulcorner \mathbf{Pr}(\overline{\ulcorner \varphi_G \urcorner}) \to \neg \varphi_G \urcorner}),$$

$$\mathrm{PA} \vdash \mathbf{Pr}(\overline{\ulcorner \mathbf{Pr}(\overline{\ulcorner \varphi_G \urcorner}) \urcorner}) \to \mathbf{Pr}(\overline{\ulcorner \neg \varphi_G \urcorner}). \qquad (2.34)$$

Consequently by (2.33) and (2.34) we have

$$\mathrm{PA} \vdash \mathbf{Pr}(\overline{\ulcorner \varphi_G \urcorner}) \to \mathbf{Pr}(\overline{\ulcorner \neg \varphi_G \urcorner}). \qquad (2.35)$$

On the other hand observe that $\mathrm{PA} \vdash \varphi_G \to (\neg \varphi_G \to \varphi_G \wedge \neg \varphi_G)$. Hence by (D1) and (D3)

$$\mathrm{PA} \vdash \mathbf{Pr}(\overline{\ulcorner \varphi_G \urcorner}) \to (\mathbf{Pr}(\overline{\ulcorner \neg \varphi_G \urcorner}) \to \mathbf{Pr}(\overline{\ulcorner \varphi_G \wedge \neg \varphi_G \urcorner})),$$

hence

$$\mathrm{PA} \vdash \mathbf{Pr}(\overline{\ulcorner \neg \varphi_G \urcorner}) \to (\mathbf{Pr}(\overline{\ulcorner \varphi_G \urcorner}) \to \mathbf{Pr}(\overline{\ulcorner \varphi_G \wedge \neg \varphi_G \urcorner})). \qquad (2.36)$$

Observe that

$$\begin{aligned} \mathrm{PA} \vdash [\mathbf{Pr}(\overline{\ulcorner \neg \varphi_G \urcorner}) \to (\mathbf{Pr}(\overline{\ulcorner \varphi_G \urcorner}) \to \mathbf{Pr}(\overline{\ulcorner \varphi_G \wedge \neg \varphi_G \urcorner}))] \to \\ \to [(\mathbf{Pr}(\overline{\ulcorner \varphi_G \urcorner}) \to \mathbf{Pr}(\overline{\ulcorner \neg \varphi_G \urcorner})) \to \\ \to (\mathbf{Pr}(\overline{\ulcorner \varphi_G \urcorner}) \to \mathbf{Pr}(\overline{\ulcorner \varphi_G \wedge \neg \varphi_G \urcorner}))]. \end{aligned}$$

Indeed, this formula can be obtained from the following tautology

$$[p \to (q \to r)] \to [(q \to p) \to (q \to r)]$$

by the following substitutions:

$$p/\mathbf{Pr}(\overline{\ulcorner\neg\varphi_G\urcorner}),\ \ q/\mathbf{Pr}(\overline{\ulcorner\varphi_G\urcorner}),\ \ r/\mathbf{Pr}(\overline{\ulcorner\varphi_G\wedge\neg\varphi_G\urcorner}).$$

By the latter formula and formulas (2.36) and (2.35) we get

$$\mathrm{PA}\vdash\mathbf{Pr}(\overline{\ulcorner\varphi_G\urcorner})\to\mathbf{Pr}(\overline{\ulcorner\varphi_G\wedge\neg\varphi_G\urcorner}). \tag{2.37}$$

But

$$\mathrm{PA}\vdash(\varphi_G\wedge\neg\varphi_G)\equiv(0=\overline{1}),$$

hence

$$\mathrm{PA}\vdash(\varphi_G\wedge\neg\varphi_G)\to(0=\overline{1})$$

and consequently by (D1) and (D3)

$$\mathrm{PA}\vdash\mathbf{Pr}(\overline{\ulcorner\varphi_G\wedge\neg\varphi_G\urcorner})\to\mathbf{Pr}(\overline{\ulcorner 0=\overline{1}\urcorner}).$$

This together with (2.37) gives

$$\mathrm{PA}\vdash\mathbf{Pr}(\overline{\ulcorner\varphi_G\urcorner})\to\mathbf{Pr}(\overline{\ulcorner 0=\overline{1}\urcorner}),$$

i.e.,

$$\mathrm{PA}\vdash\neg\mathbf{Pr}(\overline{\ulcorner 0=\overline{1}\urcorner})\to\neg\mathbf{Pr}(\overline{\ulcorner\varphi_G\urcorner}).$$

Using now the definitions of the sentence φ_G and of the formula $\mathrm{Con}_{\mathrm{PA}}$ we get

$$\mathrm{PA}\vdash\mathrm{Con}_{\mathrm{PA}}\to\varphi_G.$$

This is the required implication.

In this way we have shown that $\mathrm{PA}\vdash\varphi_G\to\mathrm{Con}_{\mathrm{PA}}$ and $\mathrm{PA}\vdash\mathrm{Con}_{\mathrm{PA}}\to\varphi_G$. So finally $\mathrm{PA}\vdash\varphi_G\equiv\mathrm{Con}_{\mathrm{PA}}$. Q.E.D.

Note that from the given proof of Gödel's Second Incompleteness Theorem it follows that a sentence stating its own unprovability in PA is equivalent in PA to a sentence $\mathrm{Con}_{\mathrm{PA}}$ stating that PA is consistent (the equivalence holds under the assumption that the conditions (D1)–(D3) are satisfied). Hence for any two sentences φ and ψ such that

$$\mathrm{PA}\vdash\varphi\equiv\neg\mathbf{Pr}(\overline{\ulcorner\varphi\urcorner})\qquad\text{and}\qquad\mathrm{PA}\vdash\psi\equiv\neg\mathbf{Pr}(\overline{\ulcorner\psi\urcorner})$$

we have

$$\mathrm{PA} \vdash \varphi \equiv \psi.$$

By consequence, any two Gödel's sentences (undecidable in PA) are provably equivalent in PA.

Note that conditions (D1)–(D3) were essential in the proof of Gödel's Second Incompleteness Theorem. They are conditions put on the formula $\mathrm{Con}_{\mathrm{PA}}$, i.e., on the formula of the language L(PA) of Peano arithmetic stating that Peano arithmetic PA is consistent. In fact the truth of Theorem 2.4.7 depends on the way in which the sentence "Peano arithmetic is consistent" is formalized in the language L(PA). In other words: the property that Peano arithmetic does not prove its own consistency depends on the way in which the indicated metamathematical sentence has been translated into the language L(PA). Indeed, one can give examples of formulas of L(PA), stating that Peano arithmetic is consistent, for which Gödel's Second Incompleteness Theorem is not true. We show two of them. The first one was given by Mostowski (1965, p. 24).

Let $n_0 = \ulcorner 0 = \overline{1} \urcorner$ and let $\mathbf{Prf}'(x, y)$ denote the formula

$$\mathbf{Prf}(x, y) \wedge \neg \mathbf{Prf}(x, n_0)$$

where **Prf** is any formula of the language L(PA) strongly representing the relation Prf in PA. It can be easily proved that for any natural numbers m and n

$$\mathrm{PA} \vdash \mathbf{Prf}'(\overline{m}, \overline{n}) \equiv \mathbf{Prf}(\overline{m}, \overline{n}).$$

Hence, in PA the formula $\mathbf{Prf}'$ strongly represents the relation Prf. Let $\mathrm{Con}'_{\mathrm{PA}}$ denote the following formula

$$\neg \exists x \mathbf{Prf}'(x, \overline{n_0}),$$

i.e., the formula

$$\neg \exists x [\mathbf{Prf}(x, \overline{\ulcorner 0 = \overline{1} \urcorner}) \wedge \neg \mathbf{Prf}(x, \overline{\ulcorner 0 = \overline{1} \urcorner})].$$

By the law of contradiction one gets

$$\mathrm{PA} \vdash \mathrm{Con}'_{\mathrm{PA}}.$$

Hence $\mathrm{Con}'_{\mathrm{PA}}$ is an example of a formula provable in PA and stating formally that Peano arithmetic PA is consistent. It implies that formula $\mathbf{Pr}'$ of the form $\exists x \mathbf{Prf}'(x, y)$ does not satisfy conditions (D1)–(D2).

Let us give one more example. Let now $\mathbf{Prf}$ be a formula strongly representing the relation Prf in PA and let $\mathbf{Prf}^R$ denote the following formula

$$\mathbf{Prf}(x, y) \wedge \forall z \leqslant x \forall w [\mathbf{Prf}(z, w) \rightarrow \neg \mathbf{Neg}(w, y) \wedge \neg \mathbf{Neg}(y, w)]$$

where $\mathbf{Neg}$ is a formula strongly representing the following recursive relation Neg in PA:

$$Neg(\ulcorner \varphi \urcorner, \ulcorner \psi \urcorner) \equiv (\varphi \text{ is of the form } \neg \psi).$$

The formula $\mathbf{Prf}^R(x, y)$ says that x is (the Gödel number of) a proof of a formula (with the Gödel number) y and that there is no proof (with a Gödel number) smaller than x of the negation of the formula (with the Gödel number) y. Finally let $\mathbf{Pr}^R(y)$ be the formula $\exists x \mathbf{Prf}^R(x, y)$ and $\mathrm{Con}^R_{\mathrm{PA}}$ be the formula $\neg \mathbf{Pr}^R(\overline{\ulcorner 0 = \overline{1} \urcorner})$. Observe that for any φ

$$\mathrm{PA} \vdash \neg[\mathbf{Pr}^R(\overline{\ulcorner \varphi \urcorner}) \wedge \mathbf{Pr}^R(\overline{\ulcorner \neg \varphi \urcorner})].$$

If φ is the sentence $\neg(0 = \overline{1})$ then we get

$$\mathrm{PA} \vdash \mathbf{Pr}^R(\overline{\ulcorner \neg(0 = \overline{1}) \urcorner}) \rightarrow \neg \mathbf{Pr}^R(\overline{\ulcorner 0 = \overline{1} \urcorner}).$$

But $\mathrm{PA} \vdash \neg(0 = \overline{1})$ (cf. axiom (A2)). Hence

$$\mathrm{PA} \vdash \mathbf{Pr}^R(\overline{\ulcorner \neg(0 = \overline{1}) \urcorner}).$$

So

$$\mathrm{PA} \vdash \neg \mathbf{Pr}^R(\overline{\ulcorner 0 = \overline{1} \urcorner}),$$

i.e.,

$$\mathrm{PA} \vdash \mathrm{Con}^R_{\mathrm{PA}}.$$

And again it is not the case that conditions (D1)–(D3) are all satisfied by the formula $\mathbf{Pr}^R$. Indeed, Jeroslov (1973) showed that $\mathbf{Pr}^R$ does not satisfy condition (D2).

Note that there is a connection between the formula $\mathbf{Prf}^R$ (and consequently $\mathbf{Pr}^R$ as well) and the Rosser's sentences considered above. Indeed, one can prove the following

Theorem 2.4.8 *Let φ_R be a sentence such that*

$$\mathrm{PA} \vdash \varphi_R \equiv \neg\mathbf{Pr}^R(\overline{\ulcorner\varphi_R\urcorner}).$$

Then

(i) $\mathrm{PA} \nvdash \varphi_R$,
(ii) $\mathrm{PA} \nvdash \neg\varphi_R$.

Proof. It can be easily seen that formulas $\mathbf{Prf}$ and $\mathbf{Prf}^R$ strongly represent the same relation in PA. Hence (D1) holds for $\mathbf{Prf}^R$, i.e., if $\mathrm{PA} \vdash \varphi$ then $\mathrm{PA} \vdash \mathbf{Pr}^R(\overline{\ulcorner\varphi\urcorner})$ for any sentence φ. Now just as in Theorem 2.4.6 we show that $\mathrm{PA} \nvdash \varphi_R$.

To prove (ii) suppose that $\mathrm{PA} \vdash \neg\varphi_R$. Hence there is a proof in PA of the sentence $\neg\varphi_R$. Let d be its Gödel number. So $Prf(d, \ulcorner\neg\varphi_R\urcorner)$ and consequently $\mathrm{PA} \vdash \mathbf{Prf}^R(\overline{d}, \overline{\ulcorner\neg\varphi_R\urcorner})$. On the other hand, by the definition of φ_R one gets

$$\mathrm{PA} \vdash \mathbf{Pr}^R(\overline{\ulcorner\varphi_R\urcorner}),$$

i.e.,

$$\begin{aligned}\mathrm{PA} \vdash \exists x[&\mathbf{Prf}(x, \overline{\ulcorner\varphi_R\urcorner})\wedge \\ &\wedge \forall z \leqslant x \forall w(\mathbf{Prf}(z,w) \to \\ &\quad \to \neg\mathbf{Neg}(w, \overline{\ulcorner\varphi_R\urcorner}) \wedge \neg\mathbf{Neg}(\overline{\ulcorner\varphi_R\urcorner}, w))].\end{aligned}$$

To simplify the notation let $A(x)$ denote the subformula of the above formula put in square brackets. Then

$$\begin{aligned}\mathrm{PA} &\vdash \exists x[(x \leqslant \overline{d} \vee x > \overline{d}) \wedge A(x)], \\ \mathrm{PA} &\vdash \exists x[(x \leqslant \overline{d} \wedge A(x)) \vee (x > \overline{d} \wedge A(x))], \\ \mathrm{PA} &\vdash \exists x[x \leqslant \overline{d} \wedge A(x)] \vee \exists x[x > \overline{d} \wedge A(x)].\end{aligned}$$

The second part of the disjuction is inconsistent since

$$\mathrm{PA} \vdash \mathbf{Prf}^R(\overline{d}, \overline{\ulcorner\neg\varphi_R\urcorner}).$$

So PA $\vdash \exists x(x \leqslant \overline{d} \wedge A(x))$. But

$$\mathrm{PA} \vdash x \leqslant \overline{d} \equiv x = 0 \vee \ldots \vee x = \overline{d}.$$

So one gets

$$\mathrm{PA} \vdash \mathbf{Prf}(0, \overline{\ulcorner \varphi_R \urcorner}) \vee \ldots \vee \mathbf{Prf}(\overline{d}, \overline{\ulcorner \varphi_R \urcorner}).$$

But this contradicts the fact that PA $\vdash \mathbf{Prf}^R(\overline{d}, \overline{\ulcorner \neg\varphi_R \urcorner})$. So finally PA $\nvdash \neg\varphi_R$. Q.E.D.

Above, we wrote (cf. the comment after the proof of Theorem 2.4.7) that any two Gödel sentences stating its own unprovability in PA are provably equivalent in PA. A similar question would be if the same holds for Rosser's sentences, i.e., for sentences stating its own unprovability in PA in the sense of the predicate $\mathbf{Pr}^R$.

Consider two further theorems in the proofs of which the Diagonalization Lemma plays an essential role. We mean here Löb's Theorem on a sentence stating its own provability and Tarski's Theorem on the undefinability of truth.

In 1952 Henkin asked if a sentence φ_0 such that

$$\mathrm{PA} \vdash \varphi_0 \equiv \mathbf{Pr}(\overline{\ulcorner \varphi_0 \urcorner})$$

is a theorem of Peano arithmetic PA. The problem was solved by Löb (1955) who proved a general theorem from which the affirmative answer to Henkin's question follows.

Theorem 2.4.9 (Löb) *For any sentence* φ *of* L(PA)

$$\mathrm{PA} \vdash \mathbf{Pr}(\overline{\ulcorner \varphi \urcorner}) \rightarrow \varphi \quad \textit{if and only if} \quad \mathrm{PA} \vdash \varphi.$$

Corollary 2.4.10 *Let* φ_0 *be Henkin's sentence, i.e., a sentence such that*

$$\mathrm{PA} \vdash \varphi_0 \equiv \mathbf{Pr}(\overline{\ulcorner \varphi_0 \urcorner}).$$

Then PA $\vdash \varphi_0$.

Two proofs of Löb's Theorem will be given: the first one comes from Löb, the author of the second is unknown.

Proof 1. The implication

$$\text{if} \quad PA \vdash \varphi \quad \text{then} \quad PA \vdash \mathbf{Pr}(\overline{\ulcorner\varphi\urcorner}) \to \varphi$$

is obvious. Hence we prove only the converse implication. So let

$$PA \vdash \mathbf{Pr}(\overline{\ulcorner\varphi\urcorner}) \to \varphi.$$

According to Diagonalization Lemma there exists a sentence ψ such that

$$PA \vdash [\psi \equiv (\mathbf{Pr}(\overline{\ulcorner\psi\urcorner}) \to \varphi)].$$

This together with (D1) and (D3) gives

$$\begin{aligned} PA &\vdash \mathbf{Pr}(\overline{\ulcorner\psi\urcorner}) \equiv \mathbf{Pr}(\overline{\ulcorner \mathbf{Pr}(\overline{\ulcorner\psi\urcorner}) \to \varphi\urcorner}), \\ PA &\vdash \mathbf{Pr}(\overline{\ulcorner\psi\urcorner}) \to [\mathbf{Pr}(\overline{\ulcorner \mathbf{Pr}(\overline{\ulcorner\psi\urcorner})\urcorner}) \to \mathbf{Pr}(\overline{\ulcorner\varphi\urcorner})]. \end{aligned}$$

By (D2)

$$PA \vdash \mathbf{Pr}(\overline{\ulcorner\psi\urcorner}) \to \mathbf{Pr}(\overline{\ulcorner \mathbf{Pr}(\overline{\ulcorner\psi\urcorner})\urcorner}).$$

So using Frege's law, i.e., the following tautology

$$(p \to (q \to r)) \to [(p \to q) \to (p \to r)],$$

one gets

$$PA \vdash \mathbf{Pr}(\overline{\ulcorner\psi\urcorner}) \to \mathbf{Pr}(\overline{\ulcorner\varphi\urcorner}).$$

By the assumption one obtains from this that

$$PA \vdash \mathbf{Pr}(\overline{\ulcorner\psi\urcorner}) \to \varphi.$$

Hence by the definition of the sentence ψ one has $PA \vdash \psi$. So by (D1) one gets

$$PA \vdash \mathbf{Pr}(\overline{\ulcorner\psi\urcorner})$$

and consequently

$$PA \vdash \varphi.$$

Proof 2. It suffices to prove the implication to the right. So assume that $PA \vdash \mathbf{Pr}(\overline{\ulcorner\varphi\urcorner}) \to \varphi$ and suppose that $PA \nvdash \varphi$. Then the theory $PA \cup \{\neg\varphi\}$ is consistent. Applying to this theory Gödel's Second Incompleteness Theorem one gets

$$PA \cup \{\neg\varphi\} \nvdash Con_{PA \cup \{\neg\varphi\}}.$$

But φ is a sentence, hence through the Deduction Theorem and the definition of the formula Con one obtains

$$PA \cup \{\neg\varphi\} \nvdash \neg\mathbf{Pr}(\overline{\ulcorner\neg\varphi \to (0 = \overline{1})\urcorner}).$$

By the law of contraposition

$$PA \vdash (\neg\varphi \to 0 = \overline{1}) \to (0 \neq \overline{1} \to \varphi).$$

Applying now conditions (D1)–(D3) one gets

$$\begin{aligned} PA \cup \{\neg\varphi\} &\nvdash \neg\mathbf{Pr}(\overline{\ulcorner 0 \neq \overline{1} \to \varphi\urcorner}), \\ PA \cup \{\neg\varphi\} &\nvdash \neg[\mathbf{Pr}(\overline{\ulcorner 0 \neq \overline{1}\urcorner}) \to \mathbf{Pr}(\overline{\ulcorner\varphi\urcorner})], \\ PA \cup \{\neg\varphi\} &\nvdash \mathbf{Pr}(\overline{\ulcorner 0 \neq \overline{1}\urcorner}) \wedge \neg\mathbf{Pr}(\overline{\ulcorner\varphi\urcorner}). \end{aligned}$$

On the other hand $PA \vdash \neg(0 = \overline{1})$, so

$$PA \vdash \mathbf{Pr}(\overline{\ulcorner 0 \neq \overline{1}\urcorner}) \quad \text{and} \quad PA \cup \{\neg\varphi\} \vdash \mathbf{Pr}(\overline{\ulcorner 0 \neq \overline{1}\urcorner}).$$

Hence

$$PA \cup \{\neg\varphi\} \nvdash \neg\mathbf{Pr}(\overline{\ulcorner\varphi\urcorner}).$$

So according to Deduction Theorem

$$PA \nvdash \neg\varphi \to \neg\mathbf{Pr}(\overline{\ulcorner\varphi\urcorner}),$$

i.e.,

$$PA \nvdash \mathbf{Pr}(\overline{\ulcorner\varphi\urcorner}) \to \varphi$$

contrary to the assumption. In this way we showed that $PA \vdash \varphi$. Q.E.D.

By Löb's Theorem any two sentences stating their own provability in PA are equivalent in PA, i.e., if φ and ψ are two sentences such that

$$\mathrm{PA} \vdash \varphi \equiv \mathbf{Pr}(\overline{\ulcorner \varphi \urcorner}) \qquad \text{and} \qquad \mathrm{PA} \vdash \psi \equiv \mathbf{Pr}(\overline{\ulcorner \psi \urcorner})$$

then

$$\mathrm{PA} \vdash \varphi \equiv \psi.$$

Indeed, for such sentences φ and ψ we have $\mathrm{PA} \vdash \varphi$ and $\mathrm{PA} \vdash \psi$ and apply now the fact that any two theorems of PA are provably equivalent in PA.

Let us turn now to the second application of the Diagonalization Lemma, i.e., to Tarski's theorem on the undefinability of truth.

Theorem 2.4.11 (Tarski, 1933) *There exists no formula* $\mathrm{St}(x)$ *of the language* L(PA) *of Peano arithmetic being the definition of truth, i.e., such a formula* $\mathrm{St}(x)$ *that for any sentence* ψ *of the language* L(PA)

$$\mathrm{PA} \vdash \psi \equiv \mathrm{St}(\overline{\ulcorner \psi \urcorner}).$$

Proof. Suppose that such a formula St does exist. By the Diagonalization Lemma there is a sentence φ_0 such that

$$\mathrm{PA} \vdash \varphi_0 \equiv \neg \mathrm{St}(\overline{\ulcorner \varphi_0 \urcorner}).$$

This together with the supposed property of the formula St gives

$$\mathrm{PA} \vdash \mathrm{St}(\overline{\ulcorner \varphi_0 \urcorner}) \equiv \neg \mathrm{St}(\overline{\ulcorner \varphi_0 \urcorner}).$$

Hence Peano arithmetic is inconsistent contrary to our general assumption. Consequently a formula St does not exist. Q.E.D.

The full sense and meaning of Tarski's Theorem as well as the justification for its name can be seen by considering a semantical interpretation of this theorem. So let $\mathfrak{N}_0$ be the standard interpretation of the language of Peano arithmetic, i.e.,

$\mathfrak{N}_0 = \langle \mathbb{N}, 0, S, +, \cdot \rangle$ where $\mathbb{N}$ is the set of natural numbers, 0 is the number zero, S is the successor function and $+$ and $\cdot$ are the addition and multiplication of natural numbers. All axioms of Peano arithmetic are satisfied in $\mathfrak{N}_0$ — the structure $\mathfrak{N}_0$ is called the standard model of PA. Tarski's Theorem states that there exists no formula St of the language L(PA) such that for any sentence ψ of L(PA), PA $\vdash \psi \equiv \mathrm{St}(\overline{\ulcorner\psi\urcorner})$, so in particular there exists no formula St such that for any sentence ψ of L(PA), $\mathfrak{N}_0 \models \psi$ if and only if $\mathfrak{N}_0 \models \mathrm{St}(\overline{\ulcorner\psi\urcorner})$, i.e., there is no definition (in the language of L(PA)) of the set of (Gödel numbers of) those sentences of L(PA) which are true in the domain of natural numbers (= in the standard model $\mathfrak{N}_0$).

Tarski's Theorem can be easily generalized to theories extending Peano arithmetic PA. In fact the following theorem holds.

Theorem 2.4.12 *Let T be any consistent first-order theory extending Peano arithmetic* PA *and let* $\mathfrak{M}$ *be any model of T. Then the set* $Th(\mathfrak{M}) = \{\ulcorner\psi\urcorner : \mathfrak{M} \models \psi\}$, *i.e., the set of Gödel numbers of all sentences true in* $\mathfrak{M}$, *is not definable in* $\mathfrak{M}$.

Proof. Suppose that there exists a formula Φ of the language L(T) of T such that Φ defines $Th(\mathfrak{M})$, i.e., such that for any sentence ψ of L(T), $\mathfrak{M} \models \psi \equiv \mathfrak{M} \models \Phi(\overline{\ulcorner\psi\urcorner})$. By the Diagonalization Lemma there exists a sentence φ_0 such that T $\vdash \varphi_0 \equiv \neg\Phi(\overline{\ulcorner\varphi_0\urcorner})$. Hence

$$\mathfrak{M} \models \varphi_0 \equiv \mathfrak{M} \models \neg\Phi(\overline{\ulcorner\varphi_0\urcorner}).$$

Consequently

$$\mathfrak{M} \models \Phi(\overline{\ulcorner\varphi_0\urcorner}) \equiv \mathfrak{M} \models \neg\Phi(\overline{\ulcorner\varphi_0\urcorner})$$

which is impossible. Hence there is no definition of the set $Th(\mathfrak{M})$. Q.E.D.

Observe that the language L(T) of the theory T from Theorem 2.4.12 can be a proper extension of the language L(PA) of

Peano arithmetic PA — in this case one uses a straightforward generalization of Diagonalization Lemma 2.4.5 for T in the proof.

In the literature one can find many other self-referential statments, i.e., sentences obtained by the Diagonalization Lemma. The reader can consult e.g., the survey paper by Smoryński (1981). We shall not give any details here. Instead we consider the (un)decidability of sentences stating their own (un)decidability or (in)consistency. So let **Prf** be a formula of the language L(PA) of Peano arithmetic strongly representing in PA the relation Prf and such that the formula **Pr** of the form $\exists x\mathbf{Prf}(x, a)$ satisfies the derivability conditions (D1)–(D3) as well as the converse of condition (D1). Then the following theorem holds.

Theorem 2.4.13 *Assume that Peano arithmetic* PA *is consistent.*

(i) *Let* φ *be a sentence such that*

$$\mathrm{PA} \vdash \varphi \equiv [\neg\mathbf{Pr}(\overline{\ulcorner\varphi\urcorner}) \wedge \neg\mathbf{Pr}(\overline{\ulcorner\neg\varphi\urcorner})],$$

i.e., a sentence stating its own undecidability. Then $\mathrm{PA} \vdash \neg\varphi$ *(and consequently* $\mathrm{PA} \nvdash \varphi$*).*

(ii) *Let* φ *be a sentence such that*

$$\mathrm{PA} \vdash \varphi \equiv [\mathbf{Pr}(\overline{\ulcorner\varphi\urcorner}) \vee \mathbf{Pr}(\overline{\ulcorner\neg\varphi\urcorner})],$$

i.e., a sentence stating its own decidability. Then $\mathrm{PA} \vdash \varphi$.

(iii) *Let* φ *be a sentence such that*

$$\mathrm{PA} \vdash \varphi \equiv \neg\mathbf{Pr}(\overline{\ulcorner\neg\varphi\urcorner}),$$

i.e., a sentence stating its own consistency with PA*. Then* $\mathrm{PA} \vdash \neg\varphi$ *(and consequently* $\mathrm{PA} \nvdash \varphi$*).*

(iv) *Let* φ *be a sentence such that*

$$\mathrm{PA} \vdash \varphi \equiv \mathbf{Pr}(\overline{\ulcorner\neg\varphi\urcorner}),$$

i.e., a sentence stating its own inconsistency with PA*. Then* φ *is undecidable in* PA*, i.e.,* $\mathrm{PA} \nvdash \varphi$ *and* $\mathrm{PA} \nvdash \neg\varphi$.

Proof. (i) By the definition of φ we have

$$\mathrm{PA} \vdash \varphi \to \neg \mathbf{Pr}(\overline{\ulcorner \neg\varphi \urcorner})$$

and consequently

$$\mathrm{PA} \vdash \mathbf{Pr}(\overline{\ulcorner \neg\varphi \urcorner}) \to \neg\varphi.$$

By Löb's Theorem 2.4.9 we get $\mathrm{PA} \vdash \neg\varphi$.

(ii) is a simple consequence of (i).

(iii) By the definition of φ we have $\mathrm{PA} \vdash \neg\varphi \equiv \mathbf{Pr}(\overline{\ulcorner \neg\varphi \urcorner})$, hence $\neg\varphi$ is the Henkin's sentence. By Corollary 2.4.10 we get $\mathrm{PA} \vdash \neg\varphi$.

(iv) Observe that by the definition of φ, the negation $\neg\varphi$ has properties of the Gödel's sentence. Hence by Theorem 2.4.6 we have $\mathrm{PA} \nvdash \varphi$ and $\mathrm{PA} \nvdash \neg\varphi$. Q.E.D.

Theorem 2.4.13 states that a sentence stating its own undecidability is inconsistent (with PA) and a sentence stating its own inconsistency is undecidable (in PA).

Gödel's theorems (and their variants) proved in this section showed the existence of sentences undecidable in Peano arithmetic PA. Hence they showed that PA is incomplete. [Recall that a theory T is complete if and only if for any sentence φ of the language of T, either $\mathrm{T} \vdash \varphi$ or $\mathrm{T} \vdash \neg\varphi$.] Note that all those theorems remain true in the case of any theory T extending Peano arithmetic PA by a recursive set of axioms and such that the set of (Gödel numbers of) nonlogical symbols of the language of T is recursive — such theories will be called 'recursive extensions of Peano arithmetic'. Indeed, the main facts used in proofs of the considered theorems were the recursiveness of the relation Prf and the strong representability of all recursive relations in PA. Hence the following theorem holds.

Theorem 2.4.14 *If Peano arithmetic* PA *is consistent then it is essentially incomplete, i.e.,* PA *and every consistent recursive extension of* PA *are incomplete.*

The assumption that the set of axioms of a theory T is recursive is natural — it means simply that one can effectively distinguish axioms and theorems which need to be proved. But it can be weakened and Gödel's theorems still remain true. In fact by a slight modification of the given proofs one can obtain the following

Theorem 2.4.15 *Any arithmetically definable (not necessarily recursive) extension T of* PA *such that* $\mathfrak{N}_0 \models \mathrm{T}$ *is undecidable.*

Explain that a theory T is an arithmetically definable extension of PA if and only if T is an extension of PA and the set of (Gödel numbers of) axioms of T is definable in the standard model $\mathfrak{N}_0$ by a formula (of any complexity) of the language L(PA) of Peano arithmetic.

Theories in which all recursive relations, and consequently all recursive functions, are representable possess another interesting property — one can show that for such a theory a set of (Gödel numbers of) its theorems and a set of (Gödel numbers of) negations of its theorems are recursively inseparable. Before we prove this let us formulate some definitions.

Let $Prf_{\mathrm{T}}(a,b)$ holds if and only if a is the Gödel number of a proof in T of a formula with the Gödel number b. This relation can be defined in a similar way as the relation Prf for Peano arithmetic (cf. Section 2.3).

Definition 2.4.16

$$\begin{aligned} a \in Thm_{\mathrm{T}} &\equiv \exists x Prf_{\mathrm{T}}(x,a), \\ a \in neg_{\mathrm{T}} &\equiv Form_{\mathrm{T}}(a) \wedge \langle SN(\neg), a\rangle \in Thm_{\mathrm{T}}. \end{aligned}$$

Hence Thm_{T} is the set of Gödel numbers of theorems of T and neg_{T} is the set of Gödel numbers of formulas whose negations are theorems of T, i.e., the set of negations of theorems of T.

Definition 2.4.17 *A set X is said to be recursively reducible to a set Y if and only if there exists a primitive recursive function f such that $X = f^{-1}(Y)$, i.e., for any n, $n \in X \equiv f(n) \in Y$.*

Definition 2.4.18 *A set Y is said to be universal for a class $\mathcal{X}$ of sets (or: $\mathcal{X}$-complete) if and only if every set $X \in \mathcal{X}$ is recursively reducible to the set Y.*

Lemma 2.4.19 *There does not exist a recursive set universal for the class of all recursive sets.*

Proof. Suppose that such a set exists and denote it by Y. Define a set X_0 as follows

$$n \in X_0 \equiv G(n,n) \notin Y$$

where $G(n,k)$ is a recursive function universal for the class $\mathcal{P}_1$ of unary primitive recursive functions (the existence of such a function was proved in fact in Theorem 1.4.15). Of course the set X_0 is recursive. Hence there exists a function $f \in \mathcal{P}_1$ such that $X_0 = f^{-1}(Y)$, i.e.,

$$n \in X_0 \equiv f(n) \in Y.$$

But by the universality of the function G for $\mathcal{P}_1$ there exists a number l_0 such that for any x, $f(x) = G(l_0, x)$. Hence for any n:

$$n \in X_0 \equiv G(l_0, n) \in Y.$$

So for $n = l_0$:

$$G(l_0, l_0) \notin Y \equiv l_0 \in X_0 \equiv G(l_0, l_0) \in Y$$

which is impossible. Consequently there exists no recursive set universal for the class of all recursive sets. Q.E.D

Theorem 2.4.20 *If all recursive sets are strongly representable in a theory* T *then the sets Thm_{T} and neg_{T} are not recursively separable, i.e., there exists no recursive set X such that*

$$Thm_{\mathrm{T}} \subseteq X \text{ and } X \cap neg_{\mathrm{T}} = \emptyset$$

(or equivalently $neg_{\mathrm{T}} \subseteq \mathbb{N} - X$).

Proof. By the assumption of the theorem, for any recursive set X there exists a formula $\varphi(x)$ of the language L(T) of the theory T such that for any n:

$$\begin{aligned} n \in X &\equiv \mathrm{T} \vdash \varphi(\overline{n}), \\ n \notin X &\equiv \mathrm{T} \vdash \neg\varphi(\overline{n}). \end{aligned}$$

Hence

$$\begin{aligned} n \in X &\equiv \ulcorner\varphi(\overline{n})\urcorner \in Thm_{\mathrm{T}}, \\ n \notin X &\equiv \ulcorner\varphi(\overline{n})\urcorner \in neg_{\mathrm{T}}. \end{aligned}$$

So

$$\begin{aligned} n \in X &\equiv Sub(\ulcorner\varphi\urcorner, \ulcorner x\urcorner, Num(n)) \in Thm_{\mathrm{T}}, \\ n \notin X &\equiv Sub(\ulcorner\varphi\urcorner, \ulcorner x\urcorner, Num(n)) \in neg_{\mathrm{T}}. \end{aligned}$$

Let $f(n) = Sub(\ulcorner\varphi\urcorner, \ulcorner x\urcorner, Num(n))$ for $n \in \mathbb{N}$. The function f is primitive recursive, hence $f \in \mathcal{P}_1$. So

$$n \in X \equiv f(n) \in Thm_{\mathrm{T}} \quad \text{and} \quad n \notin X \equiv f(n) \in neg_{\mathrm{T}}, \tag{2.38}$$

which means that X is recursively reducible (by the function f) to the set Thm_{T} and the complement of X is recursively reducible to the set neg_{T}.

Suppose now that there exists a recursive set Y separating Thm_{T} and neg_{T}, i.e., such that $Thm_{\mathrm{T}} \subseteq Y$ and $neg_{\mathrm{T}} \subseteq \mathbb{N} - Y$. Then by (2.38) for any recursive set X it would be

$$\begin{aligned} n \in X &\rightarrow f(n) \in Y, \\ n \notin X &\rightarrow f(n) \in \mathbb{N} - Y. \end{aligned}$$

Hence

$$n \in X \equiv f(n) \in Y.$$

So Y would be universal for the class of all recursive sets which is impossible by Lemma 2.4.19. In this way we showed that the sets Thm_{T} and neg_{T} are recursively inseparable. Q.E.D.

Corollary 2.4.21 *The set of (Gödel numbers of) theorems of Peano arithmetic* PA *and the set of (Gödel numbers of) negations of theorems of PA are recursively inseparable.*

Corollary 2.4.22 (i) *If all recursive relations are strongly representable in a theory* T *then the set* Thm_{T} *of (Gödel numbers of) theorems of* T *is not recursive. Similarly the set* neg_{T} *is not recursive.*

(ii) *The set of (Gödel numbers of) theorems of Peano arithmetic* PA *is not recursive. Similarly the set of (Gödel numbers of) negations of theorems of PA is not recursive.*

A theory with the indicated property, i.e., a theory such that the set of (Gödel numbers of) its theorems is not recursive, is called 'undecidable' — otherwise a theory is called 'decidable'. Problems of decidability and undecidablity of theories will be discussed in Chapter 3.

Note that the set Thm_{PA} is an example of a recursively enumerable set which is not recursive. The same holds for Thm_{T} where T is a theory in which all recursive relations are strongly representable. So the indicated sets are of the class Σ^0_1 but not of the class Σ^0_0.

It was shown above that Gödel's undecidable sentences are of the form $\forall x\psi(x)$ and have the property that all particular instances of them, i.e., sentences $\psi(\overline{n})$ for $n \in \mathbb{N}$, are theorems of Peano arithmmetic PA. Now we shall prove the same for any simple recursive extension of PA where a recursive extension T of PA is said to be simple if and only if the language L(T) is the same as L(PA).

Definition 2.4.23 BQ *is the smallest class of formulas of the language* L(T) *of a given first-order extension* T *of* PA *containing all atomic formulas and closed under connectives and bounded quantifiers.*

An analysis of the proof of Representability Theorem 2.2.12 leads to the conclusion that the following theorem holds.

Theorem 2.4.24 *Every recursive relation* $R \subseteq \mathbb{N}^k$ *is strongly represented in Peano arithmetic* PA *by a formula* $\varphi(x_1, \ldots, x_k)$ *of the form* $\exists x \psi(x, x_1, \ldots, x_k)$ *where* ψ *belongs to the class* BQ.

One can also easily prove the following

Theorem 2.4.25 *Every formula* $\varphi(x_1, \ldots, x_k)$ *of the language* L(PA) *belonging to the class* BQ *strongly represents in Peano arithmetic* PA *a certain elementary recursive relation* $R \subseteq \mathbb{N}^k$, *namely that relation which is defined by* $\varphi(x_1, \ldots, x_k)$ *in the standard model* $\mathfrak{N}_0$.

Theorem 2.4.26 *For any consistent simple recursive extension* T *of Peano arithmetic* PA *there exist undecidable sentences of the form* $\exists x \psi(x)$ *where* ψ *belongs to the class* BQ.

Proof. Let S_0 denote a set of Gödel numbers of sentences of the language L(T) of the form $\exists x \psi(x)$ and of negations of such sentences for ψ belonging to the class BQ. The set S_0 is recursive. Consider now the set $S_0 \cap Thm_{\mathrm{PA}}$. By Theorem 2.4.24 every recursive set is strongly representable in a theory whose theorems are formulas of Gödel numbers belonging to $S_0 \cap Thm_{\mathrm{PA}}$. By Corollary 2.4.22 the set $S_0 \cap Thm_{\mathrm{PA}}$ is not recursive. Similarly for any consistent recursive extension T of PA the set $S_0 \cap Thm_{\mathrm{T}}$ is not recursive.

Suppose now that there are no sentences of the form $\exists x \psi(x)$, where $\psi \in$ BQ, undecidable in the theory T. Then

$$S_0 = (S_0 \cap Thm_{\mathrm{T}}) \cup (S_0 \cap neg_{\mathrm{T}}).$$

The set neg_{T} is recursively reducible to Thm_{T} and Thm_{T} is recursively enumerable. Hence neg_{T} is also recursively enumerable. So the sets $S_0 \cap Thm_{\mathrm{T}}$ and $S_0 \cap neg_{\mathrm{T}}$ are recursively enumerable. But their sum is equal to S_0, hence it is recursive. By Negation Lemma 1.5.2 the sets $S_0 \cap Thm_{\mathrm{T}}$ and $S_0 \cap neg_{\mathrm{T}}$ are recursive. But it was shown above that the set $S_0 \cap Thm_{\mathrm{T}}$ is not recursive, a contradition. Consequently there exist sentences of the form $\exists x \psi(x)$, for $\psi \in$ BQ, undecidable in T. Q.E.D.

Theorem 2.4.27 *If a formula* φ *of the language* L(PA) *belongs to the class* BQ *then*

$$\mathrm{PA} \vdash \exists x \varphi(x) \quad \textit{if and only if} \quad \mathfrak{N}_0 \models \exists x \varphi(x).$$

Proof. The implication to the right is obvious. So to prove the implication to the left assume that $n \in \mathbb{N}$ is such that $\mathfrak{N}_0 \models \varphi(\overline{n})$. By Theorem 2.4.25 the formula φ strongly represents in PA an elementary recursive relation which it defines in $\mathfrak{N}_0$. Hence

$$\mathrm{PA} \vdash \varphi(\overline{n}).$$

So $\mathrm{PA} \vdash \exists x \varphi(x)$. Q.E.D.

Theorem 2.4.28 (Rosser, 1936) *For any consistent recursive simple extension* T *of Peano arithmetic* PA *there exist sentences* φ *of the form* $\forall x \psi(x)$ *true in* $\mathfrak{N}_0$ *but undecidable in* T *and such that for any* $n \in \mathbb{N}$, $\mathrm{PA} \vdash \psi(\overline{n})$.

Proof. By Theorem 2.4.26 there exist sentences of the form $\exists x \neg \psi(x)$, where $\psi \in \mathrm{BQ}$, undecidable in T. According to Theorem 2.4.27, if a sentence of the form $\exists x \neg \psi(x)$ for $\psi \in \mathrm{BQ}$ is undecidable in T then it is false in the model $\mathfrak{N}_0$. Consequently there are sentences φ of the form $\neg \exists x \neg \psi(x)$, i.e., of the form $\forall x \psi(x)$, where $\psi \in \mathrm{BQ}$, true in $\mathfrak{N}_0$ but undecidable in T.

By Theorem 2.4.27 for $\psi \in \mathrm{BQ}$ and any $n \in \mathbb{N}$

$$\mathrm{PA} \vdash \psi(\overline{n}) \quad \text{or} \quad \mathrm{PA} \vdash \neg \psi(\overline{n}).$$

Hence if a sentence φ of the form $\forall x \psi(x)$ is undecidable in T then $\mathrm{PA} \vdash \psi(\overline{n})$ for any $n \in \mathbb{N}$. Indeed, otherwise it would be $\mathrm{PA} \vdash \neg \psi(\overline{n})$. So $\mathrm{PA} \vdash \exists x \neg \psi(x)$. This means $\mathrm{PA} \vdash \neg \varphi$ and the sentence φ would not be undecidable in PA (and hence in T) contrary to the assumption. Q.E.D.

Note that Tarski's Theorem on the undefinability of truth (cf. Theorem 2.4.11), together with Theorems 2.4.24 and 2.4.27, implies the incompleteness of Peano arithmetic PA. Indeed, by Tarski's Theorem the set

$$S = \{\ulcorner\varphi\urcorner : \varphi \text{ is a sentence of L(PA) } \wedge \mathfrak{N}_0 \models \varphi\}$$

is not definable in $\mathfrak{N}_0$. On the other hand, the set

$$P = \{\ulcorner\varphi\urcorner : \varphi \text{ is a sentence of L(PA) } \wedge \text{PA} \vdash \varphi\}$$

is definable in $\mathfrak{N}_0$. To prove this observe that

$$\begin{aligned} \ulcorner\varphi\urcorner \in P &\equiv \text{PA} \vdash \varphi \\ &\equiv \text{there exists a proof of } \varphi \text{ in PA} \\ &\equiv \exists n \in \mathbb{N} Prf(n, \ulcorner\varphi\urcorner). \end{aligned}$$

By Theorem 2.4.24 there exists a formula $\chi \in BQ$ such that $\exists z\chi$ strongly represents the relation Prf in PA. So in particular for any $n, k \in \mathbb{N}$:

$$Prf(n, k) \equiv \text{PA} \vdash \exists z\chi(\overline{n}, \overline{k}, z).$$

Using this and Theorem 2.4.27 we obtain:

$$\begin{aligned} \exists n \in \mathbb{N} : Prf(n, \ulcorner\varphi\urcorner) &\equiv \exists n \in \mathbb{N} : \text{PA} \vdash \exists z\chi(\overline{n}, \overline{\ulcorner\varphi\urcorner}, z) \\ &\equiv \exists n \in \mathbb{N} : \mathfrak{N}_0 \models \exists z\chi(\overline{n}, \overline{\ulcorner\varphi\urcorner}, z) \\ &\equiv \mathfrak{N}_0 \models \exists x\exists z\chi(x, \overline{\ulcorner\varphi\urcorner}, z). \end{aligned}$$

Hence $\ulcorner\varphi\urcorner \in P \equiv \mathfrak{N}_0 \models \exists x\exists z\chi(x, \overline{\ulcorner\varphi\urcorner}, z)$. So the set P is definable in $\mathfrak{N}_0$. To finish the proof of the incompleteness of PA it suffices to observe that $P \subseteq S$ and $P \neq S$ — the latter follows from the fact that P is definable in $\mathfrak{N}_0$ whereas S is not definable. A a result, there exist true (in $\mathfrak{N}_0$) sentences φ of the language L(PA) of Peano arithmetic such that PA $\nvdash \varphi$. Of course PA $\nvdash \neg\varphi$ because $\mathfrak{N}_0 \models \varphi$. So φ is undecidable in PA.

Notice that the above proof of the incompleteness of Peano arithmetic PA does not give an example of an undecidable sentence whereas Gödel's and Rosser's theorems do.

After Gödel's results several other proofs of the incompleteness theorems appeared (we discussed above e.g., Rosser's theorem).[1] It is impossible to give here either a list or a classification of all of them. But we would like to present a proof given by Boolos (1989).

Gödel's proof was based on the diagonalization and on the Epimenides' paradox of the liar ("What I am now saying is not true"). Boolos' proof uses Berry's paradox. It is a paradox which B. Russell attributed to G.G. Berry, a librarian at Oxford University, in a number of writings. It is as follows: let m be *the least integer not nameable in fewer than nineteen syllables.* The paradox is that m has just been named in eighteen syllables.

Boolos formulates his theorem in terms of algorithms. By an algorithm he means "a computational (automatic, effective, mechanical) procedure or routine of the usual sort, e.g., a program in a computer language like C, basic, Lisp, ..., a Turing machine, register machine, Markov algorithm, ..., a formal system like Peano or Robinson Arithmetic, ..., or whatever" (1989, p. 389). By a statement of arithmetic a formula of the language of PA is meant. A statement φ is true if it is true in the standard model $\mathfrak{N}_0$.

Theorem 2.4.29 (Boolos, 1989) *There exists no algorithm A whose output contains all true statements of arithmetic and no false ones.*

Proof. Assume that there exists an algorithm M whose output contains no false statements of arithmetic. We show how to find a true statement of arithmetic that is not in the output of M. This will prove the theorem.

It will be said that a natural number n is named by a formula φ (or: the formula φ names the number n) if and only if the

[1]This phenomenon is known in the history of mathematics. Indeed, almost every important theorem has many proofs. As examples can serve the fundamental theorem of algebra (its discoverer C.F. Gauss gave four proofs of it) or the Pythagorean theorem which has hundreds of proofs.

following statement

$$\forall x[\varphi(x) \equiv x = \overline{n}]$$

is in the output of M. Hence this definition contains a reference to the algorithm M. It can be easily shown that no formula can name two different numbers. Moreover, for each number i there exists only a finite number of different formulas that contain i symbols. Thus for each i there are only finitely many numbers named by formulas containing i symbols. Consequently, for every m only finitely many numbers are named by formulas containing fewer than m symbols. So there exists a least number not named by any formula containing fewer than m symbols.

Let $\psi_1(x, z)$ be a formula of the language of PA that says that x is a number named by some formula containing z symbols. It can be constructed using arithmetization and coding of finite sequences of natural numbers. Let now $\psi_2(x, y)$ be the formula

$$\exists z[z < y \wedge \psi_1(x, z)].$$

So $\psi_2(x, y)$ says that x is named by some formula containing fewer than y symbols. Let $\psi_3(x, y)$ be the formula

$$\neg\psi_1(x, y) \wedge \forall z(z < x \rightarrow \psi_2(z, y)).$$

$\psi_3(x, y)$ says that x is the least number not named by any formula containing fewer than y symbols.

Let k be the number of symbols in the formula $\psi_3(x, y)$. Of course, $k > 3$. Finally, let $\varphi(x)$ be the formula

$$\exists y[y = \overline{10} \cdot \overline{k} \wedge \psi_3(x, y)].$$

$\varphi(x)$ states that x is the least number not named by any formula containing fewer than $10k$ symbols.

Observe that the formula $\varphi(x)$ contains $2k + 24$ symbols. Since $k > 3$, $2k + 24 < 10k$ and φ contains fewer than $10k$ symbols. Let n be the least number not named by any formula containing fewer than $10k$ symbols. Then n is not named by φ. In other words the formula

$$\forall x[\varphi(x) \equiv x = \overline{n}]$$

is true but not in the output of M. Q.E.D.

Note that the above proof, contrary to Gödel's proof, does not use diagonalization. Boolos (1989a) remarked that what is especially interesting is not the brevity of it, but that it "provides a *different sort of reason* for the incompleteness of algorithms".

Add also that other proofs of the incompleteness theorem are known that do not use the diagonalization — cf., e.g., the proof given by Kotlarski (1994) in which some quickly growing functions are used, or the proof given by Kikuchi (1994) where a proof of Gödel's First Theorem based on Berry's paradox is given by extending Boolos' method and Gödel's Second Theorem is derived from it by model-theoretical means (see also Kikuchi and Tanaka, 1994).

2.5 Paris-Harrington and Paris-Kirby Theorems

Undecidable sentences of Gödel and others described in the previous section are sentences of a metamathematical and not directly mathematical contents. Therefore they are not very interesting from a purely mathematical point of view. The existence of those sentences proves that arithmetic is incomplete but the examples of undecidable sentences are rather artificial from the point of view of mathematics (and in particular of number theory). Hence the question arose whether one could give examples of undecidable sentences of mathematically (in particular number-theoretically) interesting contents (whatever it could mean). This problem was even more attractive since after Gödel's results one could still cherish the hope that all sentences interesting from the point of view of mathematics are decidable in PA and that in the scope of those sentences elementary arithmetic is complete.

The above problem is connected to another problem. One can ask whether it is possible to prove the existence of sentences undecidable in arithmetic not by using the method of arithmetization but simply by indicating a certain sentence φ and two models of arithmetic such that in one of them the sentence φ would be true while in the other false (cf. Mostowski *et al.*, 1955). Note that such a method was successfully applied in studies on the foundations of set theory (cf. Gödel's proof of the consistency of the axiom of choice and the continuum hypothesis given in 1938, and Cohen's proof of their independence from 1963).

Both problems described above were solved in the late seventies by J. Paris, L. Harrington and L. Kirby. The aim of this section is to present their solutions by describing undecidable sentences of Paris-Harrington and Kirby-Paris. Those sentences are of mathematically interesting contents, more exactly, of combinatoral and number-theoretical contents, respectively.

Definition 2.5.1 (i) *Let X be a set of natural numbers. The family of all subsets of X of cardinality n will be denoted by $[X]^n$.*

(ii) *A colouring function is any function $C : [X]^n \to c$ where c is a natural number identified with the set of its predecessors, i.e., $c = \{0, 1, \ldots, c-1\}$.*

The name 'colouring function' can be explained in the following way: a function $C : [X]^n \to c$ can be interpreted as a procedure of colouring subsets of the set X of cardinality n in colours $0, 1, \ldots, c-1$.

In 1929 the English mathematician and economist F.R. Ramsey proved that if C is a function colouring $[X]^n$ and the set X is large enough with respect to c and n (the meaning of this will be explained below) then there exists a large subset Y of X such that all its subsets of cardinality n are coloured by C in one colour. Such set Y is said to be homogeneous for the function C. Ramsey proved the following two theorems.

Theorem 2.5.2 (Infinite Ramsey Theorem) *Let n and c be positive natural numbers. For any colouring function*

$$C : [\mathbb{N}]^n \to c$$

there exists an infinite set $Y \subseteq \mathbb{N}$ such that Y is homogeneous for C, i.e., the function C restricted to $[Y]^n$ is constant.

Theorem 2.5.3 (Finite Ramsey Theorem) *Let s, n and c be positive natural numbers such that $s \geqslant n+1$. Then there exists a number $R(s,n,c)$ such that for any $r \geqslant R(s,n,c)$, any set X of cardinality r and any colouring function $C : [X]^n \to c$ there exists a set homogeneous for C and of cardinality at least s.*

The above theorems are not intuitively obvious and need to be proved. They can be treated as generalizations of Dirichlet's *Schubfachprinzip* to higher dimensions. The Infinite Ramsey Theorem for $n = 1$ says that if one divides an infinite set into finitely many disjoint parts then at least one part is infinite. Finite Ramsey Theorem for $n = 1$, $s = 2$ and $R(2,1,c) = c+1$ is exactly Dirichlet's *Schubfachprinzip* and states that if a set of cardinality at least $c+1$ were divided into c parts then one of those parts should contain at least 2 elements. Finite Ramsey Theorem can be proved in Peano arithmetic PA (note that by coding finite sequences of natural numbers by natural numbers one can speak in the language L(PA) about finite sets of natural numbers).

Definition 2.5.4 *A set X of natural numbers is said to be relatively large if and only if $card(X) \geqslant \min(X)$.*

For example, the set $X = \{2,3,85,91\}$ is relatively large. Indeed, $card(X) = 4 \geqslant \min(X) = 2$. The set $\{10,13,85,91\}$ is not relatively large.

Let φ_0 denote the following sentence (called the 'Paris-Harrington sentence'):

> For any natural numbers s, n, c there exists a number $H(s, n, c)$ such that for any $h \geqslant H(s, n, c)$, any set X of cardinality h and any colouring function $C : [X]^n \to c$ there exists a relatively large set Y homogeneous for the function C and such that the cardinality of Y is at least s.

Observe that the sentence φ_0 is a modification of Finite Ramsey Theorem obtained by adding the requirement that the homogeneous set Y is relatively large.

Theorem 2.5.5 (Paris-Harrington, 1977) (i) *The sentence φ_0 is true in the standard model $\mathfrak{N}_0$. Hence* PA $\nvdash \neg\varphi_0$.

(ii) *The sentence φ_0 is independent of Peano arithmetic* PA, *i.e.,* PA $\nvdash \varphi_0$.

As a result, the Paris-Harrington sentence φ_0 is a sentence of combinatorial contents undecidable in Peano arithmetic PA. Note that it is of the form $\forall x \exists y \psi(x, y)$ where a formula ψ contains only atomic formulas, connectives and possibly bounded quantifiers. It can be shown that for any given natural number n, PA $\vdash \exists y \psi(\overline{n}, y)$, i.e., all particular instances of φ_0 are decidable (recall that the same holds in the case of Gödel's undecidable sentence).

A proof of Theorem 2.5.5 uses advanced and sophisticated methods of the theory of models of arithmetic that are beyond the framework of this book. Therefore it will not be given here. The same applies also to further theorems of this section, in particular the Kirby-Paris Theorem (cf. Theorem 2.5.8). Only some general remarks on proofs of theorems of this type will be given at the end of the section.

We now describe the Kirby-Paris undecidable sentence.

Definition 2.5.6 *Let m and n be natural numbers, $n > 1$. We define the base n representation of m as follows: first write m as the sum of powers of n, then write each exponent as the sum of powers of n, repeat with exponents of exponents and so on until the representation stabilizes.*

Example. Let $m = 266$ and $n = 2$. Then the base n representation of m is defined as follows:

$$\begin{aligned} 266 &= 2^8 + 2^3 + 2^1, \\ 266 &= 2^{2^3} + 2^{2+1} + 2^1, \\ 266 &= 2^{2^{2+1}} + 2^{2+1} + 2^1. \end{aligned}$$

Definition 2.5.7 *Let $G_n(m)$ be a function defined as follows: if $m = 0$ then $G_n(m) = 0$, if $m \neq 0$ then $G_n(m)$ is a number obtained by replacing every n in the base n representation of m by $n+1$ and then subtracting 1. The Goodstein sequence for the number m is now defined recursively in the following way:*

$$\begin{aligned} m_0 &= m, \\ m_1 &= G_2(m_0), \\ m_2 &= G_3(m_1), \\ &\dots \end{aligned}$$

i.e.,

$$\begin{aligned} m_0 &= m, \\ m_k &= G_{k+1}(m_{k-1}), \quad \textit{for } k > 0. \end{aligned}$$

Example. The Goodstein sequence for $m_0 = 266$ is defined as follows:

$$\begin{aligned} m_0 &= 266 = 2^{2^{2+1}} + 2^{2+1} + 2^1 \\ m_1 &= G_2(m_0) = 3^{3^{3+1}} + 3^{3+1} + 2 \approxeq 10^{38} \\ m_2 &= G_3(m_1) = 4^{4^{4+1}} + 4^{4+1} + 1 \approxeq 10^{616} \\ m_3 &= G_4(m_2) = 5^{5^{5+1}} + 5^{5+1} \approxeq 10^{10000} \\ &\text{etc.} \end{aligned}$$

Observe that the procedure of constructing the Goodstein sequence m_k can be described in the language L(PA) of Peano arithmetic. So let φ_1 be the following sentence of the language L(PA):

$$\forall m \exists k (m_k = 0).$$

The sentence φ_1 is called the Kirby-Paris sentence.

Theorem 2.5.8 (Kirby-Paris, 1982) (i) *The sentence* φ_1 *is true in the standard model* $\mathfrak{N}_0$. *Hence* PA $\nvdash \neg\varphi_1$.

(ii) *The sentence* φ_1 *is independent of Peano arithmetic* PA, *i.e.*, PA $\nvdash \varphi_1$.

Consequently, the Kirby-Paris sentence φ_1 is a sentence of number-theoretic contents undecidable in PA.[2] It is of the form $\forall x \exists y \psi(x, y)$ where ψ contains atomic formulas, conectives and possibly bounded quantifiers. As in the case of the Paris-Harrington sentence φ_0, for any given natural number n, PA $\vdash \exists y(\psi(\overline{n}, y)$. Hence the Kirby-Paris sentence is undecidable in Peano arithmetic PA but all particular instances of it are decidable, i.e., are provable in PA.

Theorem 2.5.8, in particular part (i), can seem to be a bit astounding when compared with the example given above. The example indicates that the Goodstein sequence increases very quickly and the theorem states that it eventually hits zero! To dispell the fears one should note that the Goodstein sequence reaches zero on very high places. Indeed, the Goodstein sequence for the number $m = 4$ stabilizes for $k = 3 \cdot 2^{402653211} - 3 \approxeq 10^{121210700}$ (to imagine how huge is this number compare it with the fact that the whole number of atoms in the Universe is estimated at 10^{80}). Just the fact that the Goodstein sequences hit zero only at very high places is the reason for the unprovability of the Kirby-Paris sentence in Peano arithmetic. We return to this problem in Chapter 4 discussing philosophical and methodological problems connected with Gödel's theorems and their generalizations.

---

[2]A proof of the above theorem slightly different from the proof by Kirby and Paris (1982) can be found in Adamowicz *et al.*, 1997.

By using methods similar to those employed by Paris and Kirby in their proof of undecidability of the sentence φ_1, one can show the undecidability in Peano arithmetic of another interesting sentence of ... mythological contents! (Cf. Kirby and Paris, 1982)

Recall that Hercules, the son of Zeus and Alcmene, had to accomplish twelve tasks for Eurystheus, his cousin and king of Mycenae, in order to gain immortality. One of those labours was the killing of the hydra, a water or marsh serpent with numerous heads, each of which, if cut off, grew back as two. What did the hydra look like? Using the language of mathematics one can describe it as a finite tree. It can be represented for example as follows:

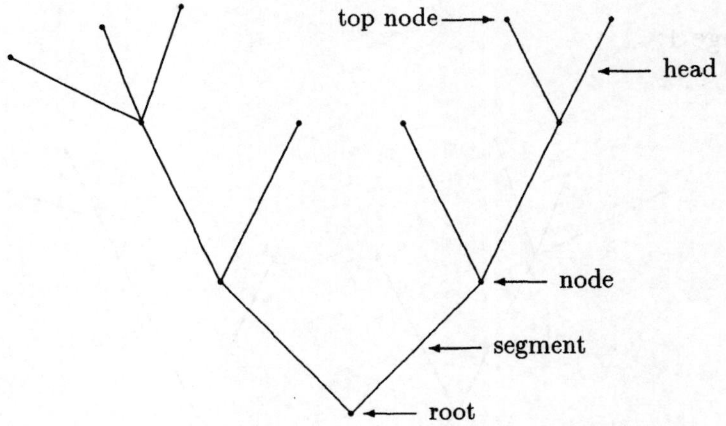

Hence a hydra is a finite tree, which may be considered as a finite collection of straight line segments, each joining two nodes, such that every node is connected by a unique path of segments to a fixed node called the 'root'. A top node of the hydra is a node of only one segment, and is not the root. A head of the hydra is a top node together with its attached segment.

A battle between Hercules and a given hydra proceeds in the following way: at stage n ($n > 1$) Hercules chopps off one head from the hydra, the hydra then grows n "new heads" in the following manner. From the node that used to be attached to the head which was just chopped off, traverse one segment towards the root until the next node is reached. From this node sproud n replicas of that part of the hydra (after decapitation) "above" the segment just traversed, i.e., those nodes and segments from which this segment would have to be traversed in order to reach the root. Where the head just chopped off had the root as one of its nodes, no new head grew. The battle might commence like this (arrows indicate which head Hercules decided to chop off at the given stage):

Stage 1:

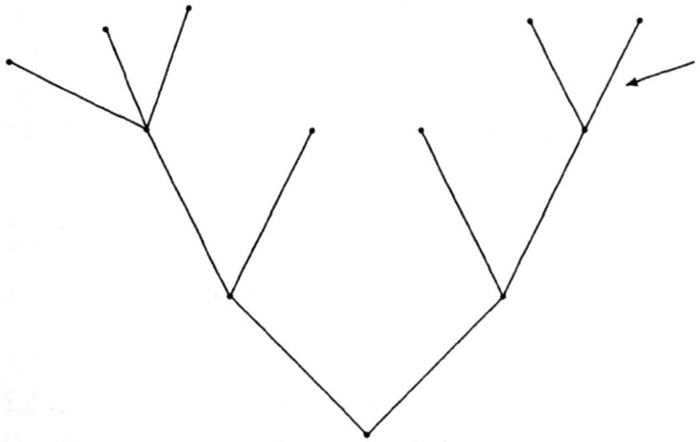

After stage 1:

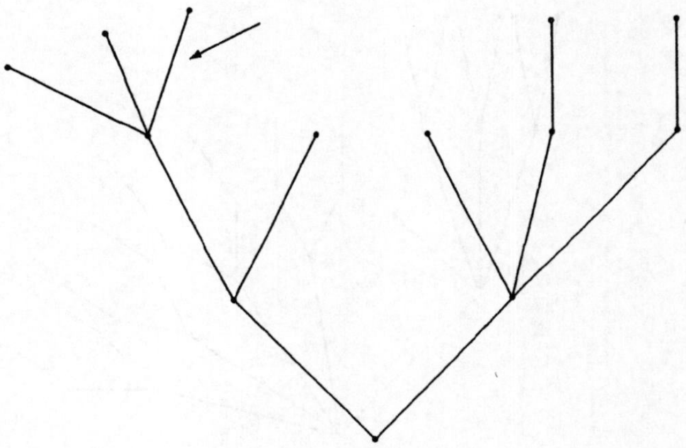

After stage 2:

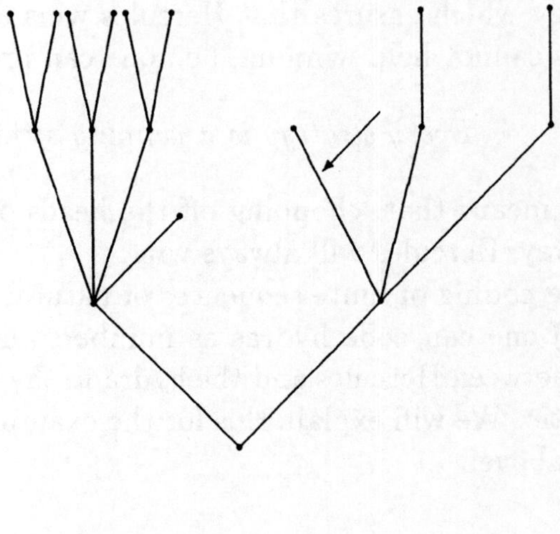

After stage 3:

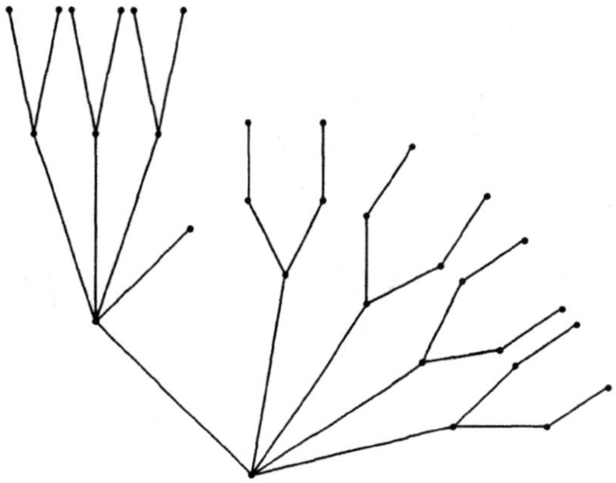

Hercules wins if after some finite number of stages nothing is left of the hydra but its root. He can of course chop off the heads in any order. A strategy is a function determining which head he chops off at each stage of any battle. A winning strategy is a strategy which ensures that Hercules wins against any hydra. Hercules cannot help winning, i.e., one can prove that

Every strategy is a winning strategy. (2.39)

This means that, chopping off the heads of any given hydra in any way, Hercules will always win.

Using coding of finite sequences of natural numbers (cf. Section 1.2) one can code hydras as numbers and thus talk about battles between Hercules and the hydra in the language of Peano arithmetic. We will explain this for the example of the hydra described above.

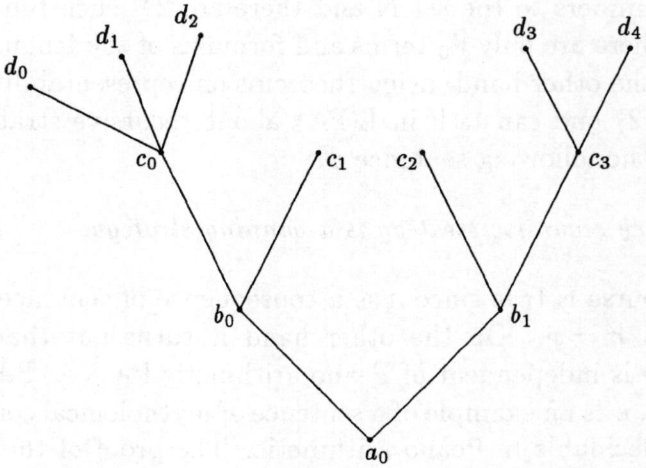

Let the correspondence between nodes and natural numbers be as follows:

$$\begin{aligned}
a_0 &\longrightarrow \langle 0 \rangle \\
b_0 &\longrightarrow \langle 0,0 \rangle \\
b_1 &\longrightarrow \langle 0,1 \rangle \\
c_0 &\longrightarrow \langle 0,0,0 \rangle \\
c_1 &\longrightarrow \langle 0,0,1 \rangle \\
c_2 &\longrightarrow \langle 0,1,0 \rangle \\
c_3 &\longrightarrow \langle 0,1,1 \rangle \\
d_0 &\longrightarrow \langle 0,0,0,0 \rangle \\
d_1 &\longrightarrow \langle 0,0,0,1 \rangle \\
d_2 &\longrightarrow \langle 0,0,0,2 \rangle \\
d_3 &\longrightarrow \langle 0,1,1,0 \rangle \\
d_4 &\longrightarrow \langle 0,1,1,1 \rangle
\end{aligned}$$

The sequence of natural numbers being codes of nodes can be coded by a single number which can be treated as the code of the hydra. Observe that the given hydra can be reconstructed in the unique way by decoding.

In the language of Peano arithmetic one cannot speak about arbitrary strategies. A strategy is a function from the set $\mathbb{N}$ of natural numbers to the set $\mathbb{N}$ and there are $2^{\aleph_0}$ such functions whereas there are only $\aleph_0$ terms and formulas of the language of PA. One the other hand, using theorems on representability (cf. Section 2.2) one can talk in L(PA) about recursive strategies. Consider the following sentence ψ:

Every recursive strategy is a winning strategy. (2.40)

This of course is true since it is a consequence of sentence 2.39. Hence PA $\nvdash \neg\psi$. On the other hand it turns out that sentence 2.40 is independent of Peano arithmetic PA, i.e., PA $\nvdash \psi$. Therefore, ψ is an example of a sentence of mythological contents being undecidable in Peano arithmetic. The proof of the independence of ψ of PA is similar to the proof of the independence of the Kirby-Paris sentence φ_1. In both proofs one uses the Cantor Normal Form Theorem, systems of notations for countable ordinals and theorems on ordinals below ε_0 proved by Ketonen and Solovay in 1981 (they developed earlier work by Paris, Kirby, Harrington, Wainer and others). Since the proofs are rather sophisticated and technically complicated, we will not give them here.

As indicated above, proofs of the Paris-Harrington and the Kirby-Paris theorems are difficult and use advanced methods of the theory of models of Peano arithmetic. Therefore they cannot be given in this book.[3] We will restrict ourselves to make only some short remarks on the proofs. Technical notions occurring in these remarks will not be explained. The reader who does not know them can simply omit this part of the book.

Proofs of theorems on undecidability of certain sentences in Peano arithmetic discussed in this section can be divided into two groups. Proofs of the first group are based on the indicator theory founded by Paris and Kirby in their studies of initial segments of countable nonstandard models of Peano arithmetic in

[3]A simple proof of the Paris-Harrington Theorem follows from the results of Mycielski (1986) if one notices that an appropriate variant of the Paris-Harrington combinatorial principle implies the consistency of FIN(PA).

the late seventies. To explain this notion consider a countable model $\mathfrak{M}$ of PA. We are interested in initial segments $\mathfrak{I}$ of the model $\mathfrak{M}$ with a given property Q (Q can be for instance the property of being a model of PA). Note that $\mathfrak{I}$ is an initial segment of the model $\mathfrak{M}$ if and only if $\mathfrak{I}$ is a submodel of $\mathfrak{M}$ and for any $x, y \in M$, if $y \in I$ and $\mathfrak{M} \models x < y$ then $x \in I$. (Latin letters denote universes of models denoted by the appropriate Gothic letters). An indicator for the property Q is a function (definable in the language L(PA) of Peano arithmetic) informing whether there exists an initial segment $\mathfrak{I}$ with the property Q or not between two given elements a and b of the model $\mathfrak{M}$.

Let $\mathfrak{M}$ be a countable nonstandard model of Peano arithmetic and assume that there exists an indicator for the family of initial segments $\mathfrak{I}$ of $\mathfrak{M}$, itself being models of PA. Hence there exists a definable function $Y(x, y)$ informing where in the model $\mathfrak{M}$ there are initial segments fulfilling all the axioms of Peano arithmetic. One proves now that there exists an initial segment $\mathfrak{I}_0$ of the model $\mathfrak{M}$ such that $\mathfrak{I}_0$ is a model of PA but the following sentence θ is false in it:

$$\forall z \forall x \exists y Y(x, y) \geqslant z.$$

On the other hand, one shows that this sentence is true in the standard model $\mathfrak{N}_0$. In this way one obtains two models of Peano arithmetic PA, namely the standard model $\mathfrak{N}_0$ and the nonstandard model $\mathfrak{I}_0$, such that the one satisfies θ and the other satisfies $\neg\theta$. Hence θ is undecidable in Peano arithmetic.

This is a general scheme of constructing undecidable sentences. The problem consists now in finding an appropriate indicator Y for models of PA, i.e., an indicator of mathematical (or, mathematically interested) contents. The indicator constructed by Paris and Harrington giving the undecidable sentence φ_0 described above was based on the Ramsey Theorem.

The second method of proving theorems on undecidable sentences is the method used by Paris and Harrington (1977). Within this method, one considers the sentence φ_0 to be a modification of the Ramsey Finite Theorem and then one constructs a theory T extending Peano arithmetic PA and proves that

$$\mathrm{PA} \vdash \mathrm{Con_T} \rightarrow \mathrm{Con_{PA}},$$

$$\mathrm{PA} \vdash \varphi_0 \rightarrow \mathrm{Con_T}.$$

This together with Gödel's Second Incompleteness Theorem gives that PA $\nvdash \varphi_0$ (since otherwise PA would prove $\mathrm{Con_{PA}}$).

We end this section by giving another example of a sentence undecidable in Peano arithmetic due to P. Pudlák.[4] The reason for this is that it has a clear contents which can be understandable to every mathematician and that one can give a relatively simple proof of its undecidability illustrating the first method described above. The sentence was given in an unpublished paper by Pudlák (1979). We describe this sentence and its properties using notes to the lectures of Smoryński given at the University of Utrecht (Smoryński, 1980).

Definition 2.5.9 *Let f be a function whose domain and co-domain are subsets of the set $\mathbb{N}$ of natural numbers. A finite set $A \subseteq \mathbb{N}$, $A = \{a_0, a_1, \ldots, a_n\}$, $a_0 < a_1 < \ldots < a_n$ is an approximation of the function f if and only if for any $i < n$ the following holds*

$$\forall x \leqslant a_i[x \in \mathrm{dom}(f) \rightarrow f(x) \leqslant a_{i+1} \vee f(x) > a_n],$$

where $\mathrm{dom}(f)$ *is the domain of f.*

For example a set $A = \{10, 12, 13, 14\}$ is an approximation of any finite restriction of the function $f(x) = x^2$. Observe that a two-element set is an approximation of any function.

---

[4]For examples of other undecidable sentences of mathematical contents see Smoryński (1982), Schütte and Simpson (1985), Simpson (1985b, 1985c and 1987a), Kanamori and McAloon (1987), Kent and Hodgson (1989) and Gordeev (1994).

Definition 2.5.10 *If X is a finite subset of the set $\mathbb{N}$ then X is said to be 0-dense if and only if X has at least three elements. The set X is said to be $(n+1)$-dense if and only if for any function f with domain and codomain contained in $\mathbb{N}$ there exists an n-dense set Y such that $Y \subseteq X$ and Y is an approximation of f.*

The set $\{0, 1, 2\}$ is 1-dense. It is not easy to give other examples of 1-dense sets. Nevertheless, one can relatively easily prove the following

Theorem 2.5.11 *For any natural numbers $a, n \in \mathbb{N}$ there exists a number $b \in \mathbb{N}$ such that $a < b$ and $[a, b]$ is n-dense.*

Both remarks made just before the theorem seem to be contradictory. The apparent inconsistency lies in the fact that the proof of the theorem is really not difficult but that it is not effective and cannot be formalized in Peano arithmetic. In fact it is based on the following combinatorial principle:

> For any infinite set $Y \subseteq \mathbb{N}$ and any function f with domain and codomain contained in $\mathbb{N}$ there exists an infinite set $X \subseteq Y$ being an approximation of f.

Add that an infinite set $X = \{x_0, x_1, \ldots\}$ where $x_0 < x_1 < \ldots$ is an approximation of a function f if and only if for any $i \in \mathbb{N}$ the following holds:

$$\forall x \leqslant x_i [x \in \mathrm{dom}(f) \rightarrow f(x) \leqslant x_{i+1}].$$

The proof of this principle is not effective.

Note that the function $F_a : \mathbb{N} \rightarrow \mathbb{N}$ defined as follows

$$F_a = \min\{b : [a, b] \text{ is } n\text{-dense}\}$$

grows rapidly for any $a \in \mathbb{N}$ — indeed, the values of F_a even for small arguments are greater than values of all rapidly growing functions known in the number theory (cf. Smoryński, 1980a).

Denote by φ the following sentence

$$\forall v_0, v_1 \exists v_2 (v_2 > v_0 \wedge [v_0, v_2] \text{ is } v_1\text{−dense}).$$

This sentence is called Pudlák's sentence. It turns out that sentence φ is undecidable in Peano arithmetic. By Theorem 2.5.11 sentence φ is true in the standard model $\mathfrak{N}_0$ of PA. Hence PA $\nvdash \neg\varphi$. It suffices now to find a (nonstandard) model $\mathfrak{M}$ of PA such that $\mathfrak{M} \models \neg\varphi$ which will imply that PA $\nvdash \varphi$.

It is a fundamental fact that Peano arithmetic PA has countable nonstandard models. One can construct a countable nonstandard model $\mathfrak{M}$ of PA being an elementary extension of the standard model $\mathfrak{N}_0$, i.e., a model in which exactly the same sentences of the language L(PA) of Peano arithmetic are true as in the model $\mathfrak{N}_0$.

Let $\mathfrak{M}$ be such a model. By Theorem 2.5.11, $\mathfrak{M} \models \varphi$. So let a and c be nonstandard elements of the model $\mathfrak{M}$. There exists an element b in M (= the universe of the model $\mathfrak{M}$) such that $a < b$ and in $\mathfrak{M}$ it is true that the set $[a, b]$ is c-dense. The model $\mathfrak{M}$ is a model of PA, hence in $\mathfrak{M}$ the minimum principle holds. So one can find the smallest (in $\mathfrak{M}$) element b with the above properties. Denote it by b_0.

The most difficult (but in fact technical) part of the proof of the independence of the sentence φ consists in showing that if a, b, c are nonstandard and in the model $\mathfrak{M}$ the set $[a, b]$ is c-dense, then there exists an initial segment $\mathfrak{M}_0$ of the model $\mathfrak{M}$ being a model of PA and such that $a \in M_0$, $c \in M_0$ but $b \notin M_0$. This means that the function

$$Y(x, y) = \max\{c : [x, y] \text{ is } c\text{−dense}\}$$

is an indicator for the family of initial segments being models of PA.

If $\mathfrak{M}_0$ is an initial segment of the model $\mathfrak{M}$ such that $a, c \in M_0$ but $b_0 \notin M_0$ then there is in M_0 no element b with the properties $a < b$ and $[a, b]$ is c-dense. Hence the model $\mathfrak{M}_0$ is a model of the negation of the sentence φ which finishes the proof. Observe that we used here the fact that the sentence "$[a, b]$ is c-dense" for $a, b, c \in M_0$ is true in the model $\mathfrak{M}_0$ if and only if it is true in the model $\mathfrak{M}$.

2.6 Satisfaction and Consistency

In Section 2.4 Tarski's theorem on undefinability of truth was proved (cf. Theorem 2.4.11). It states that the notion of truth for the language of Peano arithmetic PA is not definable in PA itself. Hence a natural question arises: where can one define truth for Peano arithmetic, and more generally, where can the notion of satisfaction for formulas of the language L(PA) be defined? The aim of this section is to answer these questions. It will be shown that the appropriate notions can be defined in a weak fragment of second-order arithmetic A_2^-. Furthermore, some connections between the definability of the notion of satisfaction and the provability of consistency will be indicated. Finally Mostowski's theorem on the reflexivity of arithmetic will be proved.

First of all we shall precisely define what it means that a given predicate is a satisfaction predicate for Peano arithmetic PA. To do this let us introduce functions

$$F: \mathbb{N} \to \mathbb{N} \text{ and } val: \mathbb{N} \times \mathbb{N} \to \mathbb{N}$$

in the following way.

$$F(t) = \begin{cases} \mu x_{x<t} \forall y < t[Fr(t, 2y) \to y < x], & \\ & \text{if } Term(t) \text{ or } Form(t), \\ 0, & \text{otherwise.} \end{cases}$$

So if α is a term then $F(\ulcorner \alpha \urcorner)$ is the smallest number i such that $j < i$ for every variable x_j occurring in α and similarly for

$F(\ulcorner\varphi\urcorner)$, φ a formula.

$$val(t,v) = \begin{cases} (v)_i, & \text{if } t = \ulcorner x_i \urcorner \text{ and } Seq(v) \text{ and } \\ & \quad lh(v) \geqslant i, \\ 0, & \text{if } t = SN(0), \\ val(t_1, v|F(t_1)) + 1, & \\ & \text{if } t = \langle SN(S), t_1 \rangle, \\ val(t_1, v|F(t_1)) + val(t_2, v|F(t_2)), & \\ & \text{if } t = \langle SN(+), t_1, t_2 \rangle, \\ val(t_1, v|F(t_1)) \cdot val(t_2, v|F(t_2)), & \\ & \text{if } t = \langle SN(\cdot), t_1, t_2 \rangle, \\ 0, & \text{otherwise.} \end{cases}$$

where $v|F(t) = In(v, F(t))$. The function val is such that if $t = \ulcorner\alpha\urcorner$ for a term α then $val(t,v)$ is the value of α when the variable x_1 is given value $(v)_1$, x_2 is given value $(v)_2$, ... and so on. The functions F and val are recursive (even primitive recursive). Hence they are representable in PA.

Let T be an extension of PA (the language L(T) can also be an extension of the language L(PA)).

Definition 2.6.1 *A binary predicate S of the language of* L(T) *is said to be a satisfaction predicate for the theory* PA *in the sense* (A) *if and only if for for every formula φ of* L(PA) *all free variables of which occur among variables $x_1, \ldots, x_n$ and any natural numbers $k_1, \ldots, k_n$:*

$$\mathrm{T} \vdash \varphi(\overline{k_1}, \ldots, \overline{k_n}) \equiv S(\overline{\ulcorner\varphi\urcorner}, \overline{\langle k_1, \ldots, k_n \rangle}).$$

Definition 2.6.2 *A binary predicate S of the language of* L(T) *is said to be a satisfaction predicate for the theory* PA *in the sense* (B) *if and only if for for every formula φ of* L(PA) *all free variables of which occur among variables $x_1, \ldots, x_n$:*

$$\mathrm{T} \vdash \forall x \{\mathbf{Seq}(x) \wedge \mathbf{lh}(x) = \overline{n} \to [\varphi((x)_1, \ldots, (x)_n) \equiv S(\overline{\ulcorner\varphi\urcorner}, x)]\}.$$

Before we formulate the next definition let us recall that if R is a recursive relation then by $\mathbf{R}$ we denote a formula of the language L(PA) strongly representing R in PA. To simplify the notation we adopt the following notational convention: if f is an n-ary recursive function and Φ is a formula of L(PA) representing f in PA then we write $y = \mathbf{f}(x)$ instead of $\forall z[\Phi(x, z) \equiv z = y]$.

Definition 2.6.3 *A binary predicate S of the language of* L(T) *is said to be a satisfaction predicate for the theory* PA *in the sense* (C) *if and only if the following formulas are provable in* T*:*

$$S(u, v) \rightarrow \mathbf{Form}(u) \wedge \mathbf{Seq}(v) \wedge \mathbf{lh}(v) = \mathbf{F}(u),$$

$$\mathbf{Term}(t_1) \wedge \mathbf{Term}(t_2) \wedge u = \langle \overline{SN(=)}, t_1, t_2 \rangle \rightarrow \\ \rightarrow [S(u, v) \equiv \mathbf{val}(t_1, v|\mathbf{F}(t_1)) = \mathbf{val}(t_2, v|\mathbf{F}(t_2))].$$

$$u = \langle \overline{SN(\neg)}, u_1 \rangle \wedge \ \mathbf{Form}(u_1) \rightarrow [S(u, v) \equiv \neg S(u_1, v)],$$

$$u = \langle \overline{SN(\vee)}, u_1, u_2 \rangle \ \wedge \ \mathbf{Form}(u_1) \wedge \mathbf{Form}(u_2) \rightarrow \\ \rightarrow [S(u, v) \equiv S(u_1, v|\mathbf{F}(u_1)) \vee S(u_2, v|\mathbf{F}(u_2))],$$

$$u = \langle \overline{SN(\exists)}, \overline{\ulcorner x_k \urcorner}, u_1 \rangle \ \wedge \ \mathbf{Form}(u_1) \wedge \neg \mathbf{Fr}(u_1, 2k) \rightarrow \\ \rightarrow [S(u, v) \equiv S(u_1, v)],$$

$$u = \langle \overline{SN(\exists)}, \overline{\ulcorner x_k \urcorner}, u_1 \rangle \ \wedge \ \mathbf{Form}(u_1) \wedge \mathbf{Fr}(u_1, 2k) \rightarrow \\ \rightarrow [S(u, v) \equiv \exists x S(u_1, v * \begin{pmatrix} k \\ x \end{pmatrix})]$$

where $v * \begin{pmatrix} k \\ x \end{pmatrix}$ *denotes a sequence number* w *such that*

$$lh(w) = max(lh(v), k),$$
$$\forall i < lh(v)[i \neq k \rightarrow (w)_i = (v)_i],$$
$$(w)_k = x,$$
$$\forall i[lh(v) < i < k \rightarrow (w)_i = 0].$$

Since other connectives (conjunction, implication and equivalence) and the universal quantifier are definable by negation, disjunction and the existential quantifier, they need not be considered in the above definition.

Note that Tarski (1933) considered the notion of a satisfaction predicate in sense (A). Observe also that if S is a satisfaction predicate in the sense (C) then it is a satisfaction predicate in the sense (B) (this follows by induction), and if S is a satisfaction predicate in the sense (B) then it is a satisfaction predicate in the sense (A) (this is obvious from the definitions). So Tarski's theorem on the undefinability of truth implies that there is no satisfaction predicate for PA in the sense (A) definable in PA. Hence there are no satisfaction predicates in the sense (B) or (C) for PA definable in PA.

Theorem 2.6.4 *Let* T *be an extension of Peano arithmetic* PA *such that induction (with respect to all formulas of the language* L(T)) *holds in* T. *If S is a satisfaction predicate for* PA *in the theory* T *in the sense* (C) *then* $\mathrm{T} \vdash \mathrm{Con}_{\mathrm{PA}}$, *where* $\mathrm{Con}_{\mathrm{PA}}$ *denotes the formula* $\neg\mathbf{Pr}(\ulcorner 0 = \overline{1}\urcorner)$.

Proof. Using induction (in T) one shows that

$$\mathrm{T} \vdash \mathbf{Ax}(x) \wedge \mathbf{Seq}(v) \wedge \mathbf{lh}(v) = \mathbf{F}(x) \to S(x, v). \qquad (2.41)$$

Consider, for example, the logical axiom $\varphi \to (\neg\varphi \to \psi)$ and assume for simplicity that the formulas φ and ψ have the same free variables. Then

$$\begin{aligned} S(\ulcorner\varphi \to (\neg\varphi \to \psi)\urcorner, v) &\equiv S(\ulcorner\neg\varphi \vee (\neg\neg\varphi \vee \psi)\urcorner, v) \equiv \\ &\equiv \neg S(\ulcorner\varphi\urcorner, v) \vee \neg\neg S(\ulcorner\varphi\urcorner, v) \vee S(\ulcorner\psi\urcorner, v). \end{aligned}$$

But the last disjunction is a theorem of T. In a similar way one proceeds in the case of other axioms.

Define the following recursive relation:

$$\begin{aligned} R(y_1, y_2, x) \quad \equiv \quad & S(y_1, x) \vee MP(y_1, y_2, x) \vee I\exists(y_1, x) \vee \\ & \vee E\exists(y_1, x) \vee I\forall(y_1, x) \vee \\ & \vee E\forall(y_1, x) \vee GR(y_1, x). \end{aligned}$$

So $R(y_1, y_2, x)$ means that the formula with the Gödel number x is obtained from formula(s) with Gödel number(s) y_1 (y_1, y_2) by one of the rules of inference. Next, using induction in T, we show that

$$\begin{aligned}\mathrm{T} \vdash [&\mathbf{Seq}(v) \wedge \mathbf{lh}(v) = max(\mathbf{F}(x), \mathbf{F}(y_1), \mathbf{F}(y_2)) \wedge \\ &\wedge \mathbf{R}(y_1, y_2, x) \wedge S(y_1, v|\mathbf{F}(y_1)) \wedge S(y_2, v|\mathbf{F}(y_2))] \rightarrow \\ &\rightarrow S(x, v|\mathbf{F}(x)). \end{aligned} \tag{2.42}$$

Using (2.41) and (2.42) one can show by induction that

$$\begin{aligned}\mathrm{T} \vdash \mathbf{Prf}(x, y) \rightarrow& \\ &\rightarrow \forall i < \mathbf{lh}(x) \forall v [\mathbf{Seq}(v) \wedge \mathbf{lh}(v) = \mathbf{F}((x)_i) \rightarrow \\ &\rightarrow S((x)_i), v)]. \end{aligned} \tag{2.43}$$

So in particular

$$\mathrm{T} \vdash \mathbf{lh}(v) = \mathbf{F}(x) \rightarrow [\mathbf{Pr}(x) \rightarrow S(x, v)]. \tag{2.44}$$

Hence

$$\mathrm{T} \vdash \mathbf{lh}(v) = \mathbf{F}(x) \rightarrow [\neg S(x, v) \rightarrow \neg \mathbf{Pr}(x)]. \tag{2.45}$$

Consider the sentence '$0 = \overline{1}$'. Since PA $\vdash 0 \neq \overline{1}$, so by the derivability condition (D1) (cf. Section 2.4) PA $\vdash \mathbf{Pr}(\ulcorner 0 \neq \overline{1} \urcorner)$. Hence T $\vdash \mathbf{Pr}(\ulcorner 0 \neq \overline{1} \urcorner)$. By (2.44) we get

$$\mathrm{T} \vdash \mathbf{lh}(v) = \mathbf{F}(\overline{\ulcorner 0 \neq \overline{1} \urcorner}) \rightarrow S(\overline{\ulcorner 0 \neq \overline{1} \urcorner}, v).$$

So

$$\mathrm{T} \vdash \mathbf{lh}(v) = \mathbf{F}(\overline{\ulcorner 0 = \overline{1} \urcorner}) \rightarrow \neg S(\overline{\ulcorner 0 = \overline{1} \urcorner}, v)$$

and consequently by (2.45)

$$\mathrm{T} \vdash \neg \mathbf{Pr}(\overline{\ulcorner 0 = \overline{1} \urcorner}),$$

$$\mathrm{T} \vdash \mathrm{Con}_{\mathrm{PA}}.$$

Q.E.D.

An analysis of the above proof shows that the assumption that induction with respect to all formulas of the language L(T) holds in T is not necessary. In fact, it suffices to assume the provability in T of induction with respect to all formulas of L(T) of the class BQ.

Note that Theorem 2.6.4 does not hold for S being a satisfaction predicate in the sense (A) or (B). In fact, property (2.43) cannot be obtained in those cases.

We now show that one can prove in second-order arithmetic A_2^- (more exactly, in its fragment) the existence of a satisfaction predicate for Peano arithmetic PA in the sense (C).

Let us describe A_2^-. This is a first-order (!) system formalized in a language with two sorts of variables: number variables x, y, z, ... and set variables X, Y, Z, ... Its nonlogical constants are those of Peano arithmetic, i.e., 0, S, $+$, $\cdot$ as well as symbols for all primitive recursive functions and the membership relation $\in$. Nonlogical axioms of A_2^- are the following:

(1) axioms of PA without the axiom scheme of induction,

(2) (extensionality) $\forall x(x \in X \equiv x \in Y) \rightarrow X = Y$,

(3) (induction axiom)

$$0 \in X \wedge \forall x(x \in X \rightarrow Sx \in X) \rightarrow \forall x(x \in X),$$

(4) recursive definitional equations for primitive recursive functions,

(5) (axiom scheme of comprehension)

$$\exists X \forall x[x \in X \equiv \varphi(x, \ldots)],$$

where φ is any formula of the language of A_2^- (possibly with free number or set variables) in which X does not occur free.

If Γ is a class of formulas of the language $L(A_2^-)$ then we denote by $A_2^-|\Gamma$ the subsystem of A_2^- obtained by restricting the comprehension axiom to formulas belonging to the class Γ. Later we shall consider in particular the system $A_2^-|\Sigma_1^1$ and the system $A_2^-|\Pi_1^1$ where Σ_1^1 is the class of formulas of the form $\exists X\varphi(X,\ldots)$ where φ is an arithmetical formula, i.e., a formula containing possibly any quantifiers bounding number variables and no quantifier over set variables, and Π_1^1 is the class of formulas of the form $\forall X\varphi(X,\ldots)$ with φ being an arithmetical formula.

To show that the existence of a satisfaction predicate for PA in sense (C) can be proved in A_2^- we present a more general fact. We show namely that the existence of a satisfaction predicate can be reduced to the existence of a certain function and that A_2^- proves the existence and uniqueness of functions defined by certain schemas using functionals. Before we do so recall that a functional is a mapping from $\mathbb{N}^\mathbb{N} \times \ldots \times \mathbb{N}^\mathbb{N} \times \mathbb{N} \times \ldots \times \mathbb{N}$ into $\mathbb{N}$ where $\mathbb{N}^\mathbb{N}$ denotes the set of all functions $f : \mathbb{N} \to \mathbb{N}$. We shall be interested in functions defined inductively by schemes of the following form

$$\begin{aligned} f(0,y) &= h(y), \\ f(x+1,y) &= \Phi((\lambda y)f(x,y),x,y), \end{aligned}$$

where $f : \mathbb{N} \to \mathbb{N}$, $\Phi : \mathbb{N}^\mathbb{N} \times \mathbb{N} \times \mathbb{N} \to \mathbb{N}$, $h : \mathbb{N} \to \mathbb{N}$ and $(\lambda y)f(x,y)$ denotes a function of x with the value $f(x,y)$ (y playes here the role of a parameter and f is in fact a function from $\mathbb{N}$ into $\mathbb{N}$).

Note that to define a relation it suffices to define its characteristic function. Observe also that the inductive definition needed to obtain a satisfaction predicate S in the sense (C) (from now on we shall consider satisfaction predicates only in this sense and the words "in the sense (C)" will be omitted) is of the following form. Let $\prec$ be a recursive well-founded partial order on a recursive subset F of $\mathbb{N}$ (in applications F will be the set $Form$) and let α and β be recursive functions from $\mathbb{N}$ into $\mathbb{N}$ such that $\alpha(x) \prec x$ and $\beta(x) \prec x$ if x has $\prec$-predecessors (in applications if, e.g., x is the Gödel number of a disjunction $\varphi \vee \psi$ then $\alpha(x)$ is the Gödel number of φ and $\beta(x)$ is the Gödel number of ψ). Let finally $h : \mathbb{N} \to \{0,1\}$ be a primitive recursive function. Define a characteristic function K_R of a relation R by the following scheme

$$\begin{aligned}K_R(x,y) &= h(x,y), \text{ if } x \text{ has no } \prec\text{-predecessors},\\ K_R(x,y) &= \Phi((\lambda y)R(\alpha(x),y),(\lambda y)R(\beta(x),y),x,y), \text{ otherwise,}\end{aligned}$$

where Φ is a given functional.

To obtain a characteristic function of a satisfaction predicate by this scheme one puts:

$$F = Form,$$

$$\prec \quad \text{is the relation of being a subformula.}$$

Functions α and β are defined in the following way:

$$\text{if } AForm(x) \text{ then } \alpha(x) = \beta(x) = x,$$

$$\text{if } x = \langle SN(\neg), x_1\rangle \wedge\ Form(x_1) \text{ then } \alpha(x) = \beta(x) = x_1,$$

$$\text{if } x = \langle SN(\vee), x_1, x_2\rangle \wedge\ Form(x_1) \wedge Form(x_2)$$
$$\text{then } \alpha(x) = x_1,\ \beta(x) = x_2,$$

$$\text{if } x = \langle SN(\exists), y, x_1\rangle \wedge\ Form(x_1) \wedge Vble(y)$$
$$\text{then } \alpha(x) = \beta(x) = x_1,$$

and otherwise the values of $\alpha(x)$ and $\beta(x)$ are 0.

The function h is defined in the following way: $h(x,y) = 0$ if and only if $AForm(x)$ [i.e., there exists $t_1 < x$ and $t_2 < x$ such that $Term(t_1)$ and $Term(t_2)$ and $x = \langle SN(=), t_1, t_2\rangle$], $Seq(y)$, $lh(y) = F(x)$, $val(t_1, y|F(t_1)) = val(t_2, y|F(t_2))$. Otherwise we put $h(x,y) = 1$.

As a functional Φ we take a functional defined by the following conditions:

1. if $x = \langle SN(\neg), x_1\rangle \wedge\ Form(x_1)$ then $\Phi(\psi_1, \psi_2, x, y) =$ $= 1 \dot{-} \psi_1(x_1, y)$,

2. if $x = \langle SN(\vee), x_1, x_2\rangle \wedge Form(x_1) \wedge Form(x_2)$ then $\Phi(\psi_1, \psi_2, x, y) = max(\psi_1(x_1, y|F(x_1)), \psi_2(x_2, y|F(x_2)))$,

3. if $x = \langle SN(\exists), \ulcorner x_k \urcorner, x_1\rangle \wedge Form(x_1) \wedge \neg Fr(x_1, 2k)$ then $\Phi(\psi_1, \psi_2, x, y) = \psi_1(x, y)$,

4. if $x = \langle SN(\exists), \ulcorner x_k \urcorner, x_1\rangle \wedge Form(x_1) \wedge Fr(x_1, 2k)$ then $\Phi(\psi_1, \psi_2, x, y) = max_{z \in \mathbb{N}}\{\psi(x_1, y * \binom{k}{z})\}$,

5. otherwise we put $\Phi(\psi_1, \psi_2, x, y) = 1$.

Using coding one can show that the scheme needed to define the characteristic function of the satisfaction predicate can be reduced to the scheme of the following form:

$$\text{(S)} \quad \begin{cases} f(0, y) = h(y), \\ f(x+1, y) = \Phi((\lambda y) f(x, y), x, y), \end{cases}$$

where $h : \mathbb{N} \to \mathbb{N}$ and $\Phi : \mathbb{N}^{\mathbb{N}} \times \mathbb{N} \times \mathbb{N} \to \mathbb{N}$.

To be able to speak in the language $L(A_2^-)$ about functions we treat them as appropriate sets, i.e., as sets of sequence numbers coding the elements of a graph of a given function. Hence we can take the following definition in the language $L(A_2^-)$:

$$\begin{aligned} Fnc_2(V) \equiv\ & \forall x[x \in V \to Seq(x) \wedge lh(x) = 3] \wedge \\ & \wedge \forall x \forall y \exists z(\langle x, y, z\rangle \in V) \wedge \\ & \wedge \forall x \forall y[x \in V \wedge y \in V \wedge (x)_0 = (y)_0 \wedge (x)_1 = (y)_1 \to \\ & \quad \to (x)_2 = (y)_2]. \end{aligned}$$

The formula $Fnc_2(V)$ means that V is a binary function.

If $f(x, y)$ is a binary function then the operation λ transforms it into a unary function $(\lambda y) f(x, y)$ of the argument y and the parameter x. This operation can be described in the language $L(A_2^-)$ in the following way:

$$\Lambda(x, V) = \{u : Seq(u) \wedge lh(u) = 2 \wedge \langle x, (u)_0, (u)_1\rangle \in V\}.$$

This formula says that $\Lambda(x, V)$ is a unary function whose value for the argument $(u)_0$ is equal to the value of the function V for the argument $\langle x, (u)_0 \rangle$. The set $\Lambda(x, V)$ exists by the comprehension axiom. From the definition of Λ it follows that

$$Fnc_2(V) \to \forall x[Fnc_1(\Lambda(x, V))],$$

where $Fnc_1(V)$ is a formula of $L(A_2^-)$ similar to $Fnc_2(V)$ stating that V is a unary function.

Let $\Phi(V, x, y, z)$ be a given formula of $L(A_2^-)$. We denote by φ the following formula of $L(A_2^-)$:

$$\forall U \forall x \forall y[Fnc_1(U) \to \exists! z \Phi(U, x, y, z)],$$

where $\exists!$ means "there exists exactly one". The formula φ says that Φ defines a functional. Since in the language of second-order arithmetic one can speak only about definable functionals we prove the following scheme of theorems.

Theorem 2.6.5 (On inductive definitions)

$$\begin{aligned} A_2^- \vdash \varphi \wedge Fnc_1(W) \to \exists! V\{&Fnc_2(V) \wedge \Lambda(0, V) = W \wedge \\ &\wedge \forall x \forall y \forall z[\langle x+1, y, z\rangle \in V \equiv \Phi(\Lambda(x, V), x, y, z)]\}. \end{aligned}$$

Note that the last condition in the curly brackets in the above formula means that the function f (V is the set of codes of elements of the graph of f) satisfies the scheme (S) defined above with h given by the sets of codes W.

Sketch of the proof. We reason in A_2^-. First we prove the existence of V. To do this consider a set X consisting of all x such that there exists a partial function V defined for arguments (s, y), $s \leqslant x$ and such that for all y:

$$V(0, y) = W(y)$$

and for $s < x$:

$$V(s+1, y) = z \equiv \Phi(\Lambda(y, V), s, y, z).$$

The set X exists by the comprehension axiom. Using induction one shows now that $X = \mathbb{N}$. Indeed, it is obvious that $0 \in X$. Assume that $x \in X$, i.e., there exists an appropriate function V_x for x. Then for $x+1$ one puts:

$$V_{x+1}(x+1, y) = z \equiv \Phi(V_x(x,y), x, y, z),$$

$$V_{x+1}(s, y) = V_x(s, y) \quad \text{for } s \leqslant x.$$

The function V_{x+1} is the needed function for $x+1$. To finish the proof of the existence of V it suffices to take $V(x,y) = V_x(x,y)$.

Now we prove the uniqueness of V. Suppose that there are two functions V_1 and V_2 fulfilling the given conditions. Hence there is a triple $\langle x, y, z\rangle$ such that

$$\neg[\langle x, y, z\rangle \in V_1 \equiv \langle x, y, z\rangle \in V_2].$$

Of course $x \neq 0$. Indeed, for $x = 0$ we have:

$$\langle 0, y, z\rangle \in V_1 \equiv \langle y, z\rangle \in W \equiv \langle 0, y, z\rangle \in V_2.$$

Hence $x = x_1 + 1$ for some x_1. One can assume that x_1 is the smallest element with the following property

$$\neg[\langle x_1+1, y, z\rangle \in V_1 \equiv \langle x_1+1, y, z\rangle \in V_2].$$

Consequently x_1 is the smallest element such that

$$\neg[\Phi(\Lambda(x_1, V_1), x_1, y, z) \equiv \Phi(\Lambda(x_1, V_2), x_1, y, z)]. \tag{2.46}$$

By the minimality of x_1 we have

$$\Lambda(x_1, V_1) = \Lambda(x_1, V_2).$$

This, however, is a contradiction because the above condition implies the negation of (2.46). Hence the uniqueness of V has been proved. Q.E.D.

Note that in the above proof only the comprehension axiom for Σ^1_1 formulas was needed. In fact, the formula describing the set X used in the proof of the existence of V was just of this form. Hence we get the following

Theorem 2.6.6 *Second order arithmetic* A_2^- *proves the existence of the satisfaction predicate for Peano arithmetic. Moreover, this can be proved in the fragment* $A_2^-|\Sigma_1^1$ *of* A_2^-.

Using this theorem and Theorem 2.6.4 we obtain

Theorem 2.6.7

$$A_2^-|\Sigma_1^1 \vdash \mathrm{Con}_{\mathrm{PA}}.$$

The next thing we want to discuss in this section is Mostowski's theorem on the reflexivity of Peano arithmetic. It states that PA, though not proving its own consistency, proves in fact the consistency of every finite fragment of itself.

Definition 2.6.8 *A first-order theory* T *containing* PA *is said to be reflexive if and only if for each finite* $\mathrm{S} \subseteq \mathrm{T}$,

$$\mathrm{T} \vdash \mathrm{Con}_{\mathrm{S}}.$$

T *is said to be essentially reflexive if and only if any consistent extension of* T *with the same constants as* T *is reflexive.*

It is quickly seen that T is reflexive just in case for each $n \in \mathbb{N}$

$$\mathrm{T} \vdash Con_{\mathrm{S}\restriction n},$$

where $\mathrm{S}\restriction n$ denotes any subtheory of S based on n axioms of S.

To show that PA is essentially reflexive we need some auxiliary facts.

Let T be any consistent first-order theory with identity. Assume that axioms of T are universal sentences. For a set A of formulas denote by $Cn_0(A)$ the smallest set of open formulas (i.e., formulas without any quantifiers) containing A, logical axioms and identity axioms and closed under substitution and *modus ponens*. Let now A be the set of all formulas obtained from the axioms of the theory T by omitting quantifiers and performing any substitutions.

Lemma 2.6.9 *If the theory* T *and the set* A *are as above then for any open sentence* φ

$$\mathrm{T} \vdash \varphi \qquad \textit{if and only if} \qquad \varphi \in Cn_0(A).$$

Proof. The implication to the left is obvious. So it suffices to prove the implication to the right. Assume that there exists an open sentence φ_0 such that $\mathrm{T} \vdash \varphi_0$ but $\varphi_0 \notin Cn_0(A)$. We construct a model of T in which φ_0 is false.

Extend $Cn_0(A)$ to a complete (with respect to open sentences) and consistent set S containing $\neg\varphi_0(t_1, \ldots, t_k)$ where $t_1, \ldots, t_k$ are any constant terms, i.e., we take a consistent set S of sentences such that $Cn_0(A) \subseteq \mathrm{S}$, $\ulcorner\neg\varphi(t_1, \ldots, t_k)\urcorner \in \mathrm{S}$ and for any open sentence ψ, either $\psi \in \mathrm{S}$ or $\neg\psi \in \mathrm{S}$. (The symbol $\ulcorner\chi\urcorner$ here does not denote the Gödel number of the formula χ but a metamathematical name of χ).

Consider the set M of all constant terms and the following relation $\sim$ on M:

$$t_1 \sim t_2 \equiv \ulcorner t_1 = t_2 \urcorner \in \mathrm{S}.$$

It can be easily seen that $\sim$ is an equivalence. It follows from the fact that S contains identity axioms and is closed under substitution. Let $M_1 = M/\!\sim$ and define functions and relations on M_1 interpreting function symbols and predicates of the language L(T). Let, for example, f be a binary function symbol belonging to L(T). The interpretation of f is a function $f^{(M_1)}$ defined as follows

$$f^{(M_1)}(t_1/\!\sim, t_2/\!\sim) = f(t_1, t_2)/\!\sim .$$

By identity axioms it holds of course that

$$t_1' \sim t_1 \wedge t_2' \sim t_2 \rightarrow f(t_1', t_2') \sim f(t_1, t_2).$$

Similarly, for a binary predicate P we define its interpretation $P^{(M_1)}$ in the following way

$$P^{(M_1)}(t_1/\!\sim, t_2/\!\sim) \equiv \ulcorner P(t_1, t_2)\urcorner \in \mathrm{S}.$$

One can prove now by structural induction that for any open formula $\varphi(x_1, \ldots, x_n)$ and any constant terms $t_1, \ldots, t_n$

$$M_1 \models \varphi[t_1/\sim, \ldots, t_n/\sim] \equiv \ulcorner \varphi(t_1, \ldots, t_n) \urcorner \in \mathrm{S}.$$

(Note that for simplicity we identify here a model with its universe). We claim that M_1 is a model of T. Indeed, consider any axiom φ of T. Let φ be of the form $\forall x_1 \ldots \forall x_n \psi(x_1, \ldots, x_n)$ for an open ψ. Let $t_1/\sim, \ldots, t_n/\sim$ be any n elements of M_1. By the definition of the set A, $\ulcorner \psi(t_1, \ldots, t_n) \urcorner \in A$. So $\ulcorner \psi(t_1, \ldots, t_n) \urcorner \in$ S. Consequently

$$M_1 \models \psi[t_1/\sim, \ldots, t_n/\sim].$$

Since the elements $t_1/\sim, \ldots, t_n/\sim$ were arbitrary, we have

$$M_1 \models \varphi.$$

This proves that $M_1 \models \mathrm{T}$. By the construction of M_1:

$$M_1 \models \neg\varphi_0[t_1/\sim, \ldots, t_k/\sim],$$

which completes the proof. Q.E.D.

In this way we proved that if there exists a proof of an open theorem on the basis of universal axioms then there exists a proof of it in which no quantifiers appear. Note also that the proof of Lemma 2.6.9 given above is not a proof within the system T — it follows from the fact that it uses semantical notions such as satisfaction and model. On the other hand, using Gentzen's formalism one can give a proof of Lemma 2.6.9 which can be formalized in T.

Extend Peano arithmetic PA by adding symbols for finitely many recursive functions to the language L(PA) and formulas defining those new symbols to the axioms of PA. In particular we add symbols for the functions $\langle \ldots \rangle$, $lh(.)$, $(x)_i$ and for the predicate Seq. Denote this extension by PA*.

Consider a theory T such that the language L(T) is an extension of L(PA*) by new function symbols (not necessarily for recursive functions). Let $\xi_1, \ldots, \xi_n$ be all function symbols of L(T) where ξ_j is a p_j-ary symbol.[5] Assume that the only predicate of L(T) is identity =. Let axioms of T be universal formulas of the form $\forall x_1 \ldots \forall x_{n_k} \varphi_k(x_1, \ldots, x_{n_k})$ for $k = 1, \ldots, N$. Let Ξ_j, $j = 1, \ldots, n$, be formulas of L(PA*) with $p_j + 1$ free variables and denote by Ξ the sequence $(\Xi_1, \ldots, \Xi_n)$. Let $Posmod_{\mathrm{T}}(\Xi)$ mean that Ξ is a possible model of T. We define it in the following way.

Definition 2.6.10

$$Posmod_{\mathrm{T}}(\Xi) \equiv \bigwedge_{j \leqslant n} \forall x_1 \ldots \forall x_{p_j} \exists ! y \Xi_j(x_1, \ldots, x_{p_j}, y).$$

For any term τ of the language L(T) define a formula τ_Ξ of L(PA*) by induction in the following way: If τ is a variable x_i then τ_Ξ is the formula $x_i = y$. If τ is of the form $\xi_j(\tau_1, \ldots, \tau_{p_j})$ then τ_Ξ is the formula

$$\exists y_1 \ldots \exists y_{p_j} [\bigwedge_{i \leqslant p_j} \tau_{i,\Xi}(\ldots, y_i) \wedge \Xi_j(y_1, \ldots, y_{p_j}, y)].$$

One can see that if a term $\tau(x_1, \ldots, x_k)$ is given then the formula $\tau_\Xi(x_1, \ldots, x_k, y)$ says that y is the value of the term τ for the arguments $x_1, \ldots, x_k$ by the interpretation of ξ_i as Ξ_i. One can also easily show that

$$\mathrm{PA}^* \vdash Posmod_{\mathrm{T}}(\Xi) \rightarrow \exists ! y \tau_\Xi(\ldots, y).$$

For any open formula φ of the language L(T) define by induction a formula φ_Ξ of L(PA*) in the following way: If φ is $\tau_1(\ldots) = \tau_2(\ldots)$ then φ_Ξ is $\exists y[\tau_{1,\Xi}(\ldots, y) \wedge \tau_{2,\Xi}(\ldots, y)]$. If φ is of the form $\psi_1 \vee \psi_2$ for some ψ_1 and ψ_2 then φ_Ξ is $\psi_{1,\Xi} \vee \psi_{2,\Xi}$. If φ is of the form $\neg\psi$ for some formula ψ then φ_Ξ is $\neg(\psi_\Xi)$.

Definition 2.6.11

$$Mod_{\mathrm{T}}(\Xi) \equiv Posmod_{\mathrm{T}}(\Xi) \wedge \bigwedge_{k \leqslant N} \forall x_1 \ldots \forall x_{n_k} \varphi_{k,\Xi}(x_1, \ldots, x_{n_k}).$$

[5] In the sequel we use the same symbols for the metamathematical description of the theory T and for its description in PA* by arithmetization.

The formula $Mod_{\mathrm{T}}(\Xi)$ says that Ξ is a model of T.

Assume from now on that $\mathrm{PA}^* \vdash Posmod_{\mathrm{T}}(\Xi)$. Then one can add to the language L(PA*) new function symbols $\tau_{\Xi}(\ldots, y)$ for functions $(\iota y)\tau_{\Xi}(\ldots, y)$ where τ is a term and $(\iota y)\varphi(y)$ denotes the unique y such that $\varphi(y)$.

Define in (PA*) a function $\mathbf{val}(t, s)$ such that the formula $\mathbf{val}(t, s) = z$ says: "z is the value of the term t of L(T) by the valuation v". So let $\mathbf{v}(t, v)$ be the following formula[6]

$$\mathbf{Seq}(v) \wedge \mathbf{lh}(v) = \mathbf{F}(t).$$

The formula $\mathbf{val}(t, v) = z$ is

$$\mathbf{Term}(t) \wedge \mathbf{v}(t, v) \wedge \exists s \exists w[\zeta(s, t) \wedge \theta(w, s, v) \wedge z = (w)_{\mathbf{lh}(w) \dot{-} 1}],$$

where $\zeta(s, t)$ is

$$\begin{aligned}&\mathbf{Seq}(s) \wedge \forall i < \mathbf{lh}(s)[\mathbf{Vble}((s)_i) \vee \\ &\qquad \vee \bigvee_{j \leqslant n} \exists i_1 < i \ldots \exists i_{p_j} < i\, (s)_i = \langle \ulcorner \xi_j \urcorner, (s)_{i_1}, \ldots, (s)_{i_{p_j}} \rangle] \wedge \\ &\qquad \wedge (s)_{\mathbf{lh}(s) \dot{-} 1} = t\end{aligned}$$

and $\theta(w, s, v)$ is

$$\begin{aligned}&\mathbf{Seq}(w) \wedge \mathbf{lh}(w) = \mathbf{lh}(s) \wedge \\ &\qquad \wedge \forall i < \mathbf{lh}(w)[\mathbf{Vble}((s)_i) \to (w)_i = (v)_i] \wedge \bigwedge_{j \leqslant n} \vartheta_j\end{aligned}$$

for ϑ_j being the formula

$$\begin{aligned}&\forall i < \mathbf{lh}(s) \forall i_1 < i \ldots \forall i_{p_j} < i[(s)_i = \langle \ulcorner \xi_j \urcorner, (s)_{i_1}, \ldots, (s)_{i_{p_j}} \rangle \to \\ &\qquad \to \Xi_j((w)_{i_1}, \ldots, (w)_{i_{p_j}}, (w)_i)].\end{aligned}$$

We have:

$$\mathrm{PA}^* \vdash \forall t \forall v[\mathbf{Term}(t) \wedge \mathbf{v}(t, v) \to \exists! z \mathbf{val}(t, v) = z],$$

[6]Symbols **Term**, etc., used here are formulas of the language L(PA*) strongly representing recursive sets $Term_{\mathrm{T}}$, etc., in PA* defined in a similar way as *Term*, etc., for L(PA). To simplify the notation we omit the subscript T. The meaning of the symbols will always be clear from the context.

$$\text{PA}^* \vdash \forall t \forall v \forall i[\mathbf{Vble}(t) \wedge t = \overline{2i} \wedge \mathbf{v}(t,v) \rightarrow \mathbf{val}(t,v) = (v)_i],$$

$$\begin{aligned}\text{PA}^* \vdash t = \langle \overline{\ulcorner \xi_i \urcorner}, t_1, \ldots, t_{p_j} \rangle \wedge \mathbf{v}(t,v) \rightarrow \\ \rightarrow \mathbf{val}(t,v) = \tau_{j,\Xi}(\mathbf{val}(t_1,v_1), \ldots, \mathbf{val}(t_{p_j}, v_{p_j}))\end{aligned}$$

where $v_i = v|\mathbf{F}(t_i)$,

$$\begin{aligned}\text{PA}^* \vdash \mathbf{v}(\overline{\ulcorner \tau \urcorner}, v) \rightarrow \\ \rightarrow \mathbf{val}(\overline{\ulcorner \tau \urcorner}, v) = \tau_{\Xi}(\mathbf{val}(\overline{\ulcorner x_1 \urcorner}, v_1), \ldots, \mathbf{val}(\overline{\ulcorner x_n \urcorner}, v_n))\end{aligned}$$

where $v_i = v|\mathbf{F}(\overline{\ulcorner x_i \urcorner})$.

Similarly as in Section 2.3 one can define a recursive relation $OForm(a)$ which means that $a = \ulcorner \varphi \urcorner$ for an open formula φ of the language L(T). Let **OForm** be a formula of L(PA) strongly representing $OForm$ in PA.

Let $Sat_0(a,v)$ be the following formula of the language L(PA*):

$$\mathbf{OForm}(a) \wedge \mathbf{v}(a,v) \wedge \exists s \exists w[\zeta'(s,a) \wedge \theta'(w,v,a)]$$

where $\zeta'(s,a)$ is

$$\begin{aligned}\mathbf{Seq}(s) \wedge \forall i < \mathbf{lh}(s)[\exists i_1 < i \exists i_2 < i(\mathbf{Term}((s)_{i_1}) \wedge \mathbf{Term}((s)_{i_2}) \wedge \\ \wedge (s)_i = \langle \overline{SN(=)}, (s)_{i_1}, (s)_{i_2} \rangle) \vee \\ \vee \exists i_1 < i((s)_i = \langle \overline{SN(\neg)}, (s)_{i_1} \rangle) \vee \\ \vee \exists i_1 < i \exists i_2 < i((s)_i = \langle \overline{SN(\vee)}, (s)_{i_1}, (s)_{i_2} \rangle)] \wedge \\ \wedge (s)_{\mathbf{lh}(s) \dot{-} 1} = a\end{aligned}$$

and $\theta'(w,v,a)$ is

$$\begin{aligned}\mathbf{Seq}(w) \wedge \mathbf{lh}(w) = \mathbf{lh}(s) \wedge \forall i < \mathbf{lh}(w)\{((w)_i = 0 \vee (w)_i = \overline{1}) \wedge \\ \wedge [\exists u, u' \leqslant s(\mathbf{Term}(u) \wedge \mathbf{Term}(u') \wedge \\ \wedge (s)_i = \langle \overline{SN(=)}, u, u' \rangle \wedge \\ \wedge ((w)_i = 1 \equiv (\mathbf{val}(u, v|\mathbf{F}(u)) = \mathbf{val}(u', v|\mathbf{F}(u'))))] \vee \\ \vee [\exists i_1 < i((s)_i = \langle \overline{SN(\neg)}, (s)_{i_1} \rangle \wedge (w)_i = 1 - (w)_{i_1}] \vee \\ \vee [\exists i_1, i_2 < i((s)_i = \langle \overline{SN(\vee)}, (s)_{i_1}, (s)_{i_2} \rangle \wedge \\ \wedge ((w)_i = 1 \equiv ((w)_{i_1} = 1 \vee (w)_{i_2} = 1))]\} \wedge \\ \wedge (w)_{\mathbf{lh}(w) \dot{-} 1} = 1.\end{aligned}$$

We have:

$$\begin{aligned} \text{PA}^* \vdash \mathbf{Term}(t_1) \wedge \mathbf{Term}(t_2) \wedge a = \langle \overline{SN(=)}, t_1, t_2\rangle \wedge \\ \wedge \mathbf{v}(a, v) \rightarrow \\ \rightarrow [Sat_0(a, v) \equiv \mathbf{val}(t_1, v|\mathbf{F}(t_1)) = \mathbf{val}(t_2, v|\mathbf{F}(t_2))]. \end{aligned} \tag{2.47}$$

$$\begin{aligned} \text{PA}^* \vdash \mathbf{OForm}(b) \wedge \mathbf{v}(b, v) \wedge a = \langle \overline{SN(\neg)}, b\rangle \rightarrow \\ \rightarrow [Sat_0(a, v) \equiv \neg Sat_0(b, v)]. \end{aligned} \tag{2.48}$$

$$\begin{aligned} \text{PA}^* \vdash \mathbf{OForm}(b) \wedge \mathbf{OForm}(c) \ \wedge a = \langle \overline{SN(\vee)}, b, c\rangle \wedge \\ \wedge \mathbf{v}(a, v) \rightarrow \\ \rightarrow [Sat_0(a, v) \equiv Sat_0(b, v|\mathbf{F}(b)) \vee Sat_0(c, v|\mathbf{F}(c))]. \end{aligned} \tag{2.49}$$

By structural induction one can prove that if φ is a formula of L(T) with free variables $x_1, \ldots, x_n$ and $\tau_1, \ldots, \tau_n$ are constant terms then

$$\text{PA}^* \vdash \varphi(\tau_1, \ldots, \tau_n)_\Xi \equiv Sat_0(\overline{\ulcorner \varphi(\tau_1, \ldots, \tau_n)\urcorner}, 0).$$

If $\forall x_1 \ldots \forall x_{n_k} \varphi_k$ is an axiom of the theory T and $\tau_1, \ldots, \tau_{n_k}$ are constant terms then

$$\text{PA}^* \vdash Mod_{\text{T}}(\Xi) \rightarrow Sat_0(\overline{\ulcorner \varphi_k(\tau_1, \ldots, \tau_{n_k})\urcorner}, 0). \tag{2.50}$$

Indeed, we have

$$\begin{aligned} \text{PA}^* \vdash Mod_{\text{T}}(\Xi) &\rightarrow \forall x_1 \ldots \forall x_{n_k} \varphi_{k,\Xi} \\ &\rightarrow \varphi_{k,\Xi}(\tau_{1,\Xi}, \ldots, \tau_{n_k,\Xi}) \\ &\rightarrow Sat_0(\overline{\ulcorner \varphi_k(\tau_1, \ldots, \tau_{n_k})\urcorner}, 0) \end{aligned}$$

where the first implication follows by logical laws and the second one can be obtained by (2.50).

$$\begin{aligned} \text{PA}^* \vdash \mathbf{OForm}(x) \wedge \mathbf{Sent}(a) \wedge (x \text{ is a Gödel number} \\ \text{of a substitution of a logical axiom}) \rightarrow \\ \rightarrow Sat_0(x, 0). \end{aligned}$$

A similar property holds for identity axioms. We also have

$$\mathbf{OForm}(x) \wedge \mathbf{OForm}(y) \wedge \mathbf{Sent}(x) \wedge \mathbf{Sent}(y) \wedge \\ \wedge Sat_0(x,0) \wedge Sat_0(\overline{\langle SN(\rightarrow), x, y\rangle}, 0) \rightarrow Sat_0(y,0).$$

and similarly for the rule of substitution.

To define now an arithmetized notion of an open proof denoted by $\mathbf{Prf}_0(x)$ we need an auxiliary notion $\mathbf{Ax}(x)$ being a substitution of an axiom of T. We define the latter as follows

$$\bigvee_{j\leqslant N} \{\exists t_1 \ldots \exists t_{n_j}[\bigwedge_{i\leqslant n_j} \mathbf{Term}(t_i) \wedge \forall v[\mathbf{Vble}(v) \rightarrow \neg\mathbf{Fr}(t_i, v)] \wedge \\ \wedge x = \overline{\ulcorner \varphi_j(x_1/t_1, \ldots, x_{n_j}/t_{n_j})\urcorner}]\}$$

where $x = \overline{\ulcorner \varphi_j(x_1/t_1, \ldots, x_{n_j}/t_{n_j})\urcorner}$ is an abbreviation for

$$\exists s\{\mathbf{Seq}(s) \wedge \mathbf{lh}(s) = n_j \wedge (s)_0 = \mathbf{Sub}(\overline{\ulcorner\varphi_j\urcorner}, \overline{\ulcorner x_1\urcorner}, t_1) \wedge \\ \wedge \forall i < \mathbf{lh}(s)[(s)_i = \mathbf{Sub}((s)_{i-1}, \overline{\ulcorner x_i\urcorner}, t_i)]\}.$$

The notion $\mathbf{Prf}_0(x)$ is now defined in the following way

$$\mathbf{Seq}(x) \wedge \forall i < \mathbf{lh}(x)\{\mathbf{OForm}((x)_i) \wedge \\ \wedge [Ax((x)_i) \vee (x)_i \text{ is an axiom of propositional calculus} \vee \\ \vee \mathbf{EAx}((x)_i) \vee \exists j,k < i\mathbf{MP}((x)_j,(x)_k,(x)_i)]\}.$$

Using induction (which can be formalized in PA*) we get

$$\mathrm{PA}^* \vdash Mod_{\mathrm{T}}(\Xi) \wedge \mathbf{Prf}_0(x) \rightarrow Sat_0((x)_{\mathbf{lh}(x)\dot{-}1}, 0).$$

Hence

$$\mathrm{PA}^* \vdash Mod_{\mathrm{T}}(\Xi) \rightarrow \neg\exists x[\mathbf{Prf}_0(x) \wedge ((x)_{\mathbf{lh}(x)\dot{-}1} = \overline{\ulcorner 0 \neq 0\urcorner}].$$

It follows from the fact that

$$\mathrm{PA}^* \vdash \neg Sat_0(\overline{\ulcorner 0 \neq 0\urcorner}, 0).$$

Observe that we assumed here that '$0 \neq 0$' is a sentence of the language L(T) and that the sentence '$0 = 0$' is an axiom of T. Otherwise one should use another axiom of T and its negation.

In this way we proved that one cannot obtain inconsistency by open proofs. But by Lemma 2.6.9

$$\mathrm{PA}^* \vdash \exists x \mathbf{Prf}(x, \overline{\ulcorner 0 \neq 0 \urcorner}) \rightarrow \exists x \mathbf{Prf}_0(x) \wedge (x)_{\mathrm{lh}-1} = \overline{\ulcorner 0 \neq 0 \urcorner}.$$

Consequently

$$\mathrm{PA}^* \vdash Mod_{\mathrm{T}}(\Xi) \rightarrow \mathrm{Con}_{\mathrm{T}}. \tag{2.51}$$

Theorem 2.6.12 (Mostowski, 1952) *Peano arithmetic* PA *is essentially reflexive.*

Proof. First show that PA is reflexive. So let $\mathrm{P}_n = \mathrm{PA}\lceil n$ be any fragment of PA based on n axioms, $n \in \mathbb{N}$. If φ is an axiom of P_n which is not an open formula then let φ' be a fomula in the prenex form provably equivalent (in P_n) to φ. Assume that φ' is of the form $\forall x \exists y \psi(x, y)$ where ψ is open and consider a formula φ^* of the form $\forall x \psi(x, f(x))$ where f is a new function symbol with the definition

$$f(x) = \mu y \psi(x, y),$$

i.e.,

$$f(x) = y \equiv \psi(x, y) \wedge \forall u < y \neg \psi(x, u).$$

In a similar way we eliminate all existential quantifiers from axioms of P_n and obtain an equivalent theory P'_n with universal axioms. We now take P'_n as the theory T from the above considerations and as PA^* a theory obtained from PA by adding (except symbols for the functions $\langle \ldots \rangle$, $lh(.)$, $(x)_i$ and for the predicate Seq also) symbols for recursive functions introduced in the process (described above) of eliminating existential quantifiers from the axioms of P_n together with their defining equations. It can easily be seen that $\mathrm{PA}^* \vdash Mod_{\mathrm{P}'_n}(\Xi)$. So by (2.51) we get

$$\mathrm{PA}^* \vdash \mathrm{Con}_{\mathrm{P}'_n}$$

and consequently

$$\mathrm{PA}^* \vdash \mathrm{Con}_{\mathrm{P}_n}.$$

Since Con_{P_n} is a formula of L(PA) and PA* is an extension by definitions of PA, we obtain

$$\mathrm{PA} \vdash \mathrm{Con}_{\mathrm{PA}\restriction n}.$$

The proof that any consistent extension of PA with the same constants as PA is reflexive is similar. Q.E.D.

Corollary 2.6.13 (Mostowski, 1952) *Peano arithmetic PA is not finitely axiomatizable.*

Proof. If PA were finitely axiomatizable then PA would prove its own consistency which contradicts Gödel's Second Incompleteness Theorem. Q.E.D.

To formulate the next results we need a hierarchy of formulas of the language L(PA) similar to the arithmetical hierarchy of relations (cf. Section 1.5). Let $\Sigma_0^0 = \Pi_0^0 = \Delta_0^0$ be BQ (cf. Definition 2.4.23) and we define Σ_{n+1}^0 to be the set of all formulas equivalent (in PA) to formulas of the form $\exists x\psi$ for $\psi \in \Pi_n^0$ and Π_{n+1}^0 to be the set of all formulas equivalent (in PA) to formulas of the form $\forall x\psi$ for $\psi \in \Sigma_n^0$. We also use Δ_n^0 as the set of all formulas equivalent (in PA) to a Σ_n^0 formula and to a Π_n^0 formula.

An analysis of the formulas **val** and Sat_0 defined above indicates that by modifying ζ and θ (instead of ξ_i one should now consider the function symbols of L(PA)) and by adding to ζ' and θ' appropriate clauses corresponding to quantifiers one can obtain formulas $Sat_{\Delta_0^0}$, $Sat_{\Sigma_n^0}$ and $Sat_{\Pi_n^0}$ of L(PA) which are definitions of satisfaction for Δ_0^0, Σ_n^0 and Π_n^0 formulas ($n \in \mathbb{N}$), respectively. Indeed, $Sat_{\Delta_0^0}(a, v)$ is the following formula

$$\mathbf{BQ}(a) \wedge \mathbf{v}(a, v) \wedge \exists s \exists t[\zeta_{\Delta_0^0}(s, a) \wedge \theta_{\Delta_0^0}(t, v, a)]$$

where **BQ** is a formula of L(PA) strongly representing in PA the recursive set of Gödel numbers of formulas of the class BQ (for the language L(PA)) and $\zeta_{\Delta_0^0}(s, a)$ is

$$\mathbf{Seq}(s) \wedge \forall i < \mathbf{lh}(s)[\exists i_1, i_2 < i(\mathbf{Term}((s)_{i_1}) \wedge \mathbf{Term}((s)_{i_2}) \wedge$$

$$
\begin{aligned}
&\quad \wedge (s)_i = \langle \overline{SN(=)}, (s)_{i_1}, (s)_{i_2} \rangle) \vee \\
&\vee \exists i_1 < i((s)_i = \langle \overline{SN(\neg)}, (s)_{i_1} \rangle) \vee \\
&\vee \exists i_1, i_2 < i((s)_i = \langle \overline{SN(\vee)}, (s)_{i_1}, (s)_{i_2} \rangle) \vee \\
&\vee \exists i_1, i_2, i_3 < i((s)_i = \langle \overline{SN(\exists)}, \overline{2i_1}, \overline{SN(<)}, \overline{2i_2}, (s)_{i_3} \rangle] \wedge \\
&\wedge (s)_{\mathbf{lh}(s) \dot{-} 1} = a
\end{aligned}
$$

and $\theta_{\Delta_0^0}(t, v, a)$ is

$$
\begin{aligned}
&\mathbf{Seq}(t) \wedge \mathbf{lh}(t) = \mathbf{lh}(s) \wedge \forall i < \mathbf{lh}(t) \{ \mathbf{Seq}((t)_i) \wedge \mathbf{lh}((t)_i) = 3 \wedge \\
&\quad \wedge (t)_1 < \mathbf{lh}(s) \wedge ((t)_3 = 0 \vee (t)_3 = \overline{1}) \wedge \\
&\quad \wedge [\exists u, u' \leqslant s(\mathbf{Term}(u) \wedge \mathbf{Term}(u') \wedge \\
&\qquad \wedge (s)_i = \langle \overline{SN(=)}, u, u' \rangle \wedge \\
&\qquad \wedge ((t)_3 = \overline{1} \equiv (\mathbf{val}(u, v | \mathbf{F}(u)) = \mathbf{val}(u', v | \mathbf{F}(u')))))] \vee \\
&\quad \vee \exists i_1 < i((s)_i = \langle \overline{SN(\neg)}, (s)_{i_1} \rangle \wedge \\
&\qquad \wedge \exists j < i \exists w_1 \leqslant \overline{1}((t)_j = \langle i_1, (t)_2, w_1 \rangle \wedge \\
&\qquad \wedge (((t)_i)_3 = \overline{1} \equiv w_1 = 0))) \vee \\
&\quad \vee \exists i_1, i_2 < i((s)_i = \langle \overline{SN(\vee)}, (s)_{i_1}, (s)_{i_2} \rangle \wedge \\
&\qquad \wedge \exists j_1, j_2 < i \exists w_1, w_2 \leqslant \overline{1}((t)_{j_1} = \langle i_1, (t)_2, w_1 \rangle \wedge \\
&\qquad \wedge (t)_{j_2} = \langle i_2, (t)_2, w_2 \rangle \wedge \\
&\qquad \wedge (((t)_i)_3 = \overline{1} \equiv w_1 = \overline{1} \vee w_2 = \overline{1}))) \vee \\
&\quad \vee \exists i_1, i_2, i_3 < i((s)_i = \langle \overline{SN(\exists)}, \overline{2i_1}, \overline{SN(<)}, \overline{2i_2}, (s)_{i_3} \rangle \wedge \\
&\qquad \wedge \forall r < \mathbf{val}((s)_{i_2}, ((t)_i)_2) \exists j < i \exists w_1 \leqslant \overline{1}((t)_j = \\
&\qquad\qquad = \langle i_3, (t)_2 * \begin{pmatrix} i_1 \\ r \end{pmatrix}, w_1 \rangle) \wedge \\
&\qquad \wedge (((t)_i)_3 = \overline{1} \equiv \\
&\qquad\qquad \equiv \exists x < \mathbf{val}((s)_{i_2}, ((t)_i)_2) \exists j < i((t)_j = \\
&\qquad\qquad\quad = \langle i_3, ((t)_i)_2 * \begin{pmatrix} i_1 \\ x \end{pmatrix}, 1 \rangle))))] \wedge \\
&\quad \wedge ((t)_{\mathbf{lh}(w) \dot{-} 1})_3 = 1 \}.
\end{aligned}
$$

In the formula $\theta_{\Delta_0^0}$, t is considered as a code for a sequence of triples $\langle i, z, w \rangle$ where $i < lh(s)$ is an index to the sequence coded by s and w is a 'truth value' (0 for false, 1 for true) of the formula $(s)_i$ when the variables $x_1, x_2, \ldots$ are interpreted by $(z)_1, (z)_2, \ldots$. The various disjuncts inside the formula correspond to the different ways of building up Δ_0^0 formulas.

To define now $Sat_{\Sigma_n^0}$ and $Sat_{\Pi_n^0}$ let us introduce the following notation: let $Form_{\Sigma_n^0}$ ($Form_{\Pi_n^0}$) be the set of Gödel numbers of Σ_n^0 (Π_n^0) formulas ($n \in \mathbb{N}$) and let $\mathbf{Form}_{\Sigma_n^0}$ ($\mathbf{Form}_{\Pi_n^0}$) be formulas of the language L(PA) strongly representing them in PA. Similarly for $Form_{\Delta_0^0}$ and $\mathbf{Form}_{\Delta_0^0}$.

Formulas $Sat_{\Sigma_0^0}$ and $Sat_{\Pi_0^0}$ are alternative names for the formula $Sat_{\Delta_0^0}$. Using induction define now $Sat_{\Sigma_{n+1}^0}(a, v)$ as the following formula

$$\exists b < a \exists i < a[a = \overline{\langle \overline{SN(\exists)}, \overline{\ulcorner x_i \urcorner}, b\rangle} \wedge \\ \wedge \mathbf{Form}_{\Pi_n^0}(b) \wedge \exists x Sat_{\Pi_n^0}(b, v * \begin{pmatrix} i \\ x \end{pmatrix})]$$

and $Sat_{\Pi_{n+1}^0}$ to be the formula

$$\exists b < a \exists i < a[a = \overline{\langle \overline{SN(\forall)}, \overline{\ulcorner x_i \urcorner}, b\rangle} \wedge \\ \wedge \mathbf{Form}_{\Sigma_n^0} \wedge \forall x Sat_{\Sigma_n^0}(b, v * \begin{pmatrix} i \\ x \end{pmatrix})].$$

From the above definitions it follows that the formula $Sat_{\Delta_0^0}$ is a Σ_1^0 formula. It can be also written as a Π_1^0 formula. Hence $Sat_{\Delta_0^0}$ is a Δ_1^0 formula of L(PA). As a consequence, $Sat_{\Sigma_n^0}$ ($Sat_{\Pi_n^0}$) is the formula of the class Σ_n^0 (Π_n^0) ($n \in \mathbb{N}$).

One can also show that appropriate properties of those satisfaction formulas, analogous to the properties (2.47)–(2.49) for Sat_0 can be proved (cf., e.g., Kaye, 1991).

In this way we come to the conclusion that for every particular $n \in \mathbb{N}$ the notion of satisfaction for formulas of the class Σ_n^0 (Π_n^0) is definable in PA. This contrasts with the fact proved at the beginning of this section which states that the notion of satisfaction for the whole language L(PA) is not definable in PA.

By arithmetizing model theory and provability (in a similar way as it has been done above in the proof of the reflexivity of arithmetic) one can show that for every $n \in \mathbb{N}$

$$I\Sigma^0_{n+1} \vdash \mathrm{Con}_{I\Sigma^0_n}$$

where $I\Sigma^0_k$ is a fragment of Peano arithmetic PA obtained by restricting the axiom scheme of induction to Σ^0_k formulas (for a proof see, e.g., Hájek *et al.*, 1993). This can be also obtained by prooftheoretical methods (cf. Takeuti, 1987).

In Section 3.3 a theory Q will be considered. It is a finitely axiomatizable and essentially undecidable subtheory of Peano arithmetic (for definitions of undecidability and essential undecidability see Section 3.1). Here we mention the fact that

$$\mathrm{PA} \vdash \mathrm{Con}_{\mathrm{Q}}.$$

It follows from the fact that $\mathrm{Q} \subseteq I\Sigma^0_1$ and from the above properties.

2.7 Historical Remarks

The first axiomatization of the arithmetic of natural numbers was given by Giuseppe Peano in his famous work *Arithmetices principia nova methodo exposita* (Bocca, Torino 1889, pp. xvi + 20). Peano wrote that Dedekind's work *Was sind und was sollen die Zahlen?* (Braunschweig 1888) was 'very useful' in preparing his paper. This remark serves as a basis for claims that Peano 'stole' in fact his axioms from Dedekind (Van Heijenoort, 1967; and Hao Wang 1957). On the other hand, Kennedy claims that, in the above sentence, Peano only acknowledged Dedekind's priority of publication and that "the reference to Dedekind's work was added to the preface of *Arithmetices principia* just before the pamphlet went to press, and we have an explanation of how Dedekind's work was 'useful'" (cf. Kennedy, 1972, p. 135). So the problem of how to interprete and understand Peano's remark and who was in fact the author of the first axiomatics for arithmetic remains and must remain unsolved (cf. Murawski, 1985 and 1987). Nevertheless, Peano's merit remains that he has presented the axioms for natural numbers in a nice, precise and clear form and — most important — has shown how to derive theorems of arithmetic from them.

Peano's system was not identical with the system called today 'Peano arithmetic'. Indeed, Peano arithmetic (described in Section 2.1) is a first-order theory based on the first-order predicate calculus — nowadays it is a standard system used in the foundational studies of arithmetic. The original system of Peano was formalized in the language using the notion of a set and the membership relation. It was a system based on set theory. It is worth describing it here.

In his system Peano used ten primitive logical notions: P (a sentence, i.e., the class of sentences), K (a class, i.e., the class of all classes), ε (is, i.e., the membership relation), $\cap$ (symbol of conjunction or intersection of classes, depending on context), $\cup$ (symbol of disjunction or the union of classes), $-$ (symbol of negation or the complement of a class), $\supset$ (symbol of implication read as "ab ... deducitur" or inclusion of classes), $=$ (symbol of equivalence or identity of classes), Λ (falsehood, absurd or the empty class), $[\ \varepsilon]$ (symbol of "inversion"; the expression $[x\varepsilon]a$ is read as "those x'ies for which a"). Arithmetical primitive notions are the following: N (a number, i.e., the class of natural numbers), 1 (one), $a+1$ (successor of a), $=$ (the identity of numbers). Peano's axioms are the following formulas (note that Peano used, following Leibniz, a system of dots as brackets):

1. $1\,\varepsilon\,N$,
2. $a\,\varepsilon\,N\,.\supset.\,a=a$,
3. $a,b\,\varepsilon\,N\,.\supset:\,a=b\,.=.\,b=a$,
4. $a,b,c\,\varepsilon\,N\,.\supset\therefore\,a=b\,.\,b=c:\supset.\,a=c$,
5. $a=b\,.\,b\,\varepsilon\,N:\supset.\,a\,\varepsilon\,N$,
6. $a\,\varepsilon\,N\,.\supset.\,a+1\,\varepsilon\,N$,

7. $a, b \,\varepsilon\, N \,.\, \supset : a = b \,.\, = \,.\, a + 1 = b + 1,$

8. $a \,\varepsilon\, N \,.\, \supset \,.\, a + 1 - = 1,$

9. $k \,\varepsilon\, K \,\therefore\, 1 \,\varepsilon\, k \,\therefore\, x \,\varepsilon\, N \,.\, x \,\varepsilon\, k : \supset_x \,.\, x + 1 \,\varepsilon\, k :: \supset \,.\, N \supset k.$

Axioms 2–5 and the implication to the right in 7 are in fact axioms of identity, other axioms are the famous Peano's axioms for natural numbers. Axiom 7 is the induction axiom. One should add here that there were no explicitly formulated rules of inference in Peano's system. Hence there were no formalized proofs. Peano was satisfied by an intuitive notion of deduction from the axioms — as all 'normal' (i.e., not being logicians) mathematicians still are. This was perhaps the consequence of the fact that he was interested in logic only as a tool in ordering mathematical theorems and clarifying them. He was not interested in logic itself. This gap was removed by subtle considerations of a German logician Gottlob Frege who introduced the first formalized axiomatic system.

Peano arithmetic described in Section 2.1 is based on an infinite set of axioms — indeed the axiom scheme of induction is a scheme of infinitely many axioms. Ryll-Nardzewski (1953) proved that it cannot be replaced by a finite set of axioms, i.e., Peano arithmetic is not finitely axiomatizable. It was also shown that Peano arithmetic cannot be based on a set of axioms of a given degree of complexity measured by the number of quantifiers occurring in them, i.e., of axioms belonging to a given level of arithmetical hierarchy (cf. Section 1.5).

Gödel's results on the incompleteness of arithmetic were announced for the first time during the Second Conference on Epistemology of Exact Sciences (organized by Die Gesellschaft für Empirische Philosophie) held in Königsberg, 5–7th September 1930. Kurt Gödel presented a twenty-minute talk devoted to the results contained in his doctoral dissertation, i.e., the completeness theorem for first-order logic (or, as it was then called 'the restricted functional calculus') (cf. Introduction). This theorem was very important from the point of view of the realization of Hilbert's program. Nobody expected that the next day Gödel would announce a result showing that the program could not be fully realized and suggesting a negative answer to the first three postulates from Hilbert's lecture held in Bologna in 1928 (cf. Introduction). Indeed, on September 7th Gödel took part in a discussion on the foundations of mathematics and told about his recent result on the incompleteness of arithmetic.

The first announcement of Gödel's results had appeared in (Gödel, 1930a) — it was an abstract of (1931) and was presented to the Vienna Academy of Sciences by Hans Hahn on 23rd October 1930. The results were published in Gödel's paper "Über formal unentscheidbare Sätze der 'Principia Mathematica' und verwandter Systeme. I" in January 1931 (the manuscript was received by the editors on 17th November 1930). In this paper the theorem today called Gödel's First Incompleteness Theorem was proved and the theorem called Gödel's Second Incompleteness Theorem was announced and promised to be published soon — in the second part of the paper — with a full proof. In fact the second part was never written and Gödel never published a proof of the Second Theorem. He explained this by saying that "the prompt acceptance of his results was one of the reasons that made him change his plan [of publishing the proof of the Second Theorem]" (cf. Heijenoort, 1967, footnote 68a, p. 616) and that there was then no need to publish it. One should also add that the hint to the proof given in (Gödel, 1931) has turned out to be incorrect (cf. Feferman, 1960). Add also here that Gödel (1931) used primitive recursive functions (calling them recursive functions) — cf. Section 1.7. The system of arithmetic considered by Gödel in this paper was not the first-order Peano arithmetic but a system of the theory of types with individual constants for natural numbers and arithmetical axioms of Peano. In 1932, Gödel gave a more general presentation of his results using Peano arithmetic instead of the simple theory of types as the basic system.

On 25th June 1932 the paper "Über formal unentscheidbare Sätze ..." was presented to the University of Vienna as a *Ha-*

bilitationsschrift. On 1st December 1932 Gödel was granted the *Habilitation*[7] and on 11th March 1933 the *venia legendi* and became *Privatdozent*.

It is interesting to follow the reception of Gödel's incompleteness theorems.[8] It seems that the participants of the conference in Königsberg in 1930 were unaware of the meaning and importance of those results. There was no discussion after Gödel's pronouncement. In the proceedings of the conference Gödel's name does not occur (cf. Dawson, 1984).

At least two persons among the participants of the conference in Königsberg should have had foreknowledge of Gödel's incompleteness result: Hans Hahn (1879–1934), the supervisor of Gödel's doctoral dissertation, and Rudolf Carnap[9] (1891–1970). It seems that they failed to understand it.[10]

John von Neumann (1903–1957) seems to be the only participant of the conference in Königsberg who grasped the meaning of Gödel's theorem. After Gödel's talk he had a long discussion with him and asked him about details of the proof. Soon after coming back from the conference to Berlin he wrote a letter to Gödel (on 20th November 1930) in which he announced that he had received a remarkable corollary from Gödel's First Theorem, namely a theorem on the unprovability of the consistency of arithmetic in arithmetic itself. In the meantime Gödel developed his Second Incompleteness Theorem and included it in his paper "Über formal unentscheidbare Sätze ..."

[7]H. Hahn wrote in his report: "... a scientific achievement of the first order ... it can be safely predicted to earn a place in the history of mathematics ... Herr Gödel is already acknowledged as the foremost authority on symbolic logic and on the foundations of mathematics."

[8]For a detailed discussion of this subject see Dawson (1985).

[9]Carnap wrote in his diary: "August 26, 1930: 6 to half past 8, in the Café Reichsrat coffee-house with Feigl, Gödel, later Waismann. Plan of travelling to Königsberg by boat. Gödel's discovery: incompleteness of the system of *Principia Mathematica*. Difficulties with the consistency proof."

[10]Compare, e.g., the following remark from Carnap's diary: "February 7, 1931: At 4 Gödel here. What concerns his work [i.e., Gödel's paper (1931) — R.M.], I must say that it is difficult to understand."

Gödel's paper "Über formal unentscheidbare Sätze ..." was published in January 1931. Already before its publication Paul Bernays (1888–1977) took interest in it. He had learned of these results from Richard Courant (cf. Moore, 1991). Around Christmas 1930 he wrote to Gödel requesting a copy of the galley proofs. The reactions of Bernays are important because he was the secretary and assistant of David Hilbert at that time. Bernays and Gödel corresponded a lot with each other discussing Gödel's results, their relation to Hilbert's program and the chances and possibilities of developing this program. Mid January 1931, Bernays wrote to Gödel that the incompleteness theorems were "an important step forward in research on foundational problems", and in a letter of mid-April he called the results "surprising and significant" (quotations after Moore (1991)).

And how did Hilbert himself react on Gödel's results? Though he took part in the conference in Königsberg, he did not attend the discussion during which Gödel announced his incompleteness results — he was simply busy preparing his speech for the local radio on the occassion of his honorary citizenship of Königsberg.[11] He learned about them from P. Bernays only in January 1931. And, as C. Reid writes in the biography (1970), was at the beginning "somewhat angry". The irritation and frustration passed in course of time and Hilbert tried to approach

[11]The honorary citizenship was presented to Hilbert at the meeting of the Society of German Scientists and Physicians. Hilbert delivered on this occassion an address "Naturerkennen und Logik" (cf. Hilbert, 1930) attended by Gödel. This was the only time Gödel ever saw Hilbert. They never met or corresponded — cf. (Wang Hao, 1987 p. 85) and (Dawson, 1985, footnote 4). Dawson writes: "In a letter to Constance Reid of March 22, 1966, Gödel stated that he "never met Hilbert ... nor [had] any correspondence with him". The stratification of the German academic system may have discouraged contact between the two men." Reidemeister and Szegő made arrangements for Hilbert to repeat a précis of the last part of his acceptance address at the Society meeting over the local radio station. There exists a 45-rpm-record of this radio broadcast — it was distributed with the book Hilbert (1971). It was published for the first time (German original and English translation) in (Vinnikov, 1999).

the new situation in a more constructive way. Reid writes (1970, pp. 198–199):

> Bernays found himself impressed that even now, at the very end of his career, Hilbert was able to make great changes in his program. It was not yet clear just what influence Gödel's work would ultimately have. Gödel himself felt — and expressed the thought in his paper — that his work did not contradict Hilbert's formalistic point of view.

Indeed it has turned out that Hilbert's proof theory can still be successfully developed and that one of the new problems that should be considered is the problem of the means necessary to prove the consistency of a given theory (cf. the proof of the consistency of arithmetic of natural numbers using the induction up to ε_0 due to Gerhard Gentzen (1909–1945) — this problem will be considered in Chapter 4).

Gödel exchanged letters (trying to explain his results) with another leading mathematician Ernst Zermelo (1871–1953) who turned out to be one of the greatest critics of Gödel. Their letters indicate that the main obstacle to understand each other were deep disagreements in their philosophical views on mathematics (cf. Dawson, 1985a; Grattan-Guinness, 1979). In particular, Zermelo saw a contradiction in the proof of Gödel's First Incompleteness Theorem (in his letter he called it "an essential gap"), namely that a certain proposition was neither true nor false. Gödel replied that the source of the error in Zermelo's argument was the fact that the notion of truth was not expressible in the formal system used in incompleteness results.[12]

Gödel's incompleteness results were presented to a popular audience for the first time in the spring of 1932 by Karl Menger (1902–1985) in his lecture "Die neue Logik" (the lecture was published by F. Deuticke in 1933 as one of "Fünf Wiener Vorträge" in

[12]This is one of the evidences that Gödel discovered the undefinability of truth independently of Tarski — for the discussion of this problem, see below Chapter 4, Section 4.1; see also Murawski (1998).

the booklet *Krise und Neuaufbau in der exakten Wissenschaften* — for an English translation cf. Menger, 1978).

Gödel was really sensitive to the fact that he was the first in discovering the incompleteness of arithmetic and reacted vividly on every attempt to deprive him of it. And there were many such attempts. For example Paul Finsler (1894–1970) from Zurich claimed that he had obtained similar results already in 1926. A careful inspection of Finsler's paper (cf. Heijenoort, 1967, p. 438–440) proved that Gödel was the first but Finsler did not give in.[13]

Quite different was the reaction of Emil Post (1897–1954). At the beginning of the twenties (hence nearly ten years before Gödel!) he tried to show that there are absolutely undecidable problems in mathematics. He realized that his method could be applied to yield a statement undecidable within *Principia Mathematica* whose truth could nevertheless be established by metamathematical considerations. His results (on the decision problem for normal systems) anticipated results by Gödel and Church on the incompleteness and undecidability of systems of first-order logic. Post knew of course that his results were, as he wrote, "fragmentary". He never published them and gave up the researches. However, twenty years later, hence already after the publication of Gödel's and Church's results, he attempted to publish results of his investigations from 1920–1921. At the beginning of the fourties he wrote a paper "Absolutely unsolvable problems and relatively undecidable propositons — account of an anticipation". It was submitted in 1941 to *American Journal of Mathematics.* In a letter to H. Weyl accompanying the manuscript Post explained why he did not publish his results twenty years earlier and wants to do it now, i.e., after the publications by Gödel and Church. Among reasons he mentions problems he had with publishing his earlier papers which did not find a recognition and appreciation by mathematicians as well as the problems with the health which delayed the preparation of full

[13]In connection with this problem see also (Ketelsen, 1994, pp. 131–133) and (Ladrière, 1957, p.95).

detailed proofs. Though the editors appreciated the significance of Post's investigations and results, the paper has been rejected. Communicating this decision H. Weyl wrote in a letter to Post of 2nd March 1942:

> ...I have little doubt that twenty years ago your work, partly because of its then revolutionary character, did not find its due recognition. However, we cannot turn the clock back; in the meantime Gödel, Church and others have done what they have done, and the American Journal is no place for historical accounts; ... (Personally, you may be comforted by the certainty that most of the leading logicians, at least in this country, know in a general way of your anticipation.)

Only a small part of Post's paper has been published, i.e., the part containg his Normal Form Theorem (cf. Post, 1944). The full version of the paper "Absolutely unsolvable problems and relatively undecidable propositions — account of an anticipation" was published posthumously in 1965 in Davis' book *The Undecidable* (cf. Post, 1965).

Post had always had great esteem of Gödel's results and expressed "the greatest admiration for them". He never sought to diminish Gödel's achievement. On the contrary, he wrote:

> The plan [i.e., his plan to prove the incompleteness of *Principia* — R.M.], however, included prior calisthenics at other mathematical and logical work, and did not count on the appearance of a Gödel! (1965, p. 418)

In a postcard to Gödel sent on 19th October 1938 Post wrote:

> I am afraid that I took advantage of you on this, I hope but our first meeting. But for fifteen years I had carried around the thought of astounding the mathematical world with my unorthodox ideas, and meeting the man chiefly responsible for the vanishing of that dream rather carried me away.

> Since you seemed interested in my way of arriving at these new developments perhaps Church can show you a long letter I wrote to him about them. As for any claims I might make perhaps the best I can say is that I would have *proved* Gödel's Theorem in 1921 — had I been Gödel.

And in the letter to Gödel of 30th October 1938 he wrote:

> ...after all it is not ideas but the execution of ideas that constitute a mark of greatness.

There also appeared some people claiming that Gödel's theorems were incorrect and false, that Gödel simply discovered a new antinomy. Such were the opinion of Charles Perelman (cf. his paper from 1936), Marcel Barzin (in a paper from 1940) or Jerzy Kuczyński (cf. his paper (1938)). All such opinions were criticized and the arguments provided by the authors turned out to be wrong.

It is also interesting to know how two leading contemporary philosophers, i.e., Bertrand Russell (1872–1970) and Ludwig Wittgenstein (1889–1951), reacted. Wittgensteins' remarks on Gödel's theorems are contained in Appendix I to his posthumous *Remarks on the Foundations of Mathematics.* They seem to be an embarrassment to his work. So M. Dummett (1978), in many respects an admirer of Wittgenstein's philosophy, writes that the remarks on Gödel and on the notion of consistency are "of poor quality and contain definite errors" (p. 166). It turns out that Wittgenstein failed to understand Gödel's results and — as Gödel himself put it in a letter to Abraham Robinson from 2nd July 1973 — "advance[d] a completely trivial and uninteresting misinterpretation" of them. On the other hand, recently some attempts were made to find a perspective from which Wittgenstein's view towards Gödel's theorems becomes more understandable (cf. Wang Hao 1987, 1991, 1991a; Shanker, 1988; see also Floyd, 1995).

Russell[14] reacted on Gödel's results in an ambiguous way. In (1959, p. 114) he seems to make the same point as Gödel (1932) made that by passing to higher types one can obtain formal systems such that the undecidable propositions constructed within each system are decidable in higher systems. Quite different are his views expressed in an unpublished letter to Leon Henkin of 1st April 1963 (now in the Russell Archive) in which he wrote:

> I realized, of course, that Gödel's work is of fundamental importance, but I was puzzled by it. It made me glad that I was no longer working at mathematical logic. If a given set of axioms leads to a contradiction, it is clear that at least one of the axioms must be false. Does this apply to school-boy's arithmetic, and, if so, can we believe anything that we were taught in youth? Are we to think that $2 + 2$ is not 4, but 4.001? Obviously this is not what is intended.

It is not clear if Russell "had recognized the futility of Hilbert's scheme for proving the consistency of arithmetic but had failed to consider the possibility of rigorously proving that futility", or "is he revealing a belief that Gödel had in fact shown arithmetic to be inconsistent" (cf. Dawson, 1985). Gödel wrote in a letter to A. Robinson of 2nd July 1973: "Russell evidently misinterprets my results; however he does so in a very interesting manner ..." (For further remarks on Russell's reception of Gödel's results cf. Rodríguez-Consuegra, 1993).

The first full proof of Gödel's Second Incompleteness Theorem on the unprovability of consistency was published in the second volume of Hilbert and Bernays' monograph *Grundlagen der Mathematik* in 1939. This book has stilled at last the opposition to Gödel's results (at least among logicians).

---

[14]Russell was, together with Gottlob Frege, the founder of logicism — one of the directions in modern philosophy of mathematics — and the co-author (with Alfred North Whitehead) of a monumental work *Principia Mathematica*. The main aim of this work was to show how the whole of mathematics can be reduced to logic.

Gödel's Second Incompleteness Theorem is in fact a formalization of his First Theorem but one proves it not simply by formalizing the proof of the latter (as is often erroneously claimed). One introduces certain properties of a formula **Pr** called derivability conditions (cf. Section 2.4) and using them one can easily reduce the Second Theorem to the First Theorem. This idea comes from Bernays and can be found in the second volume of *Grundlagen der Mathematik*. The conditions introduced by Bernays and called Hilbert-Bernays conditions were the following:

(1) if $\mathrm{PA} \vdash \varphi \to \psi$ then $\mathrm{PA} \vdash \mathbf{Pr}(\overline{\ulcorner \varphi \urcorner}) \to \mathbf{Pr}(\overline{\ulcorner \psi \urcorner})$,

(2) $\mathrm{PA} \vdash \mathbf{Pr}(\overline{\ulcorner \neg\varphi(x) \urcorner}) \to \mathbf{Pr}(\mathbf{sub}(\overline{\ulcorner \neg\varphi \urcorner}, \overline{\ulcorner x \urcorner}, y))$,

(3) $\mathrm{PA} \vdash f(x) = 0 \to \mathbf{Pr}(\mathbf{sub}(\overline{\ulcorner f(v) = 0 \urcorner}, \overline{\ulcorner v \urcorner}, x))$
for any primitive recursive term f.

Condition (1) is a generalization of the condition: if $\mathrm{PA} \vdash \varphi$ then $\mathrm{PA} \vdash \mathbf{Pr}(\overline{\ulcorner \varphi \urcorner})$. Condition (2) is a general formulation of the fact that if a formula $\neg\varphi(x)$ (with one free variable x) is provable in PA then any particular instance $\neg\varphi(\overline{n})$ is also provable. Condition (3) is a (bit strange) statement of the property that every primitive recursive function is representable in PA.

Hilbert-Bernays conditions are not elegant. A useful and elegant version of derivability conditions was given by Löb in 1954 (cf. Löb, 1955). Löb's conditions are conditions (D1)–(D3) from Section 2.4.

A deep analysis of the problems connected with the proof of Gödel's Second Incompleteness Theorem and with derivability conditions can be found in Feferman (1960).

The predicate *Pr* was also analysed by methods of the modal logic and interesting results on self-referential sentences were obtained in those studies. The reader interested in those topics can consult Boolos (1979 and 1993), Boolos and Sambin (1991) and Smoryński (1985).

Besides generalizations and strengthenings of Gödel's incompleteness results (in previous sections Rosser's theorem, Boolos' theorem as well as Paris-Harrington's and Paris-Kirby's theorems were discussed), new proofs of them were also given — e.g., proofs using certain results of recursion theory. Today a lot of proofs of Gödel's theorems are known.

The key idea used in Gödel's theorem is the diagonalization. It seems that the diagonalization procedure was applied for the first time by Paul Du Bois-Reymond (1831–1889) (cf. Thiel, 1995) and then generalized and fruitfully used by Georg Cantor (1845–1918) in 1874 (see also Cantor, 1892). At the beginning there was a resistence against this method (cf. for instance Bridgman (1934), see also Fraenkel (1935) and Thiel (1980) where this problem is discussed in details). Now the method is widely used in mathematics and logic (see Adamowicz (1995) for a survey of various applications of diagonalization in mathematical logic and Smullyan (1996) for a general account of self-reference, diagonalization and fixed points as they occur in logic, recursion theory, combinatory logic, semantics and metamathematics).

The diagonalization Lemma, on which Gödel's results are based, was not formulated in the general form right away. The form given in Section 2.4 comes essentially from John Barkley Rosser (1907–1989) (Rosser, 1939). It was a generalization of a theorem from Gödel (1931) stating that there exists a sentence φ such that $\mathrm{PA} \vdash \varphi \equiv \neg \mathbf{Pr}(\overline{\ulcorner \varphi \urcorner})$.[15] Andrzej Ehrenfeucht and Solomon Feferman (1960) proved the Diagonalization Lemma for formulas with free variables and Richard Montague (1930–1971) (1962) gave the final version of it in the form of the following theorem: for any formula $\psi(x, y)$ with two free variables x and y there exists a formula $\varphi(y)$ with one free variable (in which the variable x does not occur) such that

$$\mathrm{PA} \vdash \varphi(y) \equiv \psi(\overline{\ulcorner \varphi(x) \urcorner}, y).$$

[15]Note that in Gödel (1934, Section 7) one finds the following remark: "It is even possible, for any metamathematical property f which can be expressed in the system to construct a proposition which says of itself that it has this property". In a footnote added in (Davis, 1965) it is credited to (Carnap, 1934a).

This is an exact counterpart of Kleene's Recursion Theorem from 1938.

Alfred Tarski (1901–1983) published his theorem on the undefinability of truth in 1933, see also Tarski (1936 and 1956). He showed his indebtedness to Gödel's methods and did not claim any priority for Gödel's own results (cf. Tarski (1956), especially the bibliographical note on p. 152, footnote 1 on pp. 247–248 and the historical note on pp. 277–278 as well as footnote 2 (p. 279) to the paper (1956a)). The last reference is interesting also for another reason. Here Tarski states that he gave an example of a consistent and yet not ω-consistent system in 1927 — hence prior to Gödel (it is given again in the cited paper in a slightly altered form).

It should be added here that Gödel was aware of the formal undefinability of the notion of truth. This follows from a letter of Bernays to Gödel of 3rd May 1931 (cf. Dawson, 1985) and from a letter of Gödel to Zermelo (mentioned above). Another evidence for that is Gödel's reply to a graduate student's query from 1970. He indicated there that it was precisely his recognition of the contrast between the formal definability of demonstrability and the formal undefinability of truth that led to his discovery of incompleteness. Gödel did not bring this out in his famous paper of 1931. The reasons for that will be discussed in Chapter 4.[16]

Gödel's incompleteness theorems are still in the focus of interests of mathematicians and logicians. New interpretations of those results are searched for and their philosophical meaning is studied — we shall discuss those problems in Chapter 4. Let us here mention only a book by Hofstadter *Gödel, Escher, Bach: An Eternal Golden Briad* where Gödel's results are presented in a nonstandard way and where some connections between ... fugues of Bach, engravings of Escher and theorems of Gödel are indicated.

---

[16]Note that in (1934) Gödel explicitly wrote about the undefinability of truth. In a footnote added in Davis (1965) he credited this to Tarski (1933) and Carnap (1934). For the problem of the priority of proving the theorem on the undefinability of truth cf. Murawski (1998).

Mostowski's theorem on reflexivity of Peano arithmetic was proved in 1952. The main motive of this paper was the theorem stating that no consistent reflexive theory is finitely axiomatizable (recall that — as already mentioned above — this fact for Peano arithmetic was proved by another method by Ryll-Nardzewski (1953)). Add that the theorem on reflexivity has also been proved by Kreisel and Wang (1955). For formulation of the notion of reflexivity for a wider class of theories cf. Feferman (1960) (there one finds also a description of another class of reflexive theories discovered by Montague in (1957) and (1957a)).

The notion of satisfaction in general, especially in the context of nonstandard models of Peano arithmetic and nonstandard formulas, was studied by Abraham Robinson (1918–1974) in 1963. These researches were then rejuvenated by S. Krajewski in 1976 who introduced the notion of a satisfaction class and continued by Kossak, Kotlarski, Lachlan, Murawski, Ratajczyk and Schmerl who indicated many interesting properties and various applications of satisfaction classes in the foundations of mathematics, and in particular in the foundations of arithmetic — cf., e.g., the survey papers (Kossak, 1985), (Kotlarski, 1991) and (Murawski, 1988 and 1997).

Since satisfaction for formulas of the language L(PA) is not definable in Peano arithmetic PA (as shown in Section 2.6), one can consider an extension of PA obtained by augmenting the language L(PA) by a new binary predicate **S** and by adding to PA new nonlogical axioms stating that **S** is a satisfaction class for L(PA) with specified properties (such as, e.g., being full or inductive — for definitions see Krajewski (1976) or one of the survey papers indicated above). Extensions of this type have been studied by various authors. In particular Kotlarski and Ratajczyk (1990 and 1990a) investigated first-order part of such an extension for full satisfaction classes and of its fragments and characterized them in terms of consistency of appropriate systems of ω-logic and of transfinite induction.

Chapter 3

Decidability Theory

3.1 Basic notions and theorems

The method of arithmetization of syntax described in Secton 2.3 can be applied not only to Peano arithmetic but also to any formalized theory in a first-order language.

Let T be a first-order theory in the language L(T) such that the set of (Gödel numbers of) nonlogical symbols is recursive (observe that this assumption does not restrict essentially the class of theories — theories cosidered in mathematical practice have usually finite, hence recursive, set of nonlogical symbols). Denote by Thm_{T} the set of Gödel numbers (by a given arithmetization of syntax) of theorems of T. We will say that the theory T is recursively axiomatizable (briefly: axiomatizable) if and only if the set of (Gödel numbers of) its nonlogical axioms is recursive. Note that since the set of (Gödel numbers of) logical axioms and identity axioms of a theory T of the considered type is always recursive, we can conclude that if a theory is axiomatizable then the set of (Gödel numbers of) all its axioms is recursive. Consequently if T is axiomatizable then the set Thm_{T} is recursively enumerable (though not necessarily recursive!). To simplify the formulation of theorems we shall speak about recursiveness of sets of formulas instead of recursiveness of sets of Gödel numbers of those formulas in the sequel.

Definition 3.1.1 *A theory* T *is said to be decidable if and only if the set* Thm_T *is recursive. Otherwise* T *is called undecidable.*

Definition 3.1.2 *A theory* T_2 *in the language* $L(T_2)$ *is said to be an extension of a theory* T_1 *in the language* $L(T_1)$ *if and only if every axiom of* T_1 *is a theorem of* T_2*; under the same conditions* T_1 *is referred to as a subtheory of* T_2*. An extension* T_2 *of a theory* T_1 *is called simple if and only if* $L(T_2) = L(T_1)$.

Definition 3.1.3 (i) *A theory* T *is said to be essentially undecidable if and only if* T *is undecidable and every consistent simple extension of it is also undecidable.*

(ii) *A theory* T *is said to be hereditarily undecidable if and only if every subtheory* T_1 *of* T *such that* $L(T_1) = L(T)$ *is undecidable.*

Definition 3.1.4 *A structure* $\mathfrak{A}$ *is said to be strongly undecidable if and only if any theory* T *such that* $\mathfrak{A} \models T$ *is undecidable.*

Definition 3.1.5 *A theory* T *is said to be strongly undecidable if and only if* T *is consistent and any model of it is strongly undecidable.*

Using Church's Thesis (cf. Section 1.6), stating that the class $\mathcal{O}$ of computable functions is equal to the class $\mathcal{R}$ of recursive functions, one can say that a theory is decidable if there exists a mechanical method which permits us to decide in a finite number of steps whether a given formula of the language L(T) is a theorem of T or not. In this case it is sometimes said that the theory T admits a decision procedure.

Corollary 2.4.22(ii) implies that Peano arithmetic is essentially undecidable. Similarly any consistent first-order theory, in which every recursive set is strongly representable, is essentially undecidable. It will be shown below (cf. Corollary 3.3.22 and Theorem 3.3.23) that the structure $\mathfrak{N}_0$ is strongly undecidable and that Peano arithmetic PA is also strongly undecidable. Examples of hereditarily undecidable theories will also be given.

Now we prove a theorem indicating connections between decidability and completeness of a theory.

Theorem 3.1.6 (Janiczak, 1950) *If an axiomatizable theory* T *is consistent and complete then it is decidable.*

Proof. Define two recursive functions F and K in the following way:

$$F(0, a) = a,$$
$$F(n+1, a) = \langle SN(\forall), 2n, F(n, a)\rangle,$$
$$K(a) = F(a+1, a).$$

If $a = \ulcorner\varphi\urcorner$ for a formula φ then $K(a) = \ulcorner\forall x_a \ldots \forall x_0\varphi\urcorner$ where $x_0, \ldots, x_a$ are variables ordered alphabetically. It can be easily seen that if a variable x_i occurs as free in the formula φ then $i < \ulcorner x_i\urcorner = 2i < \ulcorner\varphi\urcorner = a$. Hence the formula $\forall x_a \ldots \forall x_0\varphi$ is closed. We have: $\mathrm{T} \vdash \varphi$ if and only if $\mathrm{T} \vdash \forall x_a \ldots \forall x_0\varphi$. By completeness of T:

$$\mathrm{T} \nvdash \varphi \quad \text{if and only if} \quad \mathrm{T} \vdash \neg\forall x_a \ldots \forall x_0\varphi.$$

Hence

$$\begin{aligned} a \notin Thm_{\mathrm{T}} &\equiv a \in \mathbb{N} - Thm_{\mathrm{T}} \\ &\equiv \neg Sent_{\mathrm{T}}(a) \vee \langle SN(\neg), K(a)\rangle \in Thm_{\mathrm{T}} \\ &\equiv \exists y[\neg Sent_{\mathrm{T}}(a) \vee Prf_{\mathrm{T}}(y, \langle SN(\neg), K(a)\rangle)]. \end{aligned}$$

Consequently the complement of the set Thm_{T} is recursively enumerable. But by the assumption that the theory T is axiomatizable, the set Thm_{T} is also recursively enumerable, hence by Negation Theorem (cf. Theorem 1.5.2) the set Thm_{T} is recursive. So the theory T is decidable. Q.E.D.

Corollary 3.1.7 *For a complete theory* T *the following conditions are equivalent:*

(i) T *is undecidable,*
(ii) T *is essentially undecidable,*
(iii) T *is not axiomatizable.*

Proof. The implication (ii) $\rightarrow$ (i) follows from the definitions of the notions involved. To prove the implication (iii) $\rightarrow$ (ii) assume that a complete theory T is not axiomatizable and that T is not essentially undecidable, i.e., either (1) T is decidable or (2) T is undecidable but there exists a consistent simple extension of T being decidable. Using in case (1) the fact that any decidable theory is (recursively) axiomatizable[1] and in case (2) the fact that any consistent simple extension of a complete theory is equivalent to it, one obtains a contradiction. Hence the implication (iii) $\rightarrow$ (ii) holds. The implication (i) $\rightarrow$ (iii) follows from Theorem 3.1.6. Q.E.D.

Corollary 3.1.8 *Let a theory* T *be consistent and axiomatizable. If* T *is undecidable then it is incomplete.*

Hence a proof of undecidability of a given consistent and axiomatizable theory yields a proof of its incompleteness. Note that this method of proving incompleteness provides no examples of undecidable formulas.

Using the above corollary and Corollary 2.4.22 we get that Peano arithmetic PA is essentially undecidable, i.e., it is incomplete and any axiomatizable and consistent extension of it (even formalized in a language extending the language L(PA)) is incomplete. Hence the incompleteness of PA cannot be removed by adding undecidable sentences as new nonlogical axioms (provided they form a recursive set).

Note also that the inverse theorem to Theorem 3.1.6 does not hold, i.e., it is not true that if an axiomatizable theory is decidable then it is complete. Indeed, consider the theory of Abelian groups (described in the next section). Szmielew proved that it is decidable. One can easily see that it is not complete. It suffices to consider the following two structures: $\mathfrak{A} = \langle \mathbb{Z}, + \rangle$ and $\mathfrak{B} = \langle \mathbb{Q}, + \rangle$, where $\mathbb{Z}$ is the set of all integers and $\mathbb{Q}$ is the set of all rationals. Both are models of the theory of Abelian groups but

$$\langle \mathbb{Z}, + \rangle \not\models \forall x \exists y (y + y = x)$$

[1]It suffices to take as axioms all theorems of the considered theory.

and

$$\langle \mathbb{Q}, + \rangle \models \forall x \exists y (y + y = x).$$

Hence $\langle \mathbb{Z}, + \rangle$ is not elementarily equivalent to $\langle \mathbb{Q}, + \rangle$ and consequently the theory of Abelian groups is not complete. As another counterexample can serve the theory of Boolean algebras (also described in the next section). Tarski proved that it is decidable. The incompleteness of it follows from the facts that (1) there exist atomic as well as nonatomic Boolean algebras and (2) the property of being (non)atomic is definable in the language of Boolean algebras.

Theorem 3.1.9 (Tarski, 1949) *Every incomplete and decidable theory* T *has a decidable, consistent and complete extension.*

Proof. By the assumption the theory T is incomplete, hence T is consistent. We define an increasing sequence T_k of extensions of T such that every T_k is consistent and decidable. Let

$$S_0 = Thm_{\mathrm{T}},$$

$$n \in S_{k+1} \equiv n \in S_k \vee \langle SN(\rightarrow), \mu m_{m \leqslant k}(Sent(m) \wedge \wedge m \notin S_k \wedge \langle SN(\neg), m \rangle \notin S_k), n \rangle \in S_k.$$

T_k is a theory such that $S_k = Thm_{\mathrm{T}_k}$. By the definition of S_k the theory T_{k+1} is obtained by adding to the theory T_k those sentences which are implied (in T_k) by the smallest sentence undecidable in T_k, i.e., by the sentence undecidable in T_k which has the smallest Gödel number. By the construction every theory T_k is consistent. We prove now that every T_k is decidable. Indeed, consider the following recursive function:

$$g(x, 0) = \begin{cases} 0, & \text{if } x \in Thm_{\mathrm{T}}, \\ 1, & \text{if } x \notin Thm_{\mathrm{T}}. \end{cases}$$

$$g(x, y+1) = g(x, y) \cdot g(\langle SN(\rightarrow), \mu m_{m \leqslant y}(Sent(m) \wedge \wedge g(m, y) \neq 0 \wedge g(\langle SN(\neg), m \rangle, y) \neq 0), x \rangle, y).$$

One has of course $n \in S_k \equiv g(n,k) = 0$. Hence the theory T_k is decidable.

Let now $\mathrm{T}' = \bigcup_{k \in \mathbb{N}} \mathrm{T}_k$. Then

$$n \in Thm_{\mathrm{T}'} \equiv \exists k [g(n,k) = 0].$$

Hence the set $Thm_{\mathrm{T}'}$ is recursively enumerable. By the construction the theory T' is consistent and complete. Using now the proof of Theorem 3.1.6 one obtains that T' is decidable. Q.E.D.

Corollary 3.1.10 *For an axiomatizable theory* T *the following conditions are equivalent:*

(i) T *is essentially undecidable,*

(ii) T *is consistent and every consistent and axiomatizable extension of* T *is incomplete,*

(iii) T *is consistent and no consistent and complete extension of* T *is axiomatizable.*

Corollary 3.1.11 *If Peano arithmetic* PA *is consistent then no consistent and complete extension of it is (recursively) axiomatizable.*

To end this section we formulate simple theorems pointing out some connections between various kinds of undecidability. Recall that for a given structure $\mathfrak{A}$ we denote by $\mathrm{Th}(\mathfrak{A})$ the theory of $\mathfrak{A}$, i.e., the set of all sentences (formulated in the language of the structure $\mathfrak{A}$) true in $\mathfrak{A}$.

Theorem 3.1.12 *A structure* $\mathfrak{A}$ *is strongly undecidable if and only if the theory* $\mathrm{Th}(\mathfrak{A})$ *is hereditarily undecidable.*

Proof. Let the structure $\mathfrak{A}$ be strongly undecidable and let T be any theory such that $\mathrm{Th}(\mathfrak{A})$ is a simple extension of T. Then $\mathfrak{A} \models \mathrm{T}$. Hence T is undecidable and consequently $\mathrm{Th}(\mathfrak{A})$ is hereditarily undecidable.

To prove the converse implication assume that the theory $\mathrm{Th}(\mathfrak{A})$ is hereditarily undecidable. Let T be any theory such that $\mathfrak{A} \models$ T. So Th($\mathfrak{A}$) is a simple extension of T. Hence T is undecidable and consequently the structure $\mathfrak{A}$ is strongly undecidable. Q.E.D.

Theorem 3.1.13 *If a theory* T *has an undecidable model then* T *is hereditarily undecidable.*

Proof. Let $\mathfrak{A}$ be a strongly undecidable model of the theory T. By Theorem 3.1.12 the theory Th($\mathfrak{A}$) is hereditarily undecidable. Hence the theory T is also hereditarily undecidable. Indeed, if T is a simple extension of a theory T_1 then Th($\mathfrak{A}$) is also a simple extension of T_1. Q.E.D.

Theorem 3.1.14 *Every strongly undecidable theory is hereditarily undecidable.*

Proof. Let T be a strongly undecidable theory and let $\mathfrak{A}$ be its model. Hence $\mathfrak{A}$ is strongly undecidable. By Theorem 3.1.13 the theory T is hereditarily undecidable. Q.E.D.

3.2 Decidable Theories

Recall that a theory T is said to be decidable if and only if the set of (Gödel numbers of) its theorems is recursive, i.e., if, using Church's Thesis, there exists an effective method which permits to decide whether any given formula of the language of T is a theorem of T or not. There are essentially three main methods of proving decidability of a theory:

(i) elimination of quantifiers,

(ii) modeltheoretic method,

(iii) method of interpretation.

All those methods will be presented in this section and some applications of them will be indicated.

In the previous section we proved the theorem stating that if a theory T is consistent, complete and axiomatizable then T is decidable (cf. Theorem 3.1.6). Hence to prove decidability of an axiomatizable theory it suffices to show its completeness. This is just the aim of the elimination of quantifiers and the modeltheoretical method.

It does not mean that only complete theories can be decidable. Indeed, there are incomplete decidable theories as well — two examples were given in the previous section. The following theorem holds.

Theorem 3.2.1 *Let* T *be a recursively axiomatizable theory. Assume that there exists a recursive sequence of sentences*

$$\alpha_1, \alpha_2, \alpha_3, \ldots$$

such that

(1) T $\cup\{\alpha_n\}$ *is consistent for every* n,

(2) *for any completion* T_1 *of the thoery* T (*i.e., for any extension* T_1 *of* T *to a complete theory*) *there exists* (*not necessarily recursive*) *set of sentences* $\Psi = \{\psi_1, \psi_2, \ldots\}$ *such that* $T_1 = Cn(\Psi)$ *and for any* k *there is an* n *such that* $T \vdash \alpha_n \equiv \psi_1 \wedge \ldots \wedge \psi_k$.

Then the theory T *is decidable.*

Proof. Let Φ be a recursive set of axioms of the theory T. Let $\Phi_n = \Phi \cup \{\alpha_n\}$ $(n = 1, 2, 3, \ldots)$. The set of consequences of a set Φ_n is of course recursively enumerable. Hence its elements can be ordered in a recursive sequence $\chi_1^n, \chi_2^n, \chi_3^n, \ldots$ (this follows from the fact that any recursively enumerable set is the image of $\mathbb{N}$ via a recursive function). Using now the dovetailing method order in a recursive sequence all the consequences of all the sets Φ_n $(n = 1, 2, 3, \ldots)$ — one obtains, e.g., the following sequence: $\chi_1 = \chi_1^1$, $\chi_2 = \chi_1^2$, $\chi_3 = \chi_2^1$, $\chi_4 = \chi_1^3$, $\chi_5 = \chi_2^2$, ... Let now $\varphi_1, \varphi_2, \varphi_3, \ldots$ be a recursive enumeration of the set of consequences of Φ, i.e., of the set of theorems of the theory T. Using again the dovetailing method order in a recursive sequence the formulas $\chi_1, \chi_2, \chi_3, \ldots$ and the formulas $\varphi_1, \varphi_2, \varphi_3, \ldots$

It can be easily seen that (1) if $\mathrm{T} \vdash \varphi$ then there exists an i such that $\varphi = \varphi_i$. We claim that (2) if $\mathrm{T} \nvdash \varphi$ then there exists a j such that $\neg\varphi$ is of the form χ_j. Indeed, assume that $\mathrm{T} \nvdash \varphi$. Then $\mathrm{T} \cup \{\neg\varphi\}$ is consistent. Hence there is a complete theory T_1 extending $\mathrm{T} \cup \{\neg\varphi\}$. Let $\Psi = \{\psi_1, \ldots, \psi_n, \ldots\}$ be such that $\mathrm{T}_1 = Cn(\Psi)$. By the finiteness property of the consequence operation Cn there exists a k such that $\{\psi_1, \ldots, \psi_k\} \vdash \neg\varphi$. By the assumption there is an n with the property $\mathrm{T} \vdash \alpha_n \equiv \psi_1 \wedge \ldots \wedge \psi_k$. Hence $\mathrm{T} \cup \{\alpha_n\} \vdash \neg\varphi$. Consequently, the formula $\neg\varphi$ occurs in the sequence (χ_l) of formulas enumerating consequences of the sets Φ_n $(n = 1, 2, 3, \ldots)$.

We can now prove the decidability of the theory T. Indeed, let φ be any formula of the language L(T). Enumerate the sequence containing all formula χ_i and all formulas φ_j. Since either $\mathrm{T} \vdash \varphi$ or $\mathrm{T} \vdash \neg\varphi$, it follows from (1) and (2) that either φ or $\neg\varphi$ must be found in this sequence. If the former holds then we conclude that $\mathrm{T} \vdash \varphi$, if the latter then we have $\mathrm{T} \nvdash \varphi$.

Observe that conditions (1) and (2) can be strengthened to equivalences and that they prove in fact that the sets Thm_{T} and $\mathbb{N} - Thm_{\mathrm{T}}$ are recursively enumerable. Using now Negation Theorem (cf. Theorem 1.5.2) we get that the set Thm_{T} is recursive, hence the theory T is decidable. Q.E.D.

Theorem 3.2.1 can be applied to the proof of the decidability of a given incomplete theory T if we are able to search all its complete extensions.

Consider now the method of elimination of quantifiers. Generally speaking, this method can be described as follows: let T be a first-order theory in a language L(T). We are looking for a set Φ of formulas in L(T), called the set of basic formulas for T, such that for every formula φ of L(T) there exists a Boolean combination φ^* of formulas from Φ having the same free variables as φ and such that $\mathrm{T} \vdash \varphi \equiv \varphi^*$. Of course one can always take as the basic set Φ the set of all formulas of the language L(T). But we are looking for 'good' basic sets consisting of 'simple' formulas. Unfortunately there is no precise criterion for being 'good' or 'simple'. Usually the following conditions are required: (1) Φ should be reasonably small and irredundant, (2) every formula in Φ should have some straightforward mathematical meaning, and (3) there should exist an algorithm for reducing every formula φ of the language L(T) to its corresponding Boolean "representation". From the point of view of the decidability problems the set Φ should have one more property: (4) there should exist an algorithm (an effective mechanical procedure) which would tell us, given any basic sentence ψ, either that ψ is a theorem of T or that ψ is refutable from T. If conditions (3) and (4) are met then we have both a completeness proof and a decision procedure for the theory T. In this case it is said that theory T admits the elimination of quantifiers. Let us add that a method fulfilling the conditions (3) and (4) is sometimes called 'effective quantifier elimination' to distinguish it from other types of the elimination of quantifiers where there are no algorithms.

As an example of application of this method let us sketch here the result of Presburger (1930). It states that the arithmetic of addition (today called 'Presburger arithmetic') is decidable. For historical accuracy it should be added here that Presburger (1930) considered the arithmetic of integers and not the arithmetic of natural numbers. The result for natural numbers was obtained in fact by Hilbert and Bernays.

Consider a first-order theory T_+ formalized in the language with the following nonlogical symbols: $S, +, 0$ and having the following nonlogical axioms:

(a) $0 \neq S(x)$,

(b) $x \neq 0 \rightarrow \exists y[x = S(y)]$,

(c) $x + y = y + x$,

(d) $x + (y + z) = (x + y) + z$,

(e) $x + z = y + z \rightarrow x = y$,

(f) $x + 0 = x$,

(g) $\exists y(x = z + y \vee z = x + y)$,

$$(\mathrm{h}_n)\quad \exists y[x = \underbrace{(y + \ldots + y)}_{n} \vee x = \overline{1} + \underbrace{(y + \ldots + y)}_{n} \vee \ldots$$
$$\ldots \vee x = (\overline{n-1}) + \underbrace{(y + \ldots + y)}_{n}] \text{ for } n > 1.$$

Note that the last axiom (being in fact a scheme of infinitely many axioms) says that for any number x either $n|x$ or the remainder of the division of x by n is 1 or 2 or ... or $n - 1$.

Define in T_+ the notion of a congruence $\equiv_n$ (for any given natural number n) in the following way:

$$(x \equiv_n y) \equiv \exists z[x = y + \underbrace{(z + \ldots + z)}_{n} \vee y = x + \underbrace{(z + \ldots + z)}_{n}].$$

Observe that the arithmetic T_+ has the following properties (the expression $x \leqslant y$ is here an abbreviation for the expression $\exists z(x + z = y)$):

(1) if $n < m$ then $\mathrm{T}_+ \vdash \overline{n} < \overline{m}$,

(2) if $n \geqslant m$ then $\mathrm{T}_+ \vdash \neg(\overline{n} < \overline{m})$,

(3) if $n \equiv m \pmod{k}$ then $\mathrm{T}_+ \vdash \overline{n} \equiv_k \overline{m}$,

(4) if $n \not\equiv m \pmod{k}$ then $\mathrm{T}_+ \vdash \neg(\overline{n} \equiv_k \overline{m})$.

Note also that the axiom (h_n) and the definition of the congruence imply that for any n:

$$\mathrm{T}_+ \vdash [x \equiv_n 0 \vee x \equiv_n 1 \vee \ldots \vee x \equiv_n n - 1]$$

and generally:

$$\mathrm{T}_+ \vdash [y \equiv_n z \vee y \equiv_n z + 1 \vee \ldots \vee y \equiv_n z + n - 1].$$

One has also:

$$\mathrm{T}_+ \vdash x < y \vee y < x \vee x = y.$$

Using appropriate tautologies of the propositional calculus one obtains:

$$\mathrm{T}_+ \vdash \neg(x < y) \equiv (y < x \vee x = y),$$

$$T_+ \vdash \neg(x = y) \equiv (x < y \lor y < x),$$

$$T_+ \vdash \neg(y \equiv_n z) \equiv (y \equiv_n z + 1 \lor \ldots \lor y \equiv_n z + n - 1).$$

We shall use in the sequel functions $E\left(\frac{a}{n}\right)$ (for any given natural number $n > 0$) defined as follows (compare those functions with the function $\mathrm{qt}(a, b)$ considered in Section 1.2):

$$E\left(\frac{a}{n}\right) = b \quad \equiv \quad [a = \underbrace{b + \ldots + b}_{n} \lor a = \underbrace{b + \ldots b}_{n} + 1 \lor \ldots$$
$$\ldots \lor a = \underbrace{b + \ldots + b}_{n} + (n-1)].$$

The existence of such a b follows from the axiom (h_n) and the uniqueness from the identity axioms and the axiom (e).

Theorem 3.2.2 *The theory* T_+ *is complete.*

Sketch of the proof. By properties (1)–(4) for sentences φ of the form

$$\overline{n} \equiv_k \overline{m}, \quad \overline{n} < \overline{m}, \quad \overline{n} = \overline{m} \tag{3.1}$$

either $T_+ \vdash \varphi$ or $T_+ \vdash \neg\varphi$. Hence such sentences are decidable in T_+. The same holds for any conjunctions and disjunctions of sentences of the indicated form. To prove the theorem it suffices to show that any sentence φ of the language $L(T_+)$ is equivalent in T_+ to a disjunction of conjunctions of sentences of the form (3.1). To do this one uses the elimination of quantifiers.

Consider a sentence φ of the language $L(T_+)$ and assume that it has the following form:

$$\ldots \exists x[(\varphi_1^1 \land \ldots \land \varphi_{n_1}^1) \lor \ldots \lor (\varphi_1^k \land \ldots \land \varphi_{n_k}^k)]$$

where φ_j^i are formulas of the form:

$$\alpha x + a = \beta x + b,$$

$$\alpha x + a < \beta x + b,$$

$$\alpha x + a \equiv_n \beta x + b$$

where a and b are terms in which the variable x does not occur, α and β are natural numbers and the expressions αx and βx are abbreviations of the expressions $\underbrace{x + \ldots + x}_{\alpha}$ and $\underbrace{x + \ldots + x}_{\beta}$, respectively. Using axioms of T_+ one can reduce φ^i_j to the form:

$$\alpha x + a = b,$$

$$\alpha x + a \leqslant b,$$

$$b < \alpha x + a,$$

$$\alpha x + a \equiv_n b.$$

Call them basic formulas. Note that we can assume that the last quantifier in φ is existental, since otherwise one can consider the negation of the sentence φ. We shall show that this quantifier (and consequently all other quantifiers occurring in φ) can be eliminated. To do this assume — for simplicity — that in the formula φ only one quantifier occurs and that it is an existential one. Using appropriate tautologies of the predicate calculus we obtain that φ is equivalent in T_+ to the disjunction:

$$\exists x(\varphi^1_1 \wedge \ldots \wedge \varphi^1_{n_1}) \vee \ldots \vee \exists x(\varphi^k_1 \wedge \ldots \wedge \varphi^k_{n_k}).$$

Consider one disjunct of φ, i.e., the formula φ_i of the form

$$\exists x(\varphi^i_1 \wedge \ldots \wedge \varphi^i_{n_i}).$$

We shall show that the existential quantifier $\exists x$ can be eliminated. Notice that the formula φ_i says that there exists a common solution of given equations and congruences lying in a given interval. The possibility of eliminating the existential quantifier means simply that the existence of such a solution can be expressed by certain equations and congruences between coefficients and that the appropriate equivalence can be proved in the theory T_+. This is given by the following

Theorem 3.2.3 (Lemma on quantifier elimination) *Every formula φ_i of the form $\exists x(\varphi^i_1 \wedge \ldots \wedge \varphi^i_{n_i})$ where φ^i_j $(j = 1, \ldots, n_i)$*

are basic formulas is equivalent in T_+ *to a formula* φ' *which contains at most the same free variables as* φ_i *and which is a conjunction (or a disjunction of conjunctions) of basic formulas.*

Sketch of the proof. We can assume that the formula φ_i is of the form:

$$\exists x(\tau_1 \wedge \ldots \wedge \tau_k \wedge \upsilon_1 \wedge \ldots \wedge \upsilon_l \wedge \chi_1 \wedge \ldots \wedge \chi_s \wedge \psi_1 \wedge \ldots \wedge \psi_t \wedge \theta)$$

where $\tau_1, \ldots, \tau_k$ are equations of the type $\alpha x + a = b$, formulas $\upsilon_1, \ldots, \upsilon_l$ are inequalities of the type $\alpha x + a \leqslant b$, formulas $\chi_1, \ldots, \chi_s$ are inequalities of the type $b < \alpha x + a$, formulas $\psi_1, \ldots, \psi_t$ are congruences of the type $\alpha x + a \equiv_n b$ and θ is the conjunction of those subformulas of φ_i in which x does not occur. The method of eliminating the quantifier depends on the number of formulas of the appropriate type occurring in the considered formula. We distinguish several cases.

Case 1. Assume that $k > 0$, i.e., at least one equation occurs in φ_i. Assume that τ_1 is of the form $\alpha x + a = b$ and denote by $\Omega(x)$ the conjunction of all other conjuncts of φ_i. We have now in T_+:

$$\begin{aligned} \varphi_i &\equiv \exists x[\alpha x + a = b \wedge \Omega(x)] \\ &\equiv \left\{\exists x(\alpha x + a = b) \wedge \Omega\left(\frac{b \dot{-} a}{\alpha}\right)\right\} \end{aligned}$$

where $\Omega\left(\frac{b \dot{-} a}{\alpha}\right)$ is a formula obtained from $\Omega(x)$ by replacing x by the term $b \dot{-} a$ and multiplying by α every member of every equation, inequality and congruence occurring in $\Omega(x)$ (in the case of congruences one multiplies also the modulus). Note that by the multiplication of X by α we mean simply the addition of X α times. Observe that one has in T_+

$$\exists x[\alpha x + a = b \wedge \Omega(x)] \equiv \left(a \equiv_\alpha b \wedge \Omega\left(\frac{b \dot{-} a}{\alpha}\right)\right).$$

In this way the existential quantifier has been eliminated. To finish the proof in this case one must show that the formula $\Omega\left(\frac{b\dot{-}a}{\alpha}\right)$ can be written using only basic formulas. But this follows from the following equivalences:

$$\alpha(b\dot{-}a) + c = d \equiv (b \leqslant a \wedge c = d) \vee (a < b \wedge \alpha b + c = d + \alpha a),$$

$$\alpha(b\dot{-}a) + c \leqslant d \equiv (b \leqslant a \wedge c \leqslant d) \vee (a < b \wedge \alpha b + c \leqslant d + \alpha a),$$

$$c < \alpha(b\dot{-}a) + d \equiv (b \leqslant a \wedge c < d) \vee (a < b \wedge \alpha a + c < \alpha b + d),$$

$$\alpha(b\dot{-}a) + c \equiv_n d \equiv (b \leqslant a \wedge c \equiv_n d) \vee (a < b \wedge \alpha b + c \equiv_n d + \alpha a).$$

Case 2. Assume that $k = 0$ but $t \neq 0$, i.e., at least one congruence occurs in φ_i. In this case one reasons as follows: if there exist roots of a congruence then there exist infinitely many roots. We are looking for roots fulfilling additionaly some inequalities of the type υ and χ. Inequalities of the type χ give lower bounds and can be easily fulfilled. Hence consider conditions of the type υ. We distinguish here two subcases.

Subcase 2a. Assume that $k = 0$, $t \neq 0$ and $l \neq 0$, i.e., at least one condition of the type υ giving an upper bound to the roots of the congruences occurs in φ_i. In this case one proceeds as follows. Let c be the least upper bound and assume that we have only one congruence and no conditions of the type χ (this means that $s = 0$). Then in T_+:

$$\begin{aligned}\exists x(\alpha x + a \equiv_n b \wedge x \leqslant c) &\equiv \\ &\equiv (\alpha c + a \equiv_n b \vee \alpha(c\dot{-}1) + a \equiv_n b \vee \ldots \\ &\ldots \vee \alpha((c\dot{-}n) + 1) + a \equiv_n b),\end{aligned}$$

what gives the elimination of the existential quantifier. In the case of more than one congruence one must write a disjunction of more disjuncts on the right hand side, more exactly the disjunction for numers $c\dot{-}[n_1, \ldots, n_t]\dot{-}1$ where $[n_1, \ldots n_t]$ is the least common multiple of moduli of all congruences for which one wants to have a common root. To prove the equivalence observe that the implication to the left is obvious. The implication to the right follows from the fact that the following formulas are theorems of T_+:

$$x \equiv_n c \vee x \equiv_n c+1 \vee \ldots \vee x \equiv_n c + (n \dot{-} 1),$$

$$x \equiv_n c + (n \dot{-} 1) \wedge x \leqslant c \rightarrow x \equiv_n c \dot{-} 1,$$

$$x \equiv_n c+1 \wedge x \leqslant c \rightarrow x \equiv_n (c \dot{-} n) + 1,$$

$$x \equiv_n y \wedge \alpha x + a \equiv_n b \rightarrow \alpha y + a \equiv_n b.$$

Note also that if m is a multiple of n than the following implication is a theorem of T_+:

$$x \equiv_m y \rightarrow x \equiv_n y.$$

Thus one obtains the following equivalence:

$$\begin{aligned} &\exists x(\psi_1(x) \wedge \ldots \wedge \psi_t \wedge x \leqslant c) \equiv \\ &\quad \equiv [(\psi_1(c) \wedge \ldots \wedge \psi_t(c)) \vee \ldots \vee (\psi_1(c \dot{-} m) \wedge \ldots \wedge \psi_t(c \dot{-} m))] \end{aligned}$$

where m is the least common multiple of the moduli $n_1, \ldots, n_t$ of the congruences $\psi_1(x), \ldots, \psi_t(x)$ and $\psi_i(x)$ denotes here the congruence $\alpha_i x + a_i \equiv_{n_i} b_i$.

In the general case when $s \neq 0$ one has in T_+:

$$\begin{aligned} &\exists x[\psi_1(x) \wedge \ldots \wedge \psi_t(x) \wedge \chi_1(x) \wedge \ldots \wedge \chi_s(x) \wedge x \leqslant c] \equiv \\ &\quad \equiv [\psi_1(c) \wedge \ldots \wedge \psi_t(c) \wedge \chi_1(c) \wedge \ldots \wedge \chi_s(c)] \vee \ldots \vee \\ &\quad \vee [\psi_1(c \dot{-} m) \wedge \ldots \wedge \psi_t(c \dot{-} m) \wedge \chi_1(c \dot{-} m) \wedge \ldots \wedge \chi_s(c \dot{-} m)] \end{aligned}$$

where χ_i denotes a condition of the form $p_i < q_i x + r_i$. The terms $c \dot{-} k$ $(k = 0, \ldots, m)$ can be eliminated in the way described above. Denote by $\zeta_{t,s}(c)$ the following formula:

$$\exists x[\psi_1(x) \wedge \ldots \wedge \psi_t(x) \wedge \chi_1(x) \wedge \ldots \wedge \chi_s(x) \wedge x \leqslant c].$$

We have just shown that the existential quantifier in the expression $\zeta_{t,s}(c)$ can be eliminated and one obtains a disjunction of conjunctions of basic formulas.

It is obvious that not all conditions of the type v can be reduced to the form $x \leqslant c$. Generally a condition of the considered type has the form $\alpha_i x + a_i \leqslant b_i$ $(i = 1, \dots, l)$. One can prove in T_+ that

$$\begin{aligned}\alpha_i x + a_i \leqslant b_i \;&\equiv\; a_i \leqslant b_i \wedge \alpha_i x \leqslant b_i \dot{-} a_i \equiv \\ &\equiv\; a_i \leqslant b_i \wedge x \leqslant E\left(\frac{b_i \dot{-} a_i}{\alpha_i}\right).\end{aligned}$$

Denote by e_i $(i = 1, \dots, l)$ the numbers $E\left(\frac{b_i \dot{-} a_i}{\alpha_i}\right)$. One can easily prove in T_+ that

$$\begin{aligned}&(e_1 \leqslant e_1 \wedge e_1 \leqslant e_2 \wedge \dots \wedge e_1 \leqslant e_l) \vee \dots \\ &\qquad \dots \vee (e_l \leqslant e_1 \wedge e_l \leqslant e_2 \wedge \dots \wedge e_l \leqslant e_l).\end{aligned}$$

Hence if in the formula φ_i there occur some conditons of the type $\alpha_i x + a_i \leqslant b_i$ then their conjunction is equivalent in T_+ to the disjunction of the form

$$(x \leqslant e_1 \wedge \vartheta_1) \vee \dots \vee (x \leqslant e_l \wedge \vartheta_l)$$

where $\vartheta_1, \dots, \vartheta_l$ are expressions in which the variable x does not occur, in fact they are conjunctions of formulas of the form $a_i \leqslant b_i$ and $e_i \leqslant e_j$.

In this way we have shown that in the considered case the formula φ_i is equivalent in T_+ to the disjunction

$$(\zeta_{t,s}(e_1) \wedge \vartheta_1) \vee \dots \vee (\zeta_{t,s}(e_l) \wedge \vartheta_l).$$

But we have shown above that the expression $\zeta_{t,s}(e_i)$ is equivalent to a formula in which the existential quantifier does not occur.

To finish the proof one must prove that expressions containing terms of the form $E\left(\frac{b_i \dot{-} a_i}{\alpha_i}\right)$ can be reduced (in T_+) to basic formulas.

Subcase 2b. Assume that $k = 0$, $t \neq 0$ and $l = 0$, i.e., in φ_i occur only congruences and possibly conditions of the type χ inducing lower bounds. In this case one simply omits those conditions because if a congruence has a root then it has arbitrarily large roots. Observe that the existence of a common root of the congruences $\psi_1 \ldots, \psi_t$ can be expressed as a statement that one of the numbers $0, 1, \ldots, m \dot{-} 1$ is such a root where m is the least common multiple of the moduli of all the congruences $\psi_1, \ldots, \psi_t$. The eliminaton of the existential quantifier is now given by he following equivalence which can be proved in T+:

$$\exists x(\psi_1(x) \wedge \ldots \wedge \psi_t(x)) \equiv$$
$$\equiv (\psi_1(0) \wedge \ldots \wedge \psi_t(0)) \vee \ldots \vee (\psi_1(m \dot{-} 1) \wedge \ldots \wedge \psi_t(m \dot{-} 1)).$$

Case 3. Assume that $k = 0$ and $t = 0$, i.e., no equation and no congruence occur in the formula φ_i. In this case one proceeds as in the subcases 2a and 2b because one can always add a trivial congruence $x \equiv_1 0$. Q.E.D.

Corollary 3.2.4 *The theory* T_+ *is decidable.*

Corollary 3.2.5 *The theory of the structure* $\langle \mathbb{N}, S, +, 0 \rangle$ *is decidable.*

Proof. It can be easily seen that all the axioms of the theory T_+ are satisfied in the structure $\langle \mathbb{N}, S, +, 0 \rangle$, hence T_+ is a subtheory of the theory $Th(\langle \mathbb{N}, S, +, 0 \rangle)$. By completeness of T_+ both theories are equal. Q.E.D.

Corollary 3.2.6 *The theory* T_+ *is equivalent to the theory with the following nonlogical axioms:*

$$0 \neq S(x),$$
$$S(x) = S(y) \rightarrow x = y,$$
$$x + 0 = x,$$
$$x + S(y) = S(x + y),$$

induction scheme for formulas φ *of the language* $\mathrm{L}(\mathrm{T}_+)$.

Proof. It suffices to show that all the axioms of the theory T_+ are theorems of the theory described in the Corollary and to use the completeness of the theory T_+. Q.E.D.

So the theory T_+ is equivalent to the fragment of Peano arithmetic obtained by restricting the language L(PA) to the language $L(T_+)$ and by taking only those nonlogical axioms of PA that are formulated in the language $L(T_+)$. Let us mention that Presburger (1930) added in his Supplement that the result on completeness of T_+ also holds for the arithmetic of integers formalized in the language with 0, 1, + and > as nonlogical symbols.

It has been shown above that the theory of the structure $\langle \mathbb{N}, S, +, 0 \rangle$ is decidable. One can ask if the theories of the structures $\langle \mathbb{N}, S, 0 \rangle$ and $\langle \mathbb{N}, S, \cdot, 0 \rangle$ are also decidable. The following theorems give an answer.

Theorem 3.2.7 (Herbrand, 1928) *The theory of the structure $\langle \mathbb{N}, S, 0 \rangle$ is decidable.*

Theorem 3.2.8 (Skolem, 1930) *The theory of the structure $\langle \mathbb{N}, S, \cdot, 0 \rangle$ is decidable.*

Both theorems were proved by the elimination of quantifiers.

As indicated in the proof of the Corollary 3.2.5, the theory T_+ is equivalent to the theory $Th(\langle \mathbb{N}, S, +, 0 \rangle)$, hence the latter is recursively axiomatizable. One can ask if there exist recursive axiomatics for the theories $Th(\langle \mathbb{N}, S, 0 \rangle)$ and $Th(\langle \mathbb{N}, S, \cdot, 0 \rangle)$. The answer is affirmative.

Consider the theory T_S in the language with 0 and S as the only nonlogical symbols and based on the following nonlogical axioms:

$$0 \neq S(x),$$

$$S(x) = S(y) \rightarrow x = y,$$

induction scheme for formulas φ of the language $L(T_S)$.

This theory is equivalent to the theory T'_S based on the following set of nonlogical axioms:

$$x \neq 0 \rightarrow \exists y(x = S(y)),$$

$$0 \neq S(x),$$

$$S(x) = S(y) \rightarrow x = y,$$

$$x \neq \underbrace{S \ldots S}_{k}(x) \quad \text{for every natural number } k > 0.$$

One can prove that the theory T'_S admits the elimination of quantifiers and that it is decidable and complete. Therefore theories T_S and T'_S are two axiomatizations of the theory of the structure $\langle \mathbb{N}, S, 0 \rangle$.

Cegielski (1981) constructed a recursively axiomatizable theory $T_{\cdot}$ admitting elimination of quantifiers and consequently complete and decidable. This theory is equivalent to the theory $Th(\langle \mathbb{N}, S, \cdot, 0 \rangle)$.

The above theorems allow us to formulate the following heuristic corollaries. The theory T_+ of the addition and the theory $T_{\cdot}$ of multiplication are decidable. The 'union' of those theories, i.e., the theory of addition and multiplication (being of course a stronger theory) is undecidable, even more, essentially undecidable. Hence the strength but also the complexity of Peano arithmetic PA (and of the theory $Th(\mathfrak{N}_0)$) have their sources in the fact that it is a theory of both addition and multiplication. Consequently, addition is not definable in the structure $\langle \mathbb{N}, S, \cdot, 0 \rangle$ since otherwise the theory of this structure would be undecidable. So the function of addition cannot be defined in terms of multiplication and successor.

Using the method of quantifier elimination the decidability of various theories has been proved. We indicate here some examples. The following theories are decidable:

1. Elementary theory of identity, i.e., the theory formalized in a language with no nonlogical symbols except identity = and based only on logical axioms and identity axioms — its decidability was shown by Löwenheim (1915).

2. Theory of finitely many sets, i.e., the theory in a language with only m unary predicates and no nonlogical axioms — its decidability was shown by Löwenheim (1915).

3. Theory of discrete order DO. This is a theory formalized in a language with two binary predicates $=$ and $\leqslant$ and based on the following nonlogical axioms:

$$x \leqslant y \wedge y \leqslant x \rightarrow x = y,$$

$$x \leqslant y \vee y \leqslant x,$$

$$x \leqslant y \wedge y \leqslant z \rightarrow x \leqslant z,$$

$$\exists x \forall y [y \leqslant x \rightarrow x = y],$$

$$\forall x \exists y [x < y \wedge \forall z (x < z \rightarrow y \leqslant z)],$$

$$\forall x \forall y [y < x \rightarrow \exists z \forall w (z < x \wedge (z < w \rightarrow x \leqslant w))].$$

Decidability of the theory DO was proved by Langford (1927) (see also Langford, 1927a). Observe that any model of DO is a linearly ordered set with the first element and such that for any element there exists its immediate successor and for any element except the first there exists its immediate predecessor. Note also that $\langle \mathbb{N}, \leqslant \rangle \models$ DO.

4. Theory of linear order in the set of rationals $\mathbb{Q}$, i.e., the theory $Th(\langle \mathbb{Q}, \leqslant \rangle)$.

5. Theory of algebraically closed fields ACF, i.e., the theory formalized in a language with symbols $+$, $\cdot$, 0, 1, -1 and based on the axioms of the theory of fields:

$$(x + y) + z = x + (y + z),$$

$$x + 0 = x,$$

$$x + (-1 \cdot x) = 0,$$

$$x + y = y + x,$$

$$(x \cdot y) \cdot z = x \cdot (y \cdot z),$$

$$x \cdot 1 = x,$$
$$x \neq 0 \to \exists y(x \cdot y = 1),$$
$$x \cdot y = y \cdot x,$$
$$x \cdot (y + z) = (x \cdot y) + (x \cdot z),$$
$$0 \neq 1,$$

and the axiom scheme stating that every polynomial of degree n ($n \geqslant 1$) has a root, i.e., the scheme

$$\forall a_0 \forall a_1 \ldots \forall a_n \exists x(a_n x^n + a_{n-1}x^{n-1} + \ldots + a_1 x + a_0 = 0).$$

One proves that any formula of the language of this theory is equivalent in ACF to an open formula (this can be done by using, e.g., the classical algebraic elimination theory). A proof of decidability of the theory ACF can be found in Kreisel and Krivine (1967, Ch. 4.IV).

6. Theory of Boolean algebras. This is a theory formalized in a language with two binary function symbols $\cup$ and $\cap$, one unary function symbol $-$ and two individual constants $\bigwedge$ and $\bigvee$, and based on the following nonlogical axioms:

$$x \cup y = y \cup x,$$
$$x \cap y = y \cap x,$$
$$x \cup (y \cup z) = (x \cup y) \cup z,$$
$$x \cap (y \cap z) = (x \cap y) \cap z,$$
$$x \cup x = x,$$
$$x \cap x = x,$$
$$x \cup (x \cap y) = x,$$
$$x \cap (x \cup y) = x,$$
$$x \cap (y \cup z) = (x \cap y) \cup (x \cap z),$$
$$x \cup (y \cap z) = (x \cup y) \cap (x \cup z),$$

$$x \cup \bigwedge = x,$$
$$x \cap \bigvee = x,$$
$$x \cup -x = \bigvee,$$
$$x \cap -x = \bigwedge.$$

Decidability of this theory was proved by Tarski (1949a).

7. Theory of real numbers, i.e. the theory of the structure $\langle \mathbb{R}, 0, 1, +, -, \cdot, \leqslant \rangle$ where $\mathbb{R}$ is the set of reals. Its decidability was proved by Tarski (1948) (cf. also Tarski, 1951; see Cohen, 1969). It is worth noting here that Tarski (1967) asked if this result could be extended to include the exponential function, i.e., he asked whether the theory of the structure $\langle \mathbb{R}, 0, 1, +, -, \cdot, \exp, \leqslant \rangle$, where exp is the exponential function, is decidable. The problem is still open, only partial answers are known (cf. Marker, 1996).

It was noted above that the method of quantifier elimination is usually applied in the discussed context as a method of proving the completeness of a given theory. If the considered theory is additionally consistent and axiomatizable then we can conclude that it is decidable. What more, if a given theory admits an effective quantifier elimination and if there exists an effective decision procedure for basic formulas, then we obtain in fact an effective decision procedure for the whole theory under consideration (and not only the conclusion that the theory is decidable). This is the case with Presburger's arithmetic of addition.

We now describe the second indicated method of proving the decidability of a theory, i.e., the modeltheoretic method. It is usually used in this context either to show (by methods of model theory) that a given theory T is complete (this fact is used then together with Theorem 3.1.6) or to systematically study all complete extensions of T. In the latter case we want to know for any given sentence φ of the language L(T) whether the theory $T \cup \{\neg\varphi\}$ is consistent and consequently whether $T \vdash \varphi$.

A combination of quantifier elimination and the modeltheoretic method is often used. It consists of choosing a set K of sentences of the language L(T) and of proving then by methods of model theory that any sentence φ of L(T) is equivalent in T to a sentence ψ belonging to K. Investigating models of the theory T one then chooses a suitable subset of the set K consisting of theorems of T.

One of the modeltheoretic methods of proving completeness of a theory is to show that it is categorical. The notion of categoricity was introduced by Vaught (1953) and independently by Łoś (1954).

Definition 3.2.9 *Let α be a cardinal. A theory* T *is said to be α-categorical if and only if any two models $\mathfrak{A}$ and $\mathfrak{B}$ of* T *with universes of the power α are isomorphic.*

The following theorem holds.

Theorem 3.2.10 (Łoś, Vaught) *If a theory* T *has infinite models and* T *is α-categorical for an infinite cardinal α then* T *is complete.*

One of the simplest examples of an application of this theorem is the proof of completeness (and consequently of decidability) of the theory DNO of dense linear order without the first and the last elements. It is a theory in a language with one binary predicate $\leqslant$ based on the following nonlogical axioms:

$$x \leqslant x,$$

$$x \leqslant y \wedge y \leqslant z \rightarrow x \leqslant z,$$

$$x \leqslant y \wedge y \leqslant x \rightarrow x = y,$$

$$x \leqslant y \vee y \leqslant x,$$

$$\forall x \forall y \exists z [x < y \rightarrow x < z < y],$$

$$\forall x \exists y \exists z [z < x < y],$$

where $x < y$ means $x \leqslant y \wedge x \neq y$. Cantor proved that any two nonempty countable sets linearly and densly ordered and having no first and no last elements are isomorphic. Hence the theory DNO is $\aleph_0$-categorical. It can also be easily seen that DNO has no finite models. Consequently by Łoś-Vaught Theorem we obtain that DNO is complete. So by Theorem 3.1.6 the theory DNO is decidable.

Another example of a proof of decidability of a theory obtained by showing that it is categorical is the proof of decidability of the theory of algebraically closed fields of a given characteristic. So let p be a prime. The theory of algebraically closed fields of a characteristic p (denoted by ACF_p) is a theory based on the axioms of the theory of algebraically closed fields ACF described above plus axioms C_p of the form

$$\underbrace{1 + \ldots + 1}_{p \text{ times}} = 0.$$

The theory of algebraically closed fields of characteristic 0 (denoted by ACF_0) is a theory ACF plus axioms $\neg C_2$, $\neg C_3$, ... By Steinitz Theorem the theories ACF_p and ACF_0 are α-categorical for any uncountable cardinal α, hence they are complete. Consequently theoreis ACF_p (p prime) and ACF_0 are decidable.

Since theories ACF_p (p prime) and ACF_0 are all complete extensions of the theory ACF, we get by Theorem 3.2.1 that the theory of algebraically closed fields ACF is decidable (it was noted above that decidability of ACF can be also proved by the method of quantifier elimination).

By using modeltheoretical methods of proving completeness (combined sometimes with the quantifier elimination), one can exhibit the decidability of various important theories. Such proofs of decidability are often based on deep mathematical results about considered structures (in this way providing interesting examples of how methods of mathematics and logic can be fruitfully combined). We give some examples.

1. The theory of p-adic fields is decidable. This result is due to Ax and Kochen (cf. Ax and Kochen, 1965, 1965a, 1966).

2. The theory of all finite fields is decidable (cf. Ax, 1968).

3. The theory of real closed fields RCF is decidable. This is a theory formalized in the language of the theory of fields extended by a binary predicate $<$ and based on the axioms of the theory of fields augmented by the following axioms:

$$\neg(x < x),$$

$$x < y \wedge y < z \rightarrow x < z,$$

$$x < y \vee x = y \vee y < x,$$

$$x < y \rightarrow x + z < y + z,$$

$$0 < x \wedge 0 < y \rightarrow 0 < x \cdot y,$$

$$0 < x \rightarrow \exists y (y \cdot y = x),$$

$$\mathrm{A}_n \quad (\text{for } n = 1, 3, 5, \ldots),$$

where A_n is a sentence stating that every polynomial of odd degree n has a root, i.e.,

$$\forall a_0 \forall a_1 \ldots \forall a_n \exists x (a_n x^n + a_{n-1} x^{n-1} + \ldots + a_1 x + a_0 = 0).$$

A proof of the decidability of RCF can be found, e.g., in Kreisel and Krivine (1967, Ch. 4.V).

It can be easily seen that the field of real numbers

$$\mathcal{R} = \langle \mathbb{R}, 0, 1, +, \cdot, < \rangle$$

is a model of RCF (it is not the unique model of this theory!). Since RCF is complete, we can conclude that $Th(\mathcal{R})$ is equivalent to RCF. In this way another proof of decidability of the theory of real numbers can be obtained.

Add that by using those results on the theory of real-closed fields Tarski (1959) showed decidability and completeness of various theories of elementary geometry (defining the latter by the folowing words: "We regard as elementary that part of Euclidean geometry which can be formulated and established without the help of any set-theoretical devices" (p. 16)).

4. The theory of linearly ordered sets is decidable. This result was announced by Ehrenfeucht (1959) in the abstract , the first full proof was given by Läuchli and Leonard (1966).

5. The theory of Abelian groups is decidable. This is a theory formalized in a language with one binary function symbol $\cdot$ and based on the following axioms of group theory

$$x \cdot (y \cdot z) = (x \cdot y) \cdot z,$$

$$\exists z(x = y \cdot z),$$
$$\exists y(x = y \cdot z),$$

and the axiom

$$x \cdot y = y \cdot x.$$

Decidability of this theory was proved by Szmielew (1949, 1949a, 1950 and 1955).

Consider now the last indicated method of proving decidability, namely the method of interpretation. It can be briefly described as follows. Let a decidable theory T_0 formalized in a language L_0 be given. We are asking if another given theory T formalized in a language L is decidable. To answer this question one defines a computable (recursive) function f mapping formulas of the language L on formulas of the language L_0 such that if φ is a sentence of L then $f(\varphi)$ is a sentence of L_0 and $T \vdash \varphi$ if and only if $T_0 \vdash f(\varphi)$. This gives us a decision procedure for the theory T. Indeed, let φ be any sentence of the language L. Compute the image $f(\varphi)$. It suffices now to check (by the assumption of decidability of the theory T_0 this can be done in a computable (recursive) way) whether $T_0 \vdash f(\varphi)$ or not. In the former case φ is a theorem of T, in the latter it is not a theorem of T.

In practice the definition of a mapping f is based on certain modeltheoretical considerations. One shows that models of T can be obtained (up to isomorphism) from models of T_0 by certain relations definable in the language L_0.

Note that the method of interpretation can be also, *mutatis mutandis*, used to prove undecidability of theories. Indeed, if an undecidable theory T is interpretable (in the above described sense) in a theory T_0 then T_0 is also undecidable.

We indicate some examples of applying the method of interpretation in proofs of decidability of theories omitting (usually complicated) technical details. In the next section applications of this method to prove undecidability of theories will be presented.

Let L_1 be a monadic language of the second order, i.e., a language containing individual variables x, y, z, ... and set variables X, Y, Z, ... (varying over sets of individuals), the predicate $\in$ and maybe other predicates and function symbols. Let $\mathfrak{A} = \langle A; R_1, R_2, \ldots \rangle$ where $R_i \subseteq A^{n_i}$. The symbol $Th_2(\mathfrak{A})$ will denote the set of all sentences of the language L_1 true in the structure $\mathfrak{A}$. This set is called 'the second-order monadic theory of the structure $\mathfrak{A}$'. If $\mathcal{K}$ is a class of structures then $Th_2(\mathcal{K})$ denotes $\bigcap_{\mathfrak{A} \in \mathcal{K}} Th_2(\mathfrak{A})$.

The first significant result on decidability of second-order theories was Büchi's theorem stating that the theory $Th_2(\langle \mathbb{N}, S \rangle)$ (denoted sometimes by S1S and called 'the monadic theory of one successor') is decidable.

Earlier a weaker result was obtained. A monadic language of the second order is called 'weak' if and only if its set variables vary only upon finite sets. A theory of a structure $\mathfrak{A}$ formalized in a weak monadic language of the second order will be denoted by $Th_w(\mathfrak{A})$ and called 'weak second-order monadic theory of the structure $\mathfrak{A}$'. Büchi (in 1960) and Elgot (in 1961) proved that the theory $Th_w(\langle \mathbb{N}, S \rangle)$ (denoted by WS1S and called 'the weak second-order monadic theory of one successor') is decidable.

Proofs of those theorems use the automata theory. In the case of the theory WS1S one uses automata on finite sequences and in the case of S1S automata on sequences of infinite length, in particular of the length ω.

In proofs of various theorems on decidability obtained by the method of interpretation Rabin's theorem on trees is applied. To describe it let $T = \{0,1\}^*$, i.e., T is the set of sequences of 0's and 1's. One can interprete T as a tree (Λ denotes the empty sequence):

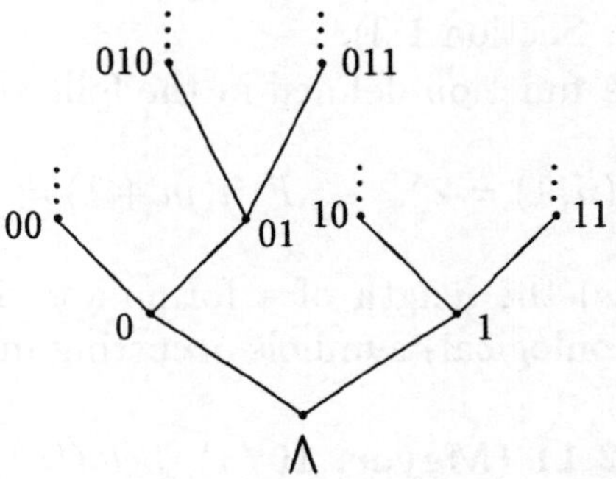

There are two successor functions in T: $r_0(x) = x0$ and $r_1(x) = x1$ for $x \in T$. Rabin's Theorem states that the second-order theory of two successors $Th_2(\langle T, r_0, r_1\rangle)$ (denoted by S2S) is decidable. The proof of this theorem uses automata on infinite trees (cf. Rabin, 1969).

Using the method of interpretation Rabin (1969) proved the decidability of the theory $Th_2(K^{\omega}_{\leqslant})$ where $K^{\omega}_{\leqslant}$ denotes the class of all countable linearly ordered sets $\langle A, \leqslant\rangle$. As a corollary one then obtains the decidability of the theory of linear order, of the second-order monadic theory of countable well-ordered sets as well as of the first-order theory of well-ordering. The latter result was also obtained by Tarski and Mostowski (1949) by quite different methods (see also Doner *et al.*, 1978).

It has been believed for a long time that a proof of decidability of a given theory T makes this theory uninteresting, since the problem whether a given sentence is provable in T becomes trivial in this way. Indeed, to solve it no creative thinking is needed — it suffices to apply the mechanical decision procedure which in a finite number of steps will give an answer. Nobody asked how big this finite number of steps can be, and in particular, if the decision procedures we have can be practically applied. In the seventies such problems of practical applicability of decision procedures were considered and it has turned out that no procedure known so far is practically applicable! We present some results of this type as examples. As a measure of complexity take the number of steps which a Turing machine needs to perform a given task (cf. Section 1.3).

Let F be a function defined in the following way:

$$F(n,1) = 2^n, \qquad F(n, m+1) = 2^{F(n,m)}.$$

Denote by $l(\varphi)$ the length of a formula φ, i.e., the number of (logical and nonlogical) symbols occurring in φ.

Theorem 3.2.11 (Meyer, 1975) *Let $f(n)$ be the function*

$$F(n, [dn]).$$

There exists a constant $d > 0$ such that for any Turing machine P deciding the theory WS1S *there exist infinitely many sentences φ with the property that the machine P needs more than $f(l(\varphi))$ steps to decide whether $\varphi \in$* WS1S *or not.*

Theorem 3.2.12 (Meyer, 1975) *The complexity of the theory of linear order is at least $F(n, [dn])$ for a certain positive constant d, i.e., to decide a formula of the length n one needs at least $F(n, [dn])$ steps.*

Theorem 3.2.13 (Fisher and Rabin, 1974) *A decision procedure for Presburger arithmetic is of the complexity at least $2^{2^{cn}}$ for a certain constant $c > 0$, i.e. to decide whether a formula φ of the length n is a theorem of Presburger arithmetic one needs at least $2^{2^{cn}}$ steps.*

Theorem 3.2.14 (Fisher and Rabin, 1974) *The complexity of the theory $Th(\langle \mathbb{N}, \cdot \rangle)$ is at least $F(cn, 3)$, i.e., $2^{2^{2^{cn}}}$ for a certain constant $c > 0$.*

So one sees that even for such simple theories as $Th(\langle \mathbb{N}, + \rangle)$ or $Th(\langle \mathbb{N}, \cdot \rangle)$ decision procedures are of exponential complexity. Add also that the decision procedure for the theory of real numbers is doubly exponential in the number of quantifier blocks (cf. Heintz *et al.*, 1989) and that even deciding first-order sentences for the ordered group $\langle \mathbb{R}, +, < \rangle$ is exponentially hard (cf. Fisher *et al.*, 1974).

Therefore one can ask whether there are decidable theories with practically applicable decision procedures. The problem is still open. It is connected with the famous problem of whether $\mathcal{P} = \mathcal{NP}$, which is nowadays the central problem of the recursion theory and of the complexity theory. Let us explain this briefly.

In Section 1.3 the notion of a Turing machine has been introduced. Various modifications of it are possible. In particular one can consider multitape or (and) nondeterministic Turing machines. A multitape Turing machine consists of a finite control with k tape read/write heads and k tapes divided into cells. Each tape is infinite in both directions. On a single move, depending on the state of the finite control and the symbol scanned by each of the tape heads, the machine can:

1. change state,
2. print a new symbol on each of the cells scanned by its tape heads,
3. move each of its tape heads, independently, one cell to the left or right, or keep it stationary.

Initially the input appears on the first tape, and the other tapes are blank.

A nondeterministic Turing machine is a device with a finite control and a single, infinite tape. For a given state and tape symbol scanned by the tape read/write head, the machine has a finite number of choices for the next move (note that in the case of Turing machines described in Section 1.4 the next move is uniquely determined — therefore they are called 'deterministic'

Turing machines). Each choice consists of a new state, a tape symbol to write, and a direction of read/write head motion.

It can be proved that those notions of a Turing machine are equivalent, i.e., if a function is computable by a multitape Turing machine then it is computable by a usual (i.e., single-tape) Turing machine, and if it is computable by a nondeterministic Turing machine then it is computable by a deterministic Turing machine (proofs can be found, e.g., in Hopcroft *et al.*, 1979).

Definition 3.2.15 *Let M be a Turing machine. The time complexity of M is the function $tc_M : \mathbb{N} \to \mathbb{N}$ such that $tc_M(n)$ is the maximum number of transitions processed by a computation of M when initiated with an input string of length n.*

It is assumed here that computations terminate for every input string. This is a natural assumption for it makes no sense to attempt to consider the efficiency or inefficiency of a computation that continues indefinitely.

Definition 3.2.16 *Let f and g be unary functions. The function f is said to be of order g, written $f = O(g)$, if and only if there exists a positive constant c and a natural number n_0 such that for all $n \geqslant n_0$*

$$f(n) \leqslant c \cdot g(n).$$

Definition 3.2.17 *A function $f : \mathbb{N} \to \mathbb{N}$ is said to be computable in polynomial time if and only if there exists a deterministic multitape Turing machine M computing it with $tc_M = O(n^r)$ where r is a natural number independent of n. The family of all functions computable in polynomial time is denoted by $\mathcal{P}$.*

Definition 3.2.18 *Let M be a nondeterministic Turing machine. The time complexity of M is the function $tc_M : \mathbb{N} \to \mathbb{N}$ such that $tc_M(n)$ is the maximum number of transitions processed by a computation, employing any choice of transitions, of an input string of length n.*

Emphasize that in the case of a nondeterministic machine an analysis of all possible computations for an input string is required.

Definition 3.2.19 *A function $f : \mathbb{N} \to \mathbb{N}$ is said to be computable in nondeterministic polynomial time if and only if there exists a nondeterministic multitape Turing machine M that computes f with $tc_M = O(n^r)$ where r is a natural number independent of n. The family of all functions computable in nondeterministic polynomial time is denoted by $\mathcal{NP}$.*

Since every deterministic machine is also nondeterministic, we can conclude that $\mathcal{P} \subseteq \mathcal{NP}$. The question if $\mathcal{NP} \subseteq \mathcal{P}$ is open. Note that nondeterministic computations utilizing a guess and check strategy are generally simpler than their deterministic counterparts. The simplicity reduces the number of transitions required for a single computation.

Definition 3.2.20 *A function f is called $\mathcal{NP}$-hard if and only if for every $g \in \mathcal{NP}$, g is reducible to f in polynomial time. An $\mathcal{NP}$-hard function that is also in $\mathcal{NP}$ is called $\mathcal{NP}$-complete.*

One can easily see that if there is an $\mathcal{NP}$-complete function f such that $f \in \mathcal{P}$ then $\mathcal{P} = \mathcal{NP}$.

Note that using characteristic functions one can say that a relation (or a decision problem) is $\mathcal{P}$ or $\mathcal{NP}$.

Many examples of $\mathcal{NP}$-complete decision problems are known. The first problem shown to be $\mathcal{NP}$-complete is the satisfiability problem. It is concerned with the truth values of formulas in propositional calculus and consists of determining whether there is an assignment of truth values to propositional variables that makes the formula true. A solution to this problem can be obtained by checking every truth assignment. The number of possible truth assignments is 2^n where n is the number of different variables occurring in a given formula, hence it is exponential. However the following theorem holds.

Theorem 3.2.21 (Cook, 1971) *The satisfiability problem for propositional calculus is $\mathcal{NP}$-complete.*

The proof of this theorem can be found of course in Cook (1971) as well as in various handbooks of the computation theory, e.g., in Hopcroft and Ullman (1979) or Sudkamp (1988). One can also find there other examples of $\mathcal{NP}$-complete problems. Instead of giving them here we shall present an example from Mycielski (cf. Mycielski, 1983; see also Mycielski, 1985) of two properties (strongly connected to mathematical logic) which are equivalent to the statement $\mathcal{P} \neq \mathcal{NP}$.

By a normal theory we mean a theory formalized in a first-order language with a finite number of nonlogical constants (including equality =) and axiomatized by a finite set of axioms and axiom schemata in which one can prove $\exists x, y(x \neq y)$.

Let Σ be a finite alphabet and Σ^* the set of all words, i.e., finite sequences of elements of Σ. For any $\xi \in \Sigma^*$, $|\xi|$ denotes the length of ξ.

A $\tau\pi$-theory is a set of pairs $T \subseteq \Sigma^* \times \Sigma^*$ such that there exists a polynomial $p(x, y)$ and a Turing machine M such that for any $(\tau, \pi) \in \Sigma^* \times \Sigma^*$, M can decide in time $\leqslant p(|\tau|, |\pi|)$ if $(\tau, \pi) \in T$. If $(\tau, \pi) \in T$ then τ is called a theorem of T and π is called a proof of τ in T. (Note that since a proof is a finite sequence of formulas satisfying appropriate conditions, one can treat it as a single formula. This justifies our definition). Observe that every normal theory defines a $\tau\pi$-theory. Indeed, it follows from the fact that the time necessary to check the correctness of a proof in a normal theory can be estimated from above by a polynomial in the length of that proof.

Call a $\tau\pi$-theory amenable (to automatization) if and only if there exists another polynomial $p_0(x, y)$ and another Turing machine M_0 such that given any word $\tau \in \Sigma^*$ and any $n \in \mathbb{N}$, $n \neq 0$, M_0 can decide in time $p_0(|\tau|, n)$ if there exists a $\pi \in \Sigma^*$ with $|\pi| \leqslant n$ and such that $(\tau, \pi) \in T$.

Theorem 3.2.22 (Mycielski, 1983) *The following conditions are equivalent:*

(i) *there exists a* $\tau\pi$*-theory which is not amenable,*

(ii) *every* $\tau\pi$*-theory which is defined by a normal theory is not amenable,*

(iii) $\mathcal{P} \neq \mathcal{NP}$.

The theorem follows from the theorem of Cook stating that the set of satisfiable formulas of propositional calculus is $\mathcal{NP}$-complete (cf. Theorem 3.2.21).

3.3 Undecidable Theories

The aim of this section is to describe basic methods of proving the undecidability of theories. The methods can be divided into two groups: (A) the direct method based on the ideas of Gödel's proof of the incompleteness of arithmetic (it can be applied to those theories within which a sufficiently large fragment of Peano arithmetic can be developed) and (B) indirect methods consisting in the reduction of the decision problem for a given theory T_1 to the decision problem for a theory T_2 whose solution is already known. For example, to prove that a theory T_1 is undecidable it suffices to show that T_1 can be obtained from an undecidable theory T_2 by removing finitely many axioms (but without changing the language). Another method of proving that T_2 is undecidable is to interprete in it a known undecidable theory (in practice Peano arithmetic or a suitable fragment of it are used here).

We begin our considerations by giving various theorems which indicate mutual interrelations between the undecidability of various theories and providing in this way methods of proving the undecidability of given theories.

Theorem 3.3.1 (Tarski, 1953) *Let* T_1 *and* T_2 *be two theories such that* T_1 *is a consistent extension of* T_2 *(the language* $L(T_1)$ *may be a proper extension of the language* $L(T_2)$*). If the theory* T_2 *is essentially undecidable then the theory* T_1 *is also essentially undecidable.*

Proof. Let T_3 be a theory in the language $L(T_2)$ defined in the following way: a set of theorems of T_3 is equal to the set of those formulas of the language $L(T_2)$ that can be proved in the theory T_1, i.e., $T_3 \vdash \varphi$ if and only if $T_1 \vdash \varphi$. Hence T_3 is a simple extension of T_2. T_3 is consistent because it is contained in the consistent theory T_1. Hence T_3 is undecidable. Consequently the theory T_1 is also undecidable since if T_1 were decidable then T_3 would also be decidable which is impossible. Using a similar reasoning one can show that any consistent simple extension of T_1 is also undecidable. Q.E.D.

Definition 3.3.2 *A theory* T_2 *is said to be an inessential extension of a theory* T_1 *if and only if every nonlogical constant of the language* $L(T_2)$ *which does not occur in the language* $L(T_2)$ *is an individual constant and if every sentence* $\varphi \in L(T_2)$ *which is a theorem of* T_2 *is derivable in* T_2 *from axioms of* T_1.

Theorem 3.3.3 (Tarski, 1953) *Let* T_2 *be an inessential extension of a theory* T_1. *Then*

(i) T_1 *is undecidable if and only if* T_2 *is undecidable,*

(ii) T_1 *is essentially undecidable if and only if* T_2 *is essentially undecidable.*

Proof. Let φ be any sentence of the language $L(T_2)$ and let $c_{m_1}, \ldots, c_{m_n}$ be all new, i.e., not belonging to the language $L(T_1)$, individual constants occurring in φ and let

$$\varphi' = \varphi(c_{m_1}/x_1, \ldots, c_{m_n}/x_n).$$

One now uses the theorem on constants stating that logic is neutral with respect to nonlogical constants, and the Closure Theorem saying that for any theory T and any formula ψ of the language L(T) the following equivalence holds:

$$T \vdash \psi(x_1, \ldots, x_n) \quad \text{if and only if} \quad T \vdash \forall x_1 \ldots \forall x_n \psi.$$

Consequently one has

$$T_1 \vdash \forall x_1 \ldots \forall x_n \varphi' \quad \text{if and only if} \quad T_2 \vdash \varphi.$$

Thus if the theory T_2 is undecidable then the theory T_1 is also undecidable. On the other hand, for any formula ψ of the language $L(T_1)$:

$$T_2 \vdash \psi \quad \text{if and only if} \quad T_1 \vdash \psi.$$

Thus, if the theory T_2 were decidable then also the theory T_1 would be decidable. As a result, if T_1 is undecidable then T_2 is also undecidable which proves (i).

To prove (ii) observe that for any extension T_1' of T_1 there exists an extension T_2' of T_2 such that T_2' is an inessential extension of T_1'. Hence if T_2 is essentially undecidable then T_1 is also essentially undecidable. The reverse implication follows from Theorem 3.3.1. In this way (ii) has been proved. Q.E.D.

Definition 3.3.4 *A theory* T_2 *is said to be a finite extension of a theory* T_1 *if and only if* T_2 *is an extension of* T_1 *and there exists a finite set* X *of sentences being theorems of* T_2 *such that for any sentence* φ, *if* $T_2 \vdash \varphi$ *then* $T_1 \cup X \vdash \varphi$.

It can be easily seen that any inessential extension is a finite extension.

Theorem 3.3.5 (Tarski, 1953) *Let* T_1 *and* T_2 *be two theories formalized in the same language and such that* T_2 *is a finite extension of* T_1. *If* T_2 *is undecidable then* T_1 *is also undecidable.*

Proof. By the definition of a finite extension, there exists a finite set $X = \{\varphi_1, \ldots, \varphi_n\}$ of sentences of the language $L(T_2)$ such that for any sentence $\varphi \in L(T_2)$:

$$T_2 \vdash \varphi \quad \text{if and only if} \quad T_1 \cup \{\varphi_1, \ldots, \varphi_n\} \vdash \varphi.$$

Hence via the Deduction Theorem:

$$T_2 \vdash \varphi \quad \text{if and only if} \quad T_1 \vdash \varphi_1 \wedge \ldots \wedge \varphi_n \to \varphi.$$

Thus if T_1 is decidable then T_2 is also decidable. Therefore, since T_2 is undecidable, T_1 is also undecidable. Q.E.D.

Definition 3.3.6 *Two theories* T_1 *and* T_2 *are said to be compatible if and only if they have a common consistent extension.*

Theorem 3.3.7 (Tarski, 1953) *Let* T_1 *and* T_2 *be two compatible theories such that the language* $L(T_2)$ *is included in the language* $L(T_1)$. *If* T_2 *is essentially undecidable and finitely axiomatizable then* T_1 *is hereditarily undecidable.*

Proof. Let T_1' be any subtheory of the theory T_1 with the same language as $L(T_1)$. It should be proved that T_1' is undecidable. So let T' be the union of the theories T_1' and T_2. By Theorem 3.3.1 the theory T' is essentially undecidable. By the construction one has that $L(T') = L(T_1')$ and T' is a finite extension of T_1'. By Theorem 3.3.5 the theory T_1' is undecidable.
Q.E.D.

A natural question arises which concerns possible improvements of Theorems 3.3.1, 3.3.3, 3.3.5 and 3.3.7. In particular, one can ask whether it is possible to remove the word 'essentially' from Theorems 3.3.1 and 3.3.7, or to weaken the assumption of Theorem 3.3.5 by taking T_2 to be an arbitrary (and not necessarily finite) extension of T_1, or to omit in Theorem 3.3.7 the assumption that T_2 is finitely axiomatizable. The answers to these questions prove to be negative (cf. Tarski, 1953).

In fact, to answer the first question assume that T_1 is an arbitrary theory in the language of which a binary predicate P occurs and such that its only nonlogical axiom is the following formula

$$\forall x \forall y (x = y)$$

(this formula states that the universe consists of only one element). Let T_2 be a theory formalized in the same language as T_1 and with no nonlogical axioms. Hence every sentence of T_2 is logically valid. The theory T_2 is a system of first-order predicate logic in the language restricted to $L(T_1)$. It can be easily seen that T_1 is a consistent extension of T_2 and hence the two theories are compatible. It is clear that T_2 is finitely axiomatizable.

By a result of Kalmár discussed at the very end of this section, it follows that T_2 is undecidable. On the other hand, using the method of quantifier elimination (described in Section 3.2) one can show that theory T_1 is decidable. Now, one sees that Theorems 3.3.1 and 3.3.7 are no longer valid if one removes the word 'essentially'.

To answer the remaining questions note that an example of a decidable and finitely axiomatizable theory T_1 is known which has uncountably many consistent and complete extensions, all in the language $L(T_1)$ — as T_1 one can take the theory of Abelian groups (cf. Szmielew, 1949, 1949a, 1950, 1955). Since there are only countably many recursive sets, one can conclude that T_1 has a consistent, complete and undecidable extension T_2. The two theories T_1 and T_2 are of course compatible, T_2 is essentially undecidable (cf. Corollary 3.1.7) and T_1, as stated above, is decidable. This example shows that one cannot extend Theorem 3.3.5 to the case where T_2 is an arbitrary (and not finite) extension of T_1. It also shows that the assumption that T_2 is finitely axiomatizable cannot be omitted in Theorem 3.3.7.

Theorem 3.3.8 *If a theory* T *is finitely axiomatizable and essentially undecidable then* T *is strongly undecidable.*

Proof. The theory T is of course consistent since otherwise it would be decidable. So let $\mathfrak{A}$ be an arbitrary model of T and let T_1 be a theory such that $\mathfrak{A} \models T_1$. So $\mathfrak{A} \models T \cup T_1$. Hence the theory $T \cup T_1$ is consistent. Since it is a simple extension of T, it is undecidable. On the other hand, $T \cup T_1$ is a finite extension of T_1. By Theorem 3.3.5, T_1 is undecidable. Consequently, the theory T is strongly undecidable. Q.E.D.

Theorem 3.3.9 *If* T_2 *is a definitional extension of* T_1 *and* T_2 *is undecidable then* T_1 *is also undecidable.*

Proof. By the theorem on elimination of defined symbols, for any formula φ of the language $L(T_2)$ there exists a formula φ^* of the language $L(T_1)$ such that

$$T_2 \vdash \varphi \quad \text{if and only if} \quad T_1 \vdash \varphi^*.$$

If T_1 were decidable then also T_2 would be decidable. Therefore, since T_2 is undecidable, T_1 is also undecidable. Q.E.D.

To widen the range of applicability of the above theorems one introduces the notion of interpretability and weak interpretability.[2]

Assume that languages of theories T_1 and T_2 have no nonlogical symbols in common.

Definition 3.3.10 *A theory* T_2 *is said to be interpretable in a theory* T_1 *if and only if there exist a theory* T *and a set* F *such that the following conditions are satisfied:*

(i) T *is a common extension of* T_1 *and* T_2, *any nonlogical symbol of the language* L(T) *is either a nonlogical symbol of* $L(T_1)$ *or a nonlogical symbol of* $L(T_2)$,

(ii) F *is a recursive set of sentences (treated as axioms of the theory* T*) which are possible definitions of nonlogical symbols of the language* $L(T_2)$ *in the theory* T_1,

(iii) *each nonlogical symbol of* $L(T_2)$ *occurs in just one sentence of the set* F,

(iv) *every theorem of* T *is derivable in* T *from a set of sentences each of which is a theorem of* T_1 *or belongs to* F.

In the general case, when languages of theories T_1 and T_2 have some common nonlogical symbols, one first replaces the

[2]Interpretability is widely used in mathematics and logic for various purposes. As examples of its application in mathematics can serve the interpretation of two-dimensional elliptic geometry in Euclidean geometry of three dimensions, the Poincaré model and the Beltrami-Klein model of hyperbolic geometry or the interpretation of arithmetic in set theory. In logic it is used, e.g., to prove the relative consistency of ZFC (Zermelo-Fraenkel set theory with the axiom scheme of choice) or ZF + CH (Zermelo-Fraenkel set theory plus the continuum hypothesis) with respect to Zermelo-Fraenkel set theory ZF (one uses here Gödel's interpretation via constructible sets) or to classify theories (cf. Mycielski *et al.* (1990) where local interpretability is applied to define chapters of mathematics — they are local interpretability types of first-order theories; chapters consitute a lattice of cardinality $2^{\aleph_0}$, its structure has been studied in the indicated paper).

nonlogical symbols of $L(T_2)$ by a new symbol not occurring in $L(T_1)$ (different symbols by different symbols) without changing the structure of T_2 in any other respect. If a theory T_2' obtained in this way is interpretable in T_1 in the sense of the above Definition 3.3.10 then one says that T_2 is interpretable in T_1. One can easily see that in the general case it is sufficient (though not necessary) to construct a common extension T of the theories T_1 and T_2, and a set F satisfying conditions (i)–(iv) of Definition 3.3.10, with the difference that F contains possible definitions of only those nonlogical symbols of the language $L(T_2)$ which do not occur in $L(T_1)$.

Definition 3.3.11 *A theory* T_2 *is said to be weakly interpretable in a theory* T_1 *if and only if the theory* T_2 *is interpretable in a consistent simple extension of the theory* T_1.

It follows from Definition 3.3.10 that (i) if a theory T_2 is interpretable in a theory T_1, then each subtheory of T_2 is interpretable in every extension of T_1, and (ii) if T_2 is interpretable in T_1 and T_3 is interpretable in T_2, then T_3 is interpretable in T_1. Observe also that if a theory T_2 is weakly interpretable in a theory T_1, every subtheory of T_2 is weakly interpretable in T_1.

Theorem 3.3.12 (Tarski, 1953) *Let* T_1 *and* T_2 *be two theories such that* T_1 *is consistent and* T_2 *is interpretable in* T_1 *or in some inessential extension of* T_1. *Then:*

(i) *if* T_2 *is essentially undecidable then* T_1 *is also essentially undecidable,*

(ii) *if* T_2 *has a finitely axiomatizable subtheory which is essentially undecidable then so has* T_1.

Proof. Assume that the theory T_2 is interpretable in the theory T_1. Without loss of generality one can suppose that the languages $L(T_1)$ and $L(T_2)$ have no nonlogical symbols in common. Let a theory T and a set F be as in Definition 3.3.10. From conditions (ii) and (iii) of this definiton it follows that for any formula φ of the language $L(T)$ one can effectively construct (by eliminating from φ nonlogical symbols of $L(T_2)$) a formula φ^* of the language $L(T_1)$ such that

$$T \vdash \varphi \quad \text{if and only if} \quad T_1 \vdash \varphi^*. \tag{3.2}$$

Hence, T_1 being consistent, T is also consistent. Thus T is a consistent extension of the essentially undecidable theory T_2. By Theorem 3.3.1 T is undecidable. On the other, hand by (3.2) any decision procedure for the theory T_1 yields a decision procedure for T. Hence the theory T_1 is undecidable.

Observe that since T_2 is interpretable in T_1, T_2 is also interpretable in any extension of T_1. Consequently, by the same argument as above, every consistent extension of the theory T_1 is undecidable and T_1 is essentially undecidable.

To prove (ii) assume now theory T_2 has a finitely axiomatizable and essentially undecidable subtheory T_2'. By the Deduction Theorem and conditions (i) and (iv) of Definition 3.3.10 there exists a finite set Y of theorems of T_1 such that every theorem of T_2' is derivable in T from $Y \cup F$. Let T_1' be a subtheory of T_1 with Y as a set of nonlogical axioms. T_1' is of course finitely axiomatizable. One can easily see that T_2' is interpretable in T_1'. Hence T_1' is essentially undecidable.

In this way the theorem has been proved by the assumption that T_2 is interpretable in T_1. Now assume that T_2 is interpretable in some inessential extension $\overline{T_1}$ of T_1. Assume additionally, to prove (i), that T_2 is essentially undecidable. Through the same argument as in the first part of the proof one can show that $\overline{T_1}$ is essentially undecidable. From Theorem 3.3.3 we conclude that T_1 is essentially undecidable.

In much the same way as above one can prove that if T_2 has a finitely axiomatizable and essentially undecidable subtheory T_2', then the theory $\overline{T_1}$ also has such a subtheory $\overline{T_1'}$. To prove (ii) it suffices to show that T_1 also has such a subtheory. Using the assumption that $\overline{T_1}$ is an inessential extension of T_1 one can eliminate from nonlogical axioms of $\overline{T_1'}$ individual constants not occurring in $L(T_1)$ without changing the structure of $\overline{T_1'}$. Let $\overline{T_1''}$ be a theory obtained in this way and let T_1' be a theory obtained from $\overline{T_1''}$ by restricting it to the language $L(T_1)$. Consequently, $\overline{T_1''}$ is an inessential extension of T_1'. By Theorem 3.3.3, T_1' is essentially undecidable. By construction the theory T_1' is finitely axiomatizable, which completes the proof. Q.E.D.

Theorem 3.3.13 (Tarski, 1953) *Let* T_1 *and* T_2 *be two such theories that* T_2 *is weakly interpretable in* T_1. *If* T_2 *is essentially undecidable and finitely axiomatizable then*

(i) T_1 *is hereditarily undecidable,*

(ii) *there exists a finite extension of* T_1 *which is formalized in the same language as* T_1 *and is essentially undecidable.*

Proof. Assume that T_2 is weakly interpretable in T_1, i.e., T_2 is interpretable in some consistent simple extension T_1^* of T_1. By Theorem 3.3.12(ii) there exists a finitely axiomatizable and essentially undecidable subtheory $\overline{T_1^*}$ of T_1^*. The theories T_1 and $\overline{T_1^*}$ are of course compatible (i.e., they have a consistent common extension) and every nonlogical symbol of the language $L(\overline{T_1^*})$ is a nonlogical symbol of the language $L(T_1)$. By Theorem 3.3.7 we obtain part (i) of our theorem.

To prove (ii) consider a theory T being a union of the theories T_1 and $\overline{T_1^*}$. Since $\overline{T_1^*}$ is finitely axiomatizable, T is a finite extension of T_1. By construction the theory T is a consistent extension of the essentially undecidable theory $\overline{T_1^*}$, so by Theorem 3.3.1 T is essentially undecidable. Q.E.D.

Analogously one proves the following

Theorem 3.3.14 (Tarski, 1953) *Let* T_1 *and* T_2 *be two such theories that* T_2 *is weakly interpretable in some inessential extension of* T_1. *If* T_2 *is essentially undecidable and finitely axiomatizable then*

(i) T_1 *is hereditarily undecidable,*

(ii) *there exists an essentially undecidable finite inessential extension of* T_1.

Theorems 3.3.12–3.3.14 proved above are generalizations of Theorems 3.3.1 and 3.3.7. Indeed, if T_1 is an extension of T_2 then

T_2 is of course interpretable in T_1. Analogously, if T_1 and T_2 are compatible and $L(T_2) \subseteq L(T_1)$, then T_2 is weakly interpretable in T_1. Notice that the assumptions of Theorems 3.3.12–3.3.14 cannot be weakened, i.e., the assumption that T_2 is essentially undecidable (in Theorems 3.3.12, 3.3.13 and 3.3.14) and that T_2 is finitely axiomatizable (in Theorems 3.3.13 and 3.3.14) are essential. It follows from the remarks we made above after Theorem 3.3.7.

The notion of interpretability will now be further generalized. The aim of this is to widen the scope of applicability of the method of interpretation.

Let a theory T and a unary predicate P be given.

Definition 3.3.15 *A relativization $\psi^{(P)}$ of a formula ψ of the language* L(T) *to the predicate P is defined inductively in the following way:*

(i) *if ψ is an atomic formula then $\psi^{(P)}$ is ψ,*

(ii) *if ψ is of the form $\neg\chi$ or $\chi_1 \vee \chi_2$ then $\psi^{(P)}$ is the formula $\neg(\chi^{(P)})$ or $(\chi_1^{(P)} \vee \chi_2^{(P)})$, resp.*

(iii) *if ψ is of the form $\forall x_i(\chi)$ or $\exists x_i(\chi)$ then $\psi^{(P)}$ is the formula $\forall x_i(P(x_i) \rightarrow \chi^{(P)})$ or $\exists x_i(P(x_i) \wedge \chi^{(P)})$, resp.*

Definition 3.3.16 *A relativization of a theory* T *to a predicate P is a theory $T^{(P)}$ defined in the following way:*

(i) $L(T^{(P)}) = L(T) \cup \{P\}$,

(ii) *φ is a theorem of the theory $T^{(P)}$ if and only if φ is a logical consequence of formulas $\psi^{(P)}$ for ψ being theorems of* T.

Definition 3.3.17 *A theory T_2 is said to be relatively interpretable (weakly relatively interpretable) in a theory T_1 if and only if there exists a unary predicate P not being a nonlogical symbol of the language $L(T_2)$ such that the theory $T_2^{(P)}$ is interpretable (weakly interpretable) in the theory T_1.*

We prove now two theorems indicating connections between axiomatizability and undecidability of theories T and $T^{(P)}$.

Theorem 3.3.18 (Tarski, 1953) *Let* T *be any theory and* P *a unary predicate which is not a nonlogical symbol of the language* L(T). *One then has:*

(i) $T^{(P)}$ *is axiomatizable if and only if* T *is axiomatizable,*

(ii) *under the assumption that only a finitely many individual constants and function symbols occur in* T, $T^{(P)}$ *is finitely axiomatizable if and only if* T *is finitely axiomatizable.*

Proof. To prove (i) assume that T is axiomatizable. Hence there exists a recursive set X of nonlogical axioms of T. Let Y be a set consisting of the following formulas:

(i) $\varphi^{(P)}$ for $\varphi \in X$,

(ii) $\exists x P(x)$,

(iii) $\forall x_1 \ldots x_n[P(x_1) \wedge \ldots \wedge P(x_n) \rightarrow P(f(x_1, \ldots, x_n))]$ for every n-ary function symbol f of L(T),

(iv) $P(e)$ for every individual constant e of L(T).

If L(T) contains at least one individual constant e then sentence (ii) is superfluous and can be omitted (it follows from any sentence of the form (iv)). The set Y is of course recursive. From the definition of the theory $T^{(P)}$, it follows that any sentence of the form (i) is a theorem of $T^{(P)}$. Similarly, the sentence $\exists x P(x)$ is provable in $T^{(P)}$. Indeed, the sentence $\exists x(x = x)$ is a theorem of T, hence its relativization $\exists x(P(x) \wedge x = x)$ is provable in $T^{(P)}$. But the latter formula is equivalent in $T^{(P)}$ to the considered sentence $\exists x P(x)$. Similarly, one can show that all sentences of the form (iii) are derivable in $T^{(P)}$. To do this observe that the sentence

$$\forall x_1 \ldots \forall x_n \exists y(y = f(x_1, \ldots, x_n))$$

is a theorem of T. Hence its relativization

$$\forall x_1 \ldots \forall x_n \exists y(P(x_1) \wedge \ldots \wedge P(x_n) \rightarrow P(y) \wedge y = f(x_1, \ldots, x_n))$$

is provable in $\mathrm{T}^{(P)}$. But this relativization is equivalent to (iii) which follows from the equivalence

$$\exists y(P(y) \wedge y = f(x_1, \ldots, x_n)) \equiv P(f(x_1, \ldots, x_n)).$$

We show now that every theorem of the theory $\mathrm{T}^{(P)}$ is derivable form the set Y. It suffices to prove that if φ is a logical consequence of X then $\varphi^{(P)}$ is a logical consequence of Y. It can be done by induction on the length of a proof. Assume that φ is a logical consequence of X and let $\varphi_1, \ldots, \varphi_n$ be a proof of it. The formula φ_1 is either a logical axiom or an axiom of identity or an element of the set X. In each case its relativization is a logical consequence of Y. So assume now that for all $i < k$, $\varphi_i^{(P)}$ is derivable from Y and prove that the same holds for $\varphi_k^{(P)}$. If φ_k is a logical axiom or an identity axiom or an element of X, one proceeds as in the case of φ_1. If φ_k is a result of applying one of the inference rules to φ_i, $i < k$, then we reason as follows. Assume, e.g., that φ_k is obtained by *modus ponens* from φ_i and φ_j, for some $i, j < k$. So the formula φ_j is of the form $\varphi_i \rightarrow \varphi_k$ and its relativization $\varphi_j^{(P)}$ is $\varphi_i^{(P)} \rightarrow \varphi_k^{(P)}$. By the inductive assumption $\varphi_i^{(P)}$ and $\varphi_j^{(P)}$ are consequences of Y. Hence $\varphi_k^{(P)}$ is also a logical consequence of Y. In the case of other inference rules (cf. Section 2.1) one proceeds in a similar way.

Assume now, conversely, that $\mathrm{T}^{(P)}$ is axiomatizable and let a recursive set Y' of sentences be the set of its nonlogical axioms. For any sentence φ of the language $\mathrm{L}(\mathrm{T}^{(P)})$ construct a sentence φ^* of the language $\mathrm{L}(\mathrm{T})$ by replacing in φ every subformula of the form $P(\alpha)$, where α is an arbitrary term, by $\alpha = \alpha$. Let X' be a recursive set of all sentences φ^* for $\varphi \in Y'$. We claim that every sentence of the set X' is a theorem of T. Indeed, observe that if $\varphi \in Y'$ then $\mathrm{T}^{(P)} \vdash \varphi$, so via the Deduction Theorem and the definition of the theory $\mathrm{T}^{(P)}$ there exist sentences $\psi_1, \ldots, \psi_n$ such that $\mathrm{T} \vdash \psi_i$ for $i = 1, \ldots, n$ and

$$\psi_1^{(P)} \wedge \ldots \wedge \psi_n^{(P)} \rightarrow \varphi$$

is a logical tautology. It can be easily seen that also the formula

$$(\psi_1^{(P)} \wedge \ldots \wedge \psi_n^{(P)} \rightarrow \varphi)^*,$$

i.e., the formula

$$(\psi_1^{(P)})^* \wedge \ldots \wedge (\psi_n^{(P)})^* \to \varphi^*$$

is a logical tautology. On the other hand, the equivalences

$$(\psi_i^{(P)})^* \equiv \psi_i \ \ (i = 1, \ldots, n)$$

are tautologies and consequently theorems of T. By the Equivalence Theorem (cf. Theorem 2.1.6) one has

$$\mathrm{T} \vdash \psi_1 \wedge \ldots \wedge \psi_n \to \varphi^*.$$

But $\mathrm{T} \vdash \psi_i$ $(i = 1, \ldots, n)$, hence $\mathrm{T} \vdash \varphi^*$.

In this way we have shown that if $\varphi^* \in X'$ then $\mathrm{T} \vdash \varphi^*$. Now we prove that X' is a set of axioms for the theory T. Assume that $\mathrm{T} \vdash \varphi$. Then of course $\mathrm{T}^{(P)} \vdash \varphi^{(P)}$, hence $\varphi^{(P)}$ is a logical consequence of Y'. By properties of the operation of consequence there exist formulas $\psi_1, \ldots, \psi_n$ such that $\psi_i \in Y'$ and the formula $\psi_1 \wedge \ldots \wedge \psi_n \to \varphi^{(P)}$ is a tautology. Consequently the formula

$$\psi_1^* \wedge \ldots \wedge \psi_n^* \to (\varphi^{(P)})^*,$$

i.e., the formula

$$\psi_1^* \wedge \ldots \wedge \psi_n^* \to \varphi$$

is also a tautology. But $\psi_i^* \in X'$, hence φ is a logical consequence of X'.

To prove part (ii) of the theorem it suffices to note that if there are only finitely many function symbols in L(T) then if X is finite then Y is also finite and if Y' is finite then the same holds also for X'. Q.E.D.

Theorem 3.3.19 (Tarski, 1953) *Let* T *be a theory and* P *a unary predicate not belonging to the language* L(T). *The theory* $\mathrm{T}^{(P)}$ *is essentially undecidable if and only if* T *is essentially undecidable.*

Proof. We show first that $\mathrm{T}^{(P)}$ is interpretable in T. For this purpose we construct an extension T^* of T in the following way: $\mathrm{L}(\mathrm{T}^*) = \mathrm{L}(\mathrm{T}) \cup \{P\}$, φ is a theorem of T^* if and only if φ is derivable from the set of theorems of T supplemented by the following sentence φ_0:

$$\forall x[P(x) \equiv x = x].$$

It is obvious that for any sentence $\varphi \in \mathrm{L}(\mathrm{T})$, $\varphi^{(P)}$ is derivable from the set $\{\varphi, \varphi_0\}$. Hence T^* is an extension of $\mathrm{T}^{(P)}$. Consequently $\mathrm{T}^{(P)}$ is interpretable in T.

Note that $\mathrm{T}^{(P)}$ is consistent if and only if T is consistent. Indeed, assume that T is consistent. By the construction of T^* every sentence of T which is a theorem of T^* is also a theorem of T. Hence the consistency of T implies that of T^*. Since T^* is an extension of $\mathrm{T}^{(P)}$, this in turn implies the consistency of $\mathrm{T}^{(P)}$. To prove the converse implication assume that $\mathrm{T}^{(P)}$ is consistent but T is inconsistent. So there exists a sentence φ such that $\mathrm{T} \vdash \varphi$ and $\mathrm{T} \vdash \neg\varphi$. Hence $\mathrm{T}^{(P)} \vdash \varphi^{(P)}$ and $\mathrm{T}^{(P)} \vdash \neg\varphi^{(P)}$. Consequently $\mathrm{T}^{(P)}$ is inconsistent contrary to the assumption. So finally, if $\mathrm{T}^{(P)}$ is consistent then T is also consistent.

To prove now the proper assertion of the theorem assume that $\mathrm{T}^{(P)}$ is essentially undecidable. It is consistent and the same holds for T. Using Theorem 3.3.12(ii) we obtain that T is essentially undecidable.

Assume, conversely, that T is essentially undecidable. So T is consistent and $\mathrm{T}^{(P)}$ is also consistent. Consider any consistent extension U of $\mathrm{T}^{(P)}$. Let

$$S = \{\varphi : \varphi \in \mathrm{L}(\mathrm{T}) \wedge \mathrm{U} \vdash \varphi^{(P)}\}.$$

The set S is closed under logical consequence. Indeed, let ψ be derivable from S. According to the Deduction Theorem there exist sentences $\chi_1, \ldots, \chi_n \in S$ such that the formula

$$\chi_1 \wedge \ldots \wedge \chi_n \to \psi$$

is a tautology. Denote it by ψ_0. Of course $\mathrm{T} \vdash \psi_0$. Therefore $\mathrm{T}^{(P)} \vdash \psi_0^{(P)}$ and $\mathrm{U} \vdash \psi_0^{(P)}$. By the definition of S, $\mathrm{U} \vdash \psi_i^{(P)}$ for $i = 1, \ldots, n$. Hence $\mathrm{U} \vdash \psi^{(P)}$ and consequently $\psi \in S$.

Construct now a theory T^* in the following way: let $L(T^*) = L(T)$ and let the set of theorems of T^* be equal to S. Of course T^* is an extension of T. The relativized theory $(T^*)^{(P)}$ is a subtheory of U. Hence it is consistent. So the theory T^* is also consistent. As a consequence, T^* is a consistent extension of T and is therefore undecidable.

The extension U is also undecidable. Indeed, if U were decidable then T^* would also be decidable contrary to our earlier assertion. As a result, $T^{(P)}$ is essentially undecidable. Q.E.D.

The notion of relativized (weak) interpretability widens the scope of applicability of the method of interpretation in proving the undecidability of theories. It is often easier to show that a given theory T_2 is relatively (weakly) interpretable in a theory T_1 than to prove its (weak) interpretability in T_1 (what is often impossible). The application of Theorem 3.3.19 (and possibly also Theorem 3.3.18) and then Theorems 3.3.12–3.3.14 enables us to solve the decision problem for T_1.

We will now give several examples of applications of the interpretability method to prove the undecidability of particular mathematical theories.

Let us begin by describing a subtheory Q of Peano arithmetic PA. The theory Q is formalized in the same language as PA and based on the following nonlogical axioms:

(Q1) $S(x) = S(y) \rightarrow x = y$,

(Q2) $0 \neq S(x)$,

(Q3) $x \neq 0 \rightarrow \exists y(x = S(y))$,

(Q4) $x + 0 = x$,

(Q5) $x + S(y) = S(x + y)$,

(Q6) $x \cdot 0 = 0$,

(Q7) $x \cdot S(y) = (x \cdot y) + x$.

To prove that Q is a subtheory of PA it suffices to show that the formula $x \neq 0 \rightarrow \exists y[x = S(y)]$ is derivable in PA. It can be easily done by applying the axiom scheme of induction to that formula (it is left to the reader as an easy exercise).

Observe that theory Q, contrary to Peano arithmetic, is finitely axiomatizable. One can also prove that every recursive function is representable in Q. This proof will not be given here because it is very long. It is in fact similar to the proof of the appropriate theorem for PA (cf. Theorem 2.2.12) — one uses a characterization of recursive functions found by J. Robinson (1950). The proof can be found in Mostowski *et al.* (1953). Hence Q is essentially undecidable (cf. comments after Definition 3.1.5). Those two properties of the theory Q, i.e., its finite axiomatizability and essential undecidability, are reasons for the fact that it turns out to be very suitable for the method of interpretation — indeed, Q became a powerful instrument in the study of the decision problem.

One can ask if there exists another theory weaker than Q with the above properties. The answer is negative, the theory Q is in a sense minimal, i.e., one can show that the following theorem holds.

Theorem 3.3.20 (Mostowski et al., 1953) *No axiomatic subtheory of the theory* Q *obtained by removing any one of the axioms* (Q1)–(Q7) *is essentially undecidable.*

Proof. Let Q_n be the axiomatic subtheory of Q obtained by omitting the axiom (Q_n) $(n = 1, \ldots, 7)$. We shall prove that none of these theories is essentially undecidable by constructing, for each of them, a consistent and decidable extension T_n. Each of the theories, with the exception of Q_6, will have the same language as Q. The consistency of the theories Q_n will follow from their descriptions. It will be easy to see that T_n is an extension of Q_n. Thus we shall show only that the theories T_n are decidable.

T_1 is defined as the theory of the structure

$$\mathfrak{M}_1 = \langle U, 0, S, +, \cdot \rangle$$

where (i) $U = \{0, 1\}$, (ii) 0 and $\cdot$ have their usual meaning, (iii) $S0 = S1 = 1$ and (iv) $x+y = 0$ if $x = y = 0$, $x+y = 1$ if $x, y \in U$ and $x = 1$ or $y = 1$. The decidability of T_1 follows from the fact that the universe U is finite. Here one uses a simple version of the method of quantifier elimination (described in general in Section 3.2). For any formula φ (in which the variable x does not occur bound) the sentences

$$\forall x \varphi(x) \equiv \varphi(0) \wedge \varphi(S0),$$

$$\exists x \varphi(x) \equiv \varphi(0) \vee \varphi(S0),$$

are theorems of T_1. Hence, with every sentence φ one can correlate in a recursive way a sentence φ^* containing no variables and no quantifiers such that $T_1 \vdash \varphi$ if and only if $T_1 \vdash \varphi^*$. The decision procedure for sentences containing no variables is based either on the direct reference to the definitions of operations which occur in the model $\mathfrak{M}_1$ (for atomic sentences) or on the reduction to the decision procedure for the propositional calculus (in the case of compound sentences).

The theory T_2 is the theory of the structure

$$\mathfrak{M}_2 = \langle U, 0, S, +, \cdot \rangle$$

where $U = \{0\}$, $S0 = 0+0 = 0 \cdot 0 = 0$. Decidability of T_2 follows from the fact that U is finite. One uses here the property that the following equivalences are theorems of T_2 for any formula φ:

$$\forall x \varphi(x) \equiv \varphi(0),$$

$$\exists x \varphi(x) \equiv \varphi(0).$$

The theory T_3 is defined as the theory of the structure

$$\mathfrak{M}_3 = \langle P, 0, S, +, \cdot \rangle$$

where P is the set of all non-negative real numbers, i.e.,

$$P = \mathbb{R}^+ \cup \{0\},$$

$Sx = x + 1$ for $x \in P$ and $0, +, \cdot$ have their usual meaning. Let $\overline{\mathrm{T}}$ be a theory with the following nonlogical symbols: **P**, **0**, **S**, $+$, $\cdot$ where **P** is a unary predicate, **0** is a constant and **S** is a binary function symbol. The theory $\overline{\mathrm{T}}$ is $Th(\langle \mathbb{R}, P, 0, S, +, \cdot \rangle$ where $\mathbb{R}$ denotes the set of all real numbers. The decidability of $\overline{\mathrm{T}}$ follows from the results in Tarski (1951) and from the fact that non-negative real numbers coincide with squares of arbitrary real numbers. With every formula φ of $\mathrm{L}(\mathrm{T}_3)$ we correlate the relativized formula $\varphi^{(P)}$. It can be easily seen that $\mathrm{T}_3 \vdash \varphi$ if and only if $\overline{\mathrm{T}} \vdash \varphi^{(P)}$. Hence a decision procedure for $\overline{\mathrm{T}}$ yields a decision procedure for T_3 and consequently T_3 is decidable.

The theory T_4 is the theory of the structure

$$\mathfrak{M}_4 = \langle \mathbb{N}, 0, S, +, \cdot \rangle$$

where $\mathbb{N}, 0, S$ have their ordinary meaning and $+$ and $\cdot$ are operations on $\mathbb{N} \times \mathbb{N}$ to $\mathbb{N}$ defined as follows

$$m + n = n, \quad m \cdot 0 = 0, \quad m \cdot n = m \quad \text{for } n \neq 0.$$

Let T^* be a theory in the language with **0** and **S** as the only nonlogical symbols and such that $\mathrm{T}^* = Th(\langle \mathbb{N}, 0, S \rangle)$. The following sentences are valid in $\mathfrak{M}_4$ and hence theorems of T_4:

$$y = x+y,$$

$$x = y \cdot z \equiv (x = \mathbf{0} \wedge z = \mathbf{0}) \vee (x = y \wedge z \neq \mathbf{0}).$$

Using this one can recursively correlate with every formula φ of $\mathrm{L}(\mathrm{T}_4)$ a formula φ^* of $\mathrm{L}(\mathrm{T}^*)$ such that $\mathrm{T}_4 \vdash \varphi \equiv \varphi^*$ and consequently $\mathrm{T}_4 \vdash \varphi$ if and only if $\mathrm{T}^* \vdash \varphi^*$. Hence the decision problem for T_4 is reduced to that for T^*. By Herbrand's Theorem (cf. Theorem 3.2.7) T^* is decidable, and therefore T_4 is also decidable.

The theory T_5 is defined as the theory of the structure

$$\mathfrak{M}_5 = \langle \mathbb{N}, 0, S, +, \cdot \rangle$$

where $\mathbb{N}, S, 0$ have their usual meaning and $+$ and $\cdot$ are defined as follows:

$$m + n = m \quad \text{and} \quad m \cdot n = 0 \quad \text{for all } m, n \in \mathbb{N}.$$

Decidability of T_5 is proved as in the case of T_4.

The theory T_6 is a theory formalized in a language with the following nonlogical symbols: $\mathbf{0}$, $\mathbf{S}$, $+$, $\cdot$, $\mathbf{0}^*$, $\mathbf{S}^*$, $+^*$. It is a theory of the structure

$$\mathfrak{M}_6 = \langle \mathbb{N}, 1, S', \oplus, \cdot, 0, S, + \rangle$$

where $\mathbb{N}, 1, 0, S, +$ have their usual meaning while $S', \oplus$ and $\cdot$ are defined by the conditions

$$S'0 = 0, \; S'n = n \quad \text{if } n \neq 0,$$

$$m \oplus n = 0 \quad \text{if } m = 0 \text{ or } n = 0,$$

$$m \oplus n = m + n - 1 \quad \text{if } m \neq 0 \text{ and } n \neq 0,$$

$$m \cdot n = 0 \quad \text{for all } m, n \in \mathbb{N}.$$

Let T^+ be a theory in a language with $\mathbf{0}^*$, $\mathbf{S}^*$, $+^*$ as the only nonlogical symbols defined as the theory of the structure

$$\langle \mathbb{N}, 0, S, + \rangle.$$

The following sentences are valid in $\mathfrak{M}_6$ and hence theorems of T_6:

$$\mathbf{0} = \mathbf{S}^*\mathbf{0}^*,$$

$$x = \mathbf{S}y \equiv (x = \mathbf{0}^* \wedge y = \mathbf{0}^*) \vee (x = \mathbf{S}^*y \wedge y \neq \mathbf{0}^*),$$

$$x = y + z \equiv (x = \mathbf{0}^* \wedge y = \mathbf{0}^*) \vee (x = \mathbf{0}^* \wedge z = \mathbf{0}^*) \vee \\ \vee (\mathbf{S}^*x = y +^* z \wedge y \neq \mathbf{0}^* \wedge z \neq \mathbf{0}^*),$$

$$y \cdot z = \mathbf{0}^*.$$

Using those formulas and arguing as in the case of T_4 one can reduce the decision problem for T_6 to that for T^+. The latter theory is decidable (cf. Presburger, 1930), hence T_6 is also decidable.

The theory T_7 is the heory of the structure

$$\mathfrak{M}_7 = \langle \mathbb{N}, 0, S, +, \cdot \rangle$$

where $m \cdot n = 0$ for all $m, n \in \mathbb{N}$ and the other operations have their ordinary meaning. The decision procedure for T_7 can be reduced to that for T^+ described above (the symbols $\mathbf{0}^*$, $\mathbf{S}^*$, $+^*$ are now replaced in T^+ by $\mathbf{0}$, $\mathbf{S}$, $+$). Hence T_7 is decidable.

Q.E.D.

We now show how the theory Q can be applied in proofs of undecidability of various theories.

Theorem 3.3.21 (Mostowski et al., 1953) *The theory of the structure $\mathfrak{N}_0$ is hereditarily undecidable. Every theory* T *compatible with* Q *and such that* $\mathrm{L(Q)} \subseteq \mathrm{L(T)}$ *is undecidable.*

Proof. Note that the theories $Th(\mathfrak{N}_0)$ and Q are compatible and that $\mathrm{L(Q)} \subseteq \mathrm{L}(Th(\mathfrak{N}_0))$. Since Q is finitely axiomatizable and essentially undecidable we conclude, using Theorem 3.3.7, that it is hereditarily undecidable. The second part of the theorem is proved in a similar way. Q.E.D.

The theory $Th(\mathfrak{N}_0)$ is complete, and so this theorem implies, together with Theorem 3.1.6, that $Th(\mathfrak{N}_0)$ is not axiomatizable (this corollary was already obtained in Chapter 2 by another method).

Corollary 3.3.22 *The structure $\mathfrak{N}_0$ is strongly undecidable.*

Theorem 3.3.23 *The theory* Q *is strongly undecidable. The same applies to Peano arithmetic* PA.

Proof. The strong undecidability of the theory Q follows from Theorem 3.3.8 and from the fact that Q is finitely axiomatizable and essentially undecidable, and the strong undecidability of Peano arithmetic PA follows from the fact that it is an extension of Q. Q.E.D.

Corollary 3.3.24 *The theory* Q *is hereditarily undecidable.*

Proof. It follows directly from Theorems 3.3.21 and 3.1.14. Q.E.D.

Corollary 3.3.25 *Every model of theory* Q *is strongly undecidable.*

Proof. The corollary follows from Theorem 3.3.7. Q.E.D.

By Theorem 3.1.13 and Corollary 3.3.22 one obtains easily the following

Corollary 3.3.26 *Peano arithmetic* PA *is hereditarily undecidable.*

Let us now turn to the decision problem of the arithmetic of integers. Consider two theories J and $\mathrm{J}^{<}$. Theory J is formalized in a first-order language with two binary function symbols $+$ and $\cdot$ and theory $\mathrm{J}^{<}$ in a language with $+$ and $\cdot$ and a binary predicate $<$. They are defined in the following way:

$$\mathrm{J} = Th(\langle \mathbb{Z}, +, \cdot \rangle) \quad \text{and} \quad \mathrm{J}^{<} = Th(\langle \mathbb{Z}, +, \cdot, < \rangle),$$

where $\mathbb{Z}$ the set of all integers, $+$ and $\cdot$ are the usual operations of addition and multiplication of integers and $<$ is the 'less than' relation on $\mathbb{Z} \times \mathbb{Z}$.

Theorem 3.3.27 (Mostowski et al., 1953) *Theory* J *and all its subtheories (in the same language as* J*) are undecidable. There are finitely axiomatizable subtheories of* J *which are essentially undecidable. The same applies to* $\mathrm{J}^{<}$.

Proof. Let $\mathbf{N}$ be a unary predicate and let $Q^{(\mathbf{N})}$ be the theory obtained from Q by relativizing the quantifiers to $\mathbf{N}$. By Theorems 3.3.18(ii) and 3.3.19, $Q^{(\mathbf{N})}$ is finitely axiomatizable and essentially undecidable.

We show now that theory $Q^{(\mathbf{N})}$ is interpretable in J. To do this consider a theory $\overline{\mathrm{J}}$. It is a first-order theory in the language with $\mathbf{N}$, 0, $S, +, \cdot$ and defined as $Th(\langle \mathbb{Z}, \mathbb{N}, 0, S, +, \cdot \rangle)$ where $\mathbb{N}$ is the set of natural numbers and other operations have their usual meaning. $\overline{\mathrm{J}}$ is obviously a common extension of J and $Q^{(\mathbf{N})}$. Consider the following formulas

$$\mathbf{N}(x) \equiv \exists y \exists z \exists u \exists v [x = y \cdot y + z \cdot z + u \cdot u + v \cdot v], \tag{3.3}$$

$$x = 0 \equiv x = x + x, \tag{3.4}$$

$$x = S(y) \equiv \exists z [x = y + z \wedge z \neq z + z \wedge z = z \cdot z]. \tag{3.5}$$

They can be treated as definitions of $\mathbf{N}$, 0, S, respectively, in J (we use here the well-known theorem of Lagrange by which natural numbers coincide with those integers which are sums of four squares). To prove that $Q^{(\mathbf{N})}$ is interpretable in J it suffices to show that for any formula φ such that $\overline{\mathrm{J}} \vdash \varphi$, φ is derivable (in $\overline{\mathrm{J}}$) from the set of sentences being either theorems of J or any of the definitions (3.3)–(3.5) — and this is obvious.

By Theorems 3.3.13(i) and 3.3.12(ii) we obtain that theory J is hereditarily undecidable and that there exists a finitely axiomatizable and essentially undecidable subtheory T of it. Nonlogical axioms of T are sentences obtained from axioms of $Q^{\mathbf{N}}$ by eliminating symbols $\mathbf{N}$, 0 and S with the help of definitions (3.3)–(3.5) (cf. the proof of Theorem 3.3.18). Since $Q^{\mathbf{N}}$ is interpretable in T, we conclude that T is essentially undecidable.

The proof for $\mathrm{J}^{<}$ is analogous but instead of (3.3) one uses the following formula

$$\mathbf{N}(x) \equiv x = x + x \vee x < x + x. \tag{3.6}$$

Q.E.D.

Among axiomatic subtheories of J and $J^<$ one finds several which are interesting from the point of view of modern algebra. All these theories are undecidable by Theorem 3.3.27. Some of them are finitely axiomatizable theories which are essentially undecidable.

To describe these theories consider the following formulas:

$$x + (y + z) = (x + y) + z, \tag{3.7}$$

$$x + y = y + x, \tag{3.8}$$

$$\exists z(x = y + z), \tag{3.9}$$

$$x \cdot (y \cdot z) = (x \cdot y) \cdot z, \tag{3.10}$$

$$x \cdot (y + z) = x \cdot y + x \cdot z, \tag{3.11}$$

$$(y + z) \cdot x = y \cdot x + z \cdot x, \tag{3.12}$$

$$x \cdot y = y \cdot x, \tag{3.13}$$

$$z = z + z \wedge z = x \cdot y \rightarrow z = x \vee z = y, \tag{3.14}$$

$$\exists u[u \neq u + u \wedge \forall x(x \cdot u = x \wedge u \cdot x = x)], \tag{3.15}$$

$$\neg(x < x), \tag{3.16}$$

$$x \neq y \rightarrow x < y \vee y < x, \tag{3.17}$$

$$x < y \wedge y < z \rightarrow x < z, \tag{3.18}$$

$$y < z \rightarrow x + y < x + z, \tag{3.19}$$

$$z = z + z \wedge z < x \wedge z < y \rightarrow z < x \cdot y, \tag{3.20}$$

$$\exists x \exists y[x < y \wedge \neg\exists z(x < z \wedge z < y)]. \tag{3.21}$$

The theory, with + and · as the only nonlogical symbols, whose nonlogical axioms are (3.7)–(3.12) is called 'the elementary theory of rings'. Clearly any ring in the sense of algebra is a model of it. If we add axiom (3.13), or axioms (3.13)–(3.14), to the theory of rings, we obtain 'the elementary theory of commutative rings', or 'of integral domains', respectively. The theory, with +,

$\cdot$ and $<$ as the only nonlogical symbols, based on axioms (3.7)–(3.12) and (3.16)–(3.20) is called 'the elementary theory of ordered rings'. If we add to it axiom (3.13) or axioms (3.13)–(3.14) then we arrive at 'the elementary theory of ordered commutative rings', or 'of ordered integral domains', respectively. The theory whose nonlogical axioms are (3.7)–(3.12) and (3.16)–(3.21) is called 'the elementary theory of non-densely ordered rings', and the theory with axioms (3.7)–(3.13) and (3.16)–(3.21) is called 'the elementary theory of non-densely ordered commutative rings'. Finally, if we include axiom (3.15) in any of the theories mentioned above then we obtain appropriate theories with unit, e.g., 'the elementary theory of rings with unit'.

Theorem 3.3.28 (Mostowski et al., 1953) *The elementary theories of rings, commutative rings, integral domains, ordered rings, and ordered commutative rings, without or with unit, are undecidable.*

Proof. The theorem follows from Theorem 3.3.27 and the fact that all mentioned theories are subtheories of J or $J^<$. Q.E.D.

Note that the theories mentioned in Theorem 3.3.28 are not essentially undecidable. In fact, there exists an extension of them which is decidable — an example of such an extension is the elementary theory of the ordered field of real numbers (cf. Section 3.2). On the other hand, the following theorem holds.

Theorem 3.3.29 (Mostowski et al., 1953) *The elementary theories of non-densely ordered rings and non-densely ordered commutative rings, without or with unit, are essentially undecidable.*

Proof. It suffices to prove that the elementary theory of non-densely ordered rings without unit T is essentially undecidable. To do this it is enough to show that the theory $Q^{(N)}$ is interpretable in T and to apply then Theorem 3.3.12(i).

Consider the relativized theory $Q^{(N)}$, however, we replace the symbol $\cdot$ by $\times$. So the nonlogical axioms of $Q^{(N)}$ are the following formulas:

$$\mathbf{N}(x) \wedge \mathbf{N}(y) \wedge S(x) = S(y) \rightarrow x = y,$$

$$\mathbf{N}(y) \rightarrow 0 \neq S(y),$$

$$\mathbf{N}(x) \wedge x \neq 0 \rightarrow \exists y[\mathbf{N}(y) \wedge x = Sy],$$

$$\mathbf{N}(x) \rightarrow x + 0 = x,$$

$$\mathbf{N}(x) \wedge \mathbf{N}(y) \rightarrow x + S(y) = S(x+y),$$

$$\mathbf{N}(x) \rightarrow x \times 0 = 0,$$

$$\mathbf{N}(x) \wedge \mathbf{N}(y) \rightarrow x \times S(y) = (x \times y) + x,$$

$$\mathbf{N}(0),$$

$$\mathbf{N}(x) \rightarrow \mathbf{N}(S(x)),$$

$$\mathbf{N}(x) \wedge \mathbf{N}(y) \rightarrow \mathbf{N}(x+y),$$

$$\mathbf{N}(x) \wedge \mathbf{N}(y) \rightarrow \mathbf{N}(x \times y).$$

In order to prove that $Q^{(N)}$ is interpretable in T one constructs an extension $\overline{\mathrm{T}}$ of T by adding $\mathbf{N}$, 0, S $\times$ to the set of nonlogical symbols and by adjoining the following formulas to the nonlogical axioms of T:

$$\begin{aligned}\mathbf{N}(u) \equiv (u = u + u \vee u < u + u) \wedge \\ \wedge \exists t[t < t + t \wedge \neg\exists x(x < t \wedge x < x + x) \wedge \\ \wedge u \cdot t = t \cdot u \wedge \forall y \exists z(y \cdot u = z \cdot t)],\end{aligned}$$

$$u = 0 \equiv u = u + u,$$

$$\begin{aligned}u = S(v) \equiv \exists t[t < t + t \wedge \\ \wedge \neg\exists x(x < t \wedge x < x + x) \wedge u = v + t].\end{aligned}$$

$$\begin{aligned}u = v \times w \equiv \exists t\{t < t + t \wedge \neg\exists x(x < t \wedge x < x + x) \wedge \\ \wedge [u \cdot t = v \cdot w \vee [u = v \wedge \forall y(y \cdot t \neq v \cdot w)]]\}.\end{aligned}$$

One now shows that these formulas are possible definitions of $\mathbf{N}$, S, $+$ and $\times$ and that $\overline{\mathrm{T}}$ is an extension of $\mathrm{Q}^{(\mathbf{N})}$, i.e., that all axioms of the latter are derivable from the axioms of T supplemented by the above possible definitions. Q.E.D.

In Section 3.2 it was stated that the theory of Abelian groups is decidable. Consider now the elementary theory of groups (without the axiom of commutativity), i.e., the theory G with the following nonlogical axioms:

$$x \cdot (y \cdot z) = (x \cdot y) \cdot z,$$

$$\exists z(x = y \cdot z),$$

$$\exists y(x = y \cdot z).$$

Theorem 3.3.30 (Tarski, 1949b and 1953a) (i) *The theory* G *is hereditarily undecidable.* (ii) *There exists a finitely axiomatizable simple extension of* G *which is essentially undecidable.*

This theorem is proved by the method of interpretation — it is shown that a certain theory of integers is weakly interpretable in some inessential extension of G.

Note that the theory G is not essentially undecidable — indeed, the theory of Abelian groups which is an extension of G is decidable. By Theorem 3.3.30(ii) there is an extension of G, i.e., the elementary theory of a particular class of groups, which is finitely axiomatizable and essentially undecidable. However, no simple and mathematically interesting theory of this type has been actually constructed so far. On the other hand, various interesting extensions of G are known to be undecidable, though not essentially undecidable. The elementary theory of centerless groups C can serve as an example. It is a theory obtained from G by adding the following nonlogical axiom:

$$\forall y(x \cdot y = y \cdot x) \rightarrow \forall y(x \cdot y = y).$$

The proof of the undecidability of C is almost the same as for G.

Among undecidable subtheories of G (cf. Theorem 3.3.30(i)) there are theories interesting from the point of view of algebra. Let us mention here the elementary theory of groupoids (associative systems) and the theory of semigroups. The former is a theory based on a single nonlogical axiom:

$$x \cdot (y \cdot z) = (x \cdot y) \cdot z,$$

and the latter is a theory based on the following axioms:

$$x \cdot (y \cdot z) = (x \cdot y) \cdot z,$$

$$x \cdot y = x \cdot z \rightarrow y = z,$$

$$x \cdot z = y \cdot z \rightarrow x = y.$$

If one supplements the theory of semigroups by the axiom:

$$\exists x \forall y (x \cdot y = y \wedge y \cdot x = y)$$

then the theory of semigroups with unit is obtained.

Corollary 3.3.31 *The theory of groupoids and the theory of semigroups (with or without unit) are undecidable.*

The method of interpretation was used to prove the (essential) undecidability of various particular mathematical theories. Let us indicate some examples.

1. J. Robinson (1949) proved that the theory

   $$Th(\langle \mathbb{Q}, +, \cdot \rangle),$$

   where $\mathbb{Q}$ is the set of rationals and $+$ and $\cdot$ are ordinary operations of addition and multiplication of rationals, is hereditarily undecidable.

2. R.M. Robinson (1951) proved the undecidability of the elementary theories of various special rings, i.e., various special complete and consistent extensions of the elementary theory of arbitrary rings.

3. Tarski (1949b and 1949c) showed that the elementary theories of lattices, modular lattices, complemented modular lattices and abstract projective geometries are undecidable. The indicated theories are not essentially undecidable, since the theories of Boolean algebras and of real projective geometry are decidable.

4. Grzegorczyk (1951) proved that: (1) there exists an elementary theory T of closure algebras that is undecidable and finitely axiomatizable and such that each theory of closure algebras consistent with T is undecidable,[3] (2) there exists an elementary theory T of the algebra of closed sets such that T is essentially undecidable and finitely axiomatizable and such that every Brouwerian algebra consistent with T is undecidable, (3) the algebra of bodies is undecidable, (4) every algebra of convexity true in a Euclidian space E_n, for $n \geqslant 2$, is undecidable, (5) every semi-projective algebra true in a Euclidean space E_n, for $n \geqslant 2$, is undecidable. (The original terminology of Grzegorczyk (1951) is used here. By an algebra of convexity (similarly for a semi-projective algebra) true in a Euclidean space, a (syntactically given) theory is meant whose axioms are true in a Euclidian space). The main idea of the proofs of the above theorems is that theory Q can be interpreted as an arithmetic of finite sets.

5. Janiczak (1953) proved the undecidability of some theories of relations and functions. In particular, he proved the undecidability of the theory of two equivalence relations, of the theory of two equivalence relations whose intersection is the identity relation, of the theory of one equivalence relation and one bijection, and of the theory of one 1–1 relation and one function (many–one relation). Those results were obtained by using the undecidability of the

[3]The undecidability of the closure algebras has also been proved by another method by S. Jaśkowski in 1939 (cf. Jaśkowski, 1948).

theory of non-densely ordered rings. Janiczak (1953) also mentioned that the theory of one equivalence relation is decidable.

6. Szmielew and Tarski (1952) showed (cf. also Tarski, 1953) that a small fragment S of set theory is essentially undecidable. The theory S is formalized in the language with two nonlogical symbols: E (= being a set) and the membership relation $\in$, and based on the set of axioms stating that: (i) any two sets with the same elements are identical (extensionality axiom), (ii) there is a set with no elements (the axiom of the empty set), (iii) for any two sets x and y there is a set z consisting of those and only those elements that are elements of a or are identical with b. The proof uses the interpretability of Q in S. From this theorem of Tarski and Szmielew it follows that every consistent theory that is an extension of S is essentially undecidable, hence all axiomatic systems of set theory (with E and $\in$ as nonlogical symbols) that are known from the literature are essentially undecidable. The result can be extended to systems of set theory formulated in the language with $\in$ alone.

Applications of the method of interpretation in proofs of the undecidability of theories usually employ some finitely axiomatizable and essentially undecidable theory. Rabin (1965) proposed a new principle for establishing the undecidability of theories which can use any undecidable theory. Rabin's method is semantical rather than syntactical, and turned out particularly convenient for applications. Using it Rabin (1965) gave short proofs of some known results concerning the undecidability of the theory of a symmetric irreflexive relation (this was proved earlier in Church and Quine (1952)), theory of distributive atomistic lattices (proved earlier in Grzegorczyk (1951)) and of the theory of groups (Rabin's result is in fact a generalization of Tarski (1953a)). He also proved a new result stating that the theory of finite commutative rings is undecidable.

To end this section consider the decision problem for the classical predicate calculus.

Theorem 3.3.32 (Church, 1936) *The classical first-order predicate calculus is hereditarily undecidable.*

Proof. The theory Q and the classical first-order predicate calculus are compatible. Observe also that each nonlogical symbol of the language L(Q) is a nonlogical symbol of the language of predicate calculus. Since Q is finitely axiomatizable and essentially undecidable, we conclude by Theorem 3.3.7 that the classical first-order predicate calculus is hereditarily undecidable. Q.E.D.

Classical first-order predicate calculus, being hereditarily undecidable, possesses decidable fragments. Consider classical monadic first-order predicate calculus. This is the fragment of the predicate calculus obtained by restricting the language to the use of unary predicates as only nonlogical symbols.

Theorem 3.3.33 (Löwenheim, 1915) *The classical monadic first-order predicate calculus is decidable.*

Proof. We describe a decision procedure for the monadic predicate calculus. Let φ be a formula of the language of this calculus and let $P_1, \ldots, P_k$ be all predicates occurring in it. We claim that φ is a theorem of classical monadic first-order predicate calculus if and only if φ is true in every interpretation with at most 2^k elements. Since there are only finitely many (up to isomorphism) interpretations with at most 2^k elements, this yields a decision procedure.

To prove the above equivalence observe that the implication to the right is obvious. So assume that φ is true in every interpretation with at most 2^k elements and let $\mathfrak{M}$ be any interpretation. For any elements b and c of the domain M of $\mathfrak{M}$ define a relation

$$b \sim c \quad \text{if and only if} \quad \mathfrak{M} \models [P_i(x) \equiv P_i(y)][b, c]$$

for every $i = 1, \ldots, k$, i.e.,

$$b \sim c \quad \text{if and only if} \quad \mathfrak{M} \models P_i(x)[b] \equiv \mathfrak{M} \models P_i(y)[c]$$

for $i = 1, \ldots, k$. It can be easily seen that $\sim$ is an equivalence relation on M. So let $\mathfrak{M}' = \mathfrak{M}/\sim$. The structure $\mathfrak{M}$ has at most 2^k elements. Indeed, every P_i determines two elements in $\mathfrak{M}'$, i.e., the equivalence class consisting of elements b of M for which $\mathfrak{M} \models P_i(x)[b]$ and the class consisting of elements b for which $\mathfrak{M} \models \neg P_i(x)[b]$ and there are k different predicates P_i (the word 'at most' is justified because some of the equivalence classes may be empty or coincide). By induction on the structure of φ one can show that

$$\mathfrak{M} \models \varphi \quad \text{if and only if} \quad \mathfrak{M}' \models \varphi.$$

Q.E.D.

Note that Theorem 3.3.33 is the best one possible. For Kalmár (1936) proved that there is an effective procedure producing, for each formula φ of the language of predicate calculus without function symbols, another formula φ^* of this language such that φ^* contains only one predicate symbol, a binary one, and such that φ is a theorem of the classical first-order predicate calculus if and only if φ^* is a theorem of this calculus. On the other hand, there exists a recursive function F such that if n is a Gödel number of a formula φ of the language of predicate calculus, then $F(n)$ is a Gödel number of a formula φ' of the language of predicate calculus without function symbols and such that φ is a theorem of the predicate calculus if and only if φ' is a theorem. Both facts together with Church's Theorem (cf. Theorem 3.3.32) imply that there exists no decision procedure for formulas containing only binary predicates.

3.4 Historical Remarks

Researches on decidability problems have arisen from Hilbert's program (cf. Introduction). The formalization of mathematics (of mathematical theories) proposed by Hilbert yielded a problem closely connected to completeness, called the 'decision problem', 'decidability problem' or '*Entscheidungsproblem*': one could ask whether a given formalized theory is decidable, i.e., whether there exists a uniform mechanical method which enables us to decide in a finite number of steps whether a given formula in the language of the theory under consideration is a theorem. Note that Hilbert and Ackermann (1928, pp. 73–74) explicitly spoke about the decision problem as about one of the most important problems that should be solved. They wrote:

> Das Entscheidungsproblem ist gelöst, wenn man ein Verfahren kennt, das bei einem vorgelegten logischen Ausdruck durch endlich viele Operationen die Entscheidung über die Allgemeingültigkeit bzw. Erfüllbarkeit erlaubt. (...) Das Entscheidungsproblem muss als das Hauptproblem der mathematischen Logik bezeichnet werden.

They were not alone in their evaluation of the importance of the decision problem. Bernays and Schönfinkel wrote:

> Das zentrale Problem der mathematischen Logik, welches auch mit den Fragen der Axiomatik im engsten Zusammenhang steht, ist das *Entscheidungsproblem.* (1928)

Herbrand wrote in (1930):

> We consider the fundamental problem of mathematics to be the following: Problem A: What is the necessary and sufficient condition for a theorem to be true having only a finite number of hypotheses?

In (1931a) he added:

> In a sense it [i.e. the classical decision problem — R.M.] is the most general problem of mathematics.

Ramsey wrote in (1929) that his paper was

> concerned with a special case of one of the leading problems in mathematical logic, the problem of finding a regular procedure to determine the truth or falsity of any given logical formula.

Using notions from recursion theory, one can make the definition of decidability more precise — cf. definitions from Section 3.1. It should be noticed here that these definitions, completely standard today, come from Alfred Tarski and were formulated in 1953.

In the twenties and thirties, in the framework of attempts to realize Hilbert's program, many results on the decidability and completeness of theories were obtained. Several general methods of proving the completeness and decidability of theories were proposed.

Incompleteness results of Gödel published in 1931 and the undecidability result of Church published in 1936 indicated some difficulties and obstacles in attempts to realize Hilbert's program. One consequence was the fact that since the end of the thirties logicians' and mathematicians' interests concentrated on the proofs of undecidability rather than of the decidability of theories. General methods of proving undecidability were investigated and particular mathematical theories were shown to be undecidable.

The method of eliminating quantifiers was initiated by Leopold Löwenheim (1878–1957) (cf. Löwenheim, 1915) and used in fully-developed form by Thoralf Skolem (1887–1963) (cf. Skolem, 1919) and by Cooper Harold Langford (1895–1964) (cf. Langford, 1927 and 1927a). It should be noted here that they did not speak about decision procedures. Langord, for instance, formulated his results in the form: "if a sentence φ is given then either φ follows from the given set of axioms or $\neg\varphi$ follows from it". The method of quantifier elimination was also studied at the seminar led by Alfred Tarski (1901–1983) at Warsaw University in 1927–29. Tarski and his students used this method to characterize definability and to prove the decidability of particular mathematical theories (cf. Murawski, 1996; Doner and Hodges,

1988). It was used also to describe and classify all complete extensions of a given theory. In relation to this, Tarski introduced some key notions of model theory in his seminar, in particular the notion of elementary equivalence. In this way, the elimination of quantifiers became *the* method in Tarski's school and a paradigm of how logicians should study an axiomatic theory. The very name of the method comes from Tarski.

The theorem on the decidability of the first-order theory of well-ordering mentioned in Section 3.2 has an interesting and dramatic history. In the late thirties Alfred Tarski and his student Andrzej Mostowski (1913–1975) developed the outline of a proof of the decidability of this theory described semantically as the set of sentences true in all well-ordered structures. They used the method of the elimination of quantifiers and by the summer of 1939 they reached a clear idea of the basic formulas. Their main aim was to prove the adequacy of certain axiom systems and to classify the complete extensions of the theory (decidability would be a by-product of it). Many technical details remained to be worked out. The work was interrupted by World War II. Tarski escaped the German occupation of Poland and stayed in the USA, Mostowski spent the war time in Poland. Each began to work out the technical details. Unfortunately, Mostowski's notes were destroyed in Warsaw in 1944 and Tarski's ones were lost in the course of his many moves. Hence the work had to be restarted from the very beginning. They published the abstract (1949) and made plans to reconstruct the proof. This was not realized and nothing more was done than writing down the specification of the basic formulas. In 1964 Tarski assigned his student John Doner the task to work out some of the details. This was done in a rough form and again nothing happened till 1975. In this year Mostowski and Tarski met and made plans to finish the paper. Unfortunately, Mostowski's unexpected death in August 1975 upset the plans. In this situation Tarski invited Doner to take up the work. The final result was the joint paper of Doner, Mostowski and Tarski (1978). One finds there a number of results not mentioned in Tarski and Mostowski (1949).

In the school year 1927/28 Tarski gave a course of lectures on first-order theories. During the course he presented a set of axioms for the theory of addition of natural numbers and formulated a problem of showing that the axioms are complete. This was solved by Mojżesz Presburger (1904–1943(?)) in May 1928 and published in Presburger (1930). It is worth noting here that Tarski did not accept this result as a sufficient basis for a doctoral dissertation and it was included into Presburger's master's thesis (cf. Crossley, 1974; Woleński, 1989; Zygmunt, 1991). Note also that Presburger, like Tarski at that time, formulated his results in terms of completeness and did not mention decidability.

The family $\mathcal{P}$ was introduced by Cobham (1964). The class $\mathcal{NP}$ was first studied by Edmonds in 1965. The foundations of the theory of $\mathcal{NP}$-completeness were presented in Cook (1971). Examples of $\mathcal{NP}$-complete problems can be found in Karp (1972), Garey and Johnson (1979), Hopcroft and Ullman (1979) and Sudkamp (1988). For recent results on interconnections between the problem of whether $\mathcal{P} = \mathcal{NP}$ on the one hand and weaker arithmetics (bounded arithmetic) as well as logic (propositional calculus) on the other see, e.g., survey papers by Krajíček (1996) and Pudlák (1995) and the book Krajíček (1995).

The theorem on decidability of the classical monadic first-order predicate calculus was proved independently by Löwenheim (1915), Skolem (1919) and Heinrich Behmann (1891–1970) (1922).

Tarski contributed very much to the study of the decidability of theories (as indicated above) as well as to the development of various general methods of proving the undecidability. His approach was very original: he applied methods of recursion theory together with ideas of model theory. The notion of interpretation (weak interpretation, relative interpretation, etc.) introduced by him in 1953 has turned out to be a powerful tool in proving the undecidability of particular theories. It should be added that the results on undecidability were very often only one kind of consequence of Tarski's results on definability and interpretability (cf. McNulty, 1986 and Murawski, 1996).

To apply the method of interpretation one has to have a finitely axiomatizable and essentially undecidable theory which is weak enough to be interpreted in theories even quite distant from it. The problem was solved by Mostowski, Tarski and Raphael M. Robinson (1911–1995). In 1939, Mostowski and Tarski constructed a finitely axiomatizable and essentially undecidable subtheory $\overline{Q}$ of the arithmetic of natural numbers. It was closely related to the theory of non-densely ordered rings (cf. Mostowski and Tarski, 1949) where an analogous theory is described). Around 1949/50 R.M. Robinson and Tarski simplified this system and finally a simple, finitely axiomatizable and essentially axiomatizable theory Q arose (Mostowski, Robinson and Tarski, 1953).

Another example of a theory with properties similar to those of Q, i.e., being finitely axiomatizable and essentially undecidable was constructed by Andrzej Grzegorczyk (b. 1922) in 1962. In fact two theories F and F^* are given there.

The first result on undecidability was in fact Church's theorem on the undecidability of the classical first-order predicate calculus from 1936 — cf. Theorem 3.3.32. One should also mention here works of the Polish logician Józef Pepis, a mathematician active at the University of Lvov, killed by the Gestapo in August 1941. His works were devoted to the problem of mutual reducibility of the decision problem for the predicate calculus for various classes of formulas (cf. Pepis, 1936, 1937, 1938, 1938a; see also Murawski (1996) for the discussion of Pepis' results).[4]

[4]Comprehensive information on decidability and undecidability of theories as well as on the classical decision problem (*Entscheidungsproblem*) can be found, e.g., in Ershov *et al.* (1965) and in the monograph by Börger *et al.* (1997).

Chapter 4

Philosophical Comments

In this chapter various philosophical and methodological consequences of Gödel's incompleteness theorems and their generalizations (presented in Chapter 2) as well as of theorems on undecidability of theories (Chapter 3) will be considered. In Section 4.1 we discuss the direct implications of Gödel's results. In Section 4.2 the problem of connections between incompleteness theorems and Hilbert's program will be discussed. Next, in Sections 4.3 and 4.4, generalizations of the program of Hilbert, i.e., generalized and relativized Hilbert's program will be presented and the influence of results of the so-called reverse mathematics on the latter will be studied. Section 4.5 deals with Hilbert's tenth problem and Matiyasevich's theorem. Further connections between Gödel's theorems and computer science (Section 4.6) and general philosophy (Section 4.7) are discussed.

4.1 Direct Consequences of Gödel's Results

First note that Gödel's First Incompleteness Theorem indicates that the cognitive power of the axiomatic-deductive method is in fact fundamentally and immanently bounded. The theorem states that every first-order theory formalized in a language with a recursive set of nonlogical symbols, based on a recursive set of nonlogical axioms and extending Peano arithmetic PA is incomplete. Furthermore, this incompleteness is essential; it cannot be removed by adding new sentences (for example undecidable sentences) as new axioms (provided that the set of nonlogical axioms remains recursive, i.e., that it can be effectively decided whether a given sentence is an axiom or not). Hence the whole of mathematics cannot be reduced to a single axiomatic system (even if everything can be expressed in a universal language of the first-order predicate calculus). Even more, a single theory cannot incorporate the whole truth about natural numbers. There will always be undecidable sentences of the form $\forall x \psi(x)$ such that all particular instances of ψ, i.e., sentences of the form $\psi(0)$, $\psi(\overline{1})$, $\psi(\overline{2})$, ... are provable. So one cannot construct a consistent and complete system of arithmetic, and all the more of the whole of mathematics (cf. Hilbert's program presented in the Introduction). The meaning of those results is increased by the fact that the particular formalization which is used is in fact irrelevant — Gödel's incompleteness results hold for any formalization provided the condition of recursiveness of the set of axioms as well as of the inference rules is satisfied.

Gödel's incompleteness theorems also revealed the distinction between demonstrability and validity (truth). They showed that formalized theories with their inference rules are not (and cannot be) adequate with respect to the usual research practice of mathematicians. A mathematician's aim is to decide whether a given property of objects studied in a considered domain (e.g., in the domain of natural numbers) does hold or does not hold, if the sentence stating that a given property holds is true or false. To answer those questions any correct methods can be used (and the scope of admissible methods is not fixed or bounded beforehand). Methods are chosen according to the needs and problems that appear. Gödel's results indicated that this practice cannot be adequately represented and expressed by a formal system (with fixed axioms and rules of inference).

This pragmatic aspect is connected with (and is a consequence of) another one, namely the problem of the distinction between syntactical notions (provability) and semantical ones (validity, truth). Gödel's theorems showed that provability is not the same as validity. Indeed, every theorem of Peano arithmetic PA is true (in the standard model $\mathfrak{N}_0$) but there exist true sentences which are not theorems of PA, i.e., there exist undecidable sentences (the truth of which can be decided by methods going beyond methods admissible in PA). Hence the semantic truth cannot be adequately expressed by syntactical truth (provability).

In connection with those issues it is worth noting that Gödel himself avoided the usage of semantical terms 'true' and 'truth'. Indeed, in Gödel (1931) one finds the word 'true' only in the introductory Section 1 (the situation is a bit different in (1934) where the term is used). What were the reasons for that? It seems that "despite his conviction of the objectivity of the concept of mathematical truth, Gödel feared that work assuming such a concept would be rejected by the foundational establishment, dominated as it was by Hilbert's ideas" (Feferman, 1984, p. 108). In a draft of a reply to a graduate student's query in 1970, in a crossed-out passage Gödel wrote:

> However in consequence of the philosophical prejudices of our time ... a concept of objective mathematical truth as opposed to demonstrability was viewed with greatest suspicion and widely rejected as meaningless.

On the other hand, as already mentioned above, Gödel was convinced of the objectivity of the concept of mathematical truth. In a letter to Hao Wang (cf. Wang Hao, 1974, p. 9) he wrote:

> I may add that my objectivist conception of mathematics and metamathematics in general, and of transfinite reasoning in particular, was fundamental also to my other work in logic. How indeed could one think of *expressing* metamathematics *in* the mathematical systems them-

> selves, if the latter are considered to consist of meaningless symbols which acquire some substitute of meaning only *through* metamathematics ... it should be noted that the heuristic principle of my construction of undecidable number theoretical propositions in the formal systems of mathematics is the highly transfinite concept of 'objective mathematical truth' as *opposed* to that of 'demonstrability' (cf. M. Davis, *The Undecidable*, New York 1965, p. 64 where I explain the heuristic argument by which I arrive at the incompleteness results), with which it was generally confused before my own and Tarski's work.

And later he adds (p. 10)

> ... formalists considered formal demonstrability to be an *analysis* of the concept of mathematical truth and, therefore were of course not in a position to *distinguish* the two.

Indeed, only after Gödel's results the semantical concepts (validity, truth) were properly defined and carefully studied (especially by Carnap and Tarski).

To illustrate the 'gap' between syntactic and semantical approach to mathematics consider the set V of Gödel numbers of sentences of the language L(PA) of Peano arithmetic which are true in the standard model $\mathfrak{N}_0$, i.e.,

$$V = \{\ulcorner\varphi\urcorner : \varphi \text{ is a sentence of } L(PA) \wedge \mathfrak{N}_0 \models \varphi\}.$$

The set V belongs to no arithmetical class, i.e., there is no $n \in \mathbb{N}$ such that $V \in \Sigma^0_n$ or $V \in \Pi^0_n$. On the other hand, the set of Gödel numbers of theorems of PA is recursively enumerable, so it is of the class Σ^0_1 (cf. Chapter 2). To prove that V is not of an arithmetical class assume that $V \in \Sigma^0_n$ for some $n \in \mathbb{N}$. By the definition of the arithmetical hierarchy (cf. Section 1.5) there exists a recursive relation R such that for any k:

$$k \in V \equiv \exists x_1 \forall x_2 \ldots Q x_n R(k, x_1, \ldots, x_n),$$

where Q is the universal quantifier $\forall$ if n is even and the existential quantifier $\exists$ if n is odd. By Theorem 2.4.24 there is a formula ψ of the class BQ such that the formula $\exists y\psi$ strongly represents in PA the relation R, i.e., for any $k, k_1, \ldots, k_n$:

$$R(k, k_1, \ldots, k_n) \text{ if and only if } \mathrm{PA} \vdash \exists y\psi(y, \overline{k}, \overline{k_1}, \ldots, \overline{k_n}),$$

$$\neg R(k, k_1, \ldots, k_n) \text{ if and only if } \mathrm{PA} \vdash \neg\exists y\psi(y, \overline{k}, \overline{k_1}, \ldots, \overline{k_n}).$$

Hence in particular

$$R(k, k_1, \ldots, k_n) \text{ if and only if } \mathfrak{N}_0 \models \exists y\psi(y, \overline{k}, \overline{k_1}, \ldots, \overline{k_n}).$$

But for any formula φ of the language L(PA):

$$\mathfrak{N}_0 \models \forall x\varphi(x) \text{ if and only if for every } k \in \mathbb{N}: \ \mathfrak{N}_0 \models \varphi(\overline{k}),$$

$$\mathfrak{N}_0 \models \exists x\varphi(x) \text{ if and only if there exists a } k \in \mathbb{N}: \ \mathfrak{N}_0 \models \varphi(\overline{k}).$$

So consequently

$$\begin{aligned} k \in V \ &\equiv \ \exists k_1 \forall k_2 \ldots Q k_n \mathfrak{N}_0 \models \exists y\psi(y, \overline{k}, \overline{k_1}, \ldots, \overline{k_n}) \\ &\equiv \ \mathfrak{N}_0 \models \exists x_1 \forall x_2 \ldots Q x_n \exists y\psi(y, \overline{k}, x_1, \ldots, x_n). \end{aligned}$$

In this way we have shown that the set V is definable in $\mathfrak{N}_0$ by the formula χ of the form

$$\exists x_1 \forall x_2 \ldots Q x_n \exists y\psi(y, x, x_1, \ldots, x_n)$$

which contradicts Tarski's theorem on undefinability of truth (cf. Theorem 2.4.11). We have proved that the set V does not belong to the arithmetical hierarchy.

To see the meaning of this corollary more clearly recall that the aim of formulating Peano's axioms was to select those facts about natural numbers from which all other truths about those numbers could be derived. Gödel's theorem showed that the proposed set of axioms does not (and cannot!) possess the required property of completeness, even more, no other recursive set of axioms can have this property. Futhermore, one can show that

Peano arithmetic has a high degree of incompleteness. Its axioms do not describe adequately natural numbers (in the sense that they are true in the domain of natural numbers $\mathfrak{N}_0$ and only in it) but arithmetic has a great variety of different models. In all of them Peano's axioms are true but for any two such models there exists a property of natural numbers expressible in the language of L(PA) such that it is true in one model and false in the other. To describe this phenomenon more exactly recall that two models $\mathfrak{A}$ and $\mathfrak{B}$ of a theory T are said to be elementarily equivalent if and only if they satisfy the same sentences, i.e., if they possess the same elementary (expressible in the language L(T)) properties. Note that there can be two different (even nonisomorphic) models different which are elementarily equivalent.

Theorem 4.1.1 *There exist* $2^{\aleph_0}$ *countable elementarily nonequivalent models of Peano arithmetic* PA.

Proof. Observe that from considerations of Section 2.4 it follows that every consistent finite extension (i.e., an extension obtained by adding finitely many new nonlogical axioms) of Peano arithmetic PA is incomplete. We construct $2^{\aleph_0}$ extensions of PA in the following way: let $T_0 = PA$, let ψ_0 be any sentence undecidable in T_0 and let $T_{00} = PA \cup \{\psi_0\}$, $T_{01} = PA \cup \{\neg\psi_0\}$. Let σ be a finite sequence of 0's and 1's and let ψ_σ be any sentence undecidable in the theory T_σ. We put $T_{\sigma * \langle 0 \rangle} = T_\sigma \cup \{\psi_\sigma\}$ and $T_{\sigma * \langle 1 \rangle} = T_\sigma \cup \{\neg\psi_\sigma\}$. In this way we constructed the binary tree of extensions of PA shown in Diagram 4.1 below. This tree has $2^{\aleph_0}$ paths and the sum of theories on every path is an extension of PA. Futhermore, it is a consistent extension of PA – indeed, every finite fragment of such an extension is consistent by the construction, so by the Compactness Theorem the whole extension is also consistent. By Gödel's Completeness Theorem every such extension has a countable model which is of course also a model of PA. No two such models can be elementarily equivalent because one of them satisfies certain undecidable sentence and the other its negation. Q.E.D.

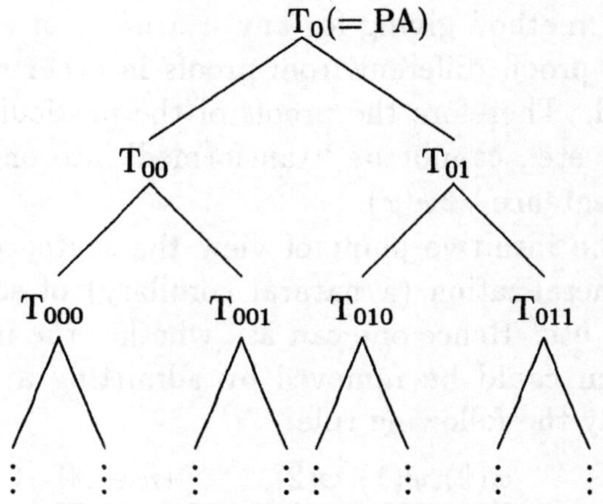

Diagram 4.1

Consequently, Peano arithmetic PA has a variety of different models. So therefore the description of natural numbers contained in the axioms of PA is incomplete in a high degree and, even more, it must remain incomplete provided we require that the set of axioms is effective (recursive) and that the notion of proof — and consequently the scope of admissible inference methods — is precisely given and also effective (recursive), i.e., that we can effectively distinguish what is and what is not an axiom and effectively check whether a given formula was obtained with the help of a given inference rule or not.

Peano arithmetic PA is based on an infinite set of axioms. On the other hand, there exist examples of theories weaker than PA, based on a finite set of axioms and such that counterparts of Gödel's incompleteness theorems hold for them. The best known example of such a theory is the Mostowski-Robinson-Tarski's theory Q discussed in Section 3.3.

What are the sources of the undecidability of Gödel's sentence? Remind that it is of the form $\forall x\psi(x)$ and that PA $\vdash \psi(0)$, PA $\vdash \psi(\overline{1})$, PA $\vdash \psi(\overline{2})$, etc. The source of undecidability of the universal sentence $\forall x\psi(x)$ is the fact that proofs of the particular instances $\psi(0)$, $\psi(\overline{1})$, $\psi(\overline{2})$, etc., are different, i.e., there is no uniform method giving for any n a proof of $\psi(\overline{n})$. In every case a new proof, different from proofs in other cases, must be constructed. Therefore the proofs of the particular cases $\psi(0)$, $\psi(\overline{1})$, $\psi(\overline{2})$, etc., cannot be 'transformed' into one proof of the universal sentence $\forall x\psi(x)$.

From the intuitive point of view the sentence $\forall x\psi(x)$ is a natural generalization (a natural corollary) of sentences $\psi(0)$, $\psi(\overline{1})$, $\psi(\overline{2})$, etc. Hence one can ask whether the incompleteness phenomenon could be removed by admitting a new inference rule, namely the following rule

$$\frac{\varphi(0), \varphi(\overline{1}), \varphi(\overline{2}), \ldots \quad (n \in \mathbb{N})}{\forall x\varphi(x)},$$

called ω-rule.[1] Notice that the ω-rule has (in contrast with rules accepted so far) infinitely many premises. Since there is *a priori*

---

[1]It seems that Tarski has been the first person to formulate the ω-rule (cf. Feferman, 1986; Isaacson, 1992). He discussed it in 1926 and presented in a lecture in 1927, though he did not publish the idea until 1933 (cf. Tarski, 1933a, p. 294; see also p. 279, footnote 2; Tarski, 1933, pp. 259–261). He called it 'the rule of infinite induction'.

The ω-rule was published for the first time in Ajdukiewicz (1928) where it was called 'the rule of complete quasi-induction'.

Tarski knew quite well the Ajdukiewicz's book (1928) and quoted it several times. However, he never refered to it when writing about the ω-rule. He refered to Hilbert who formulated this rule in an implicitly constructive form in (1931, pp. 491–492) and used it further in (1931a). In (1931) he gave no characteristic name of the rule and called it merely 'a new rule of inference'. Later (1931a) he designated it as 'transfinite axiom and inference schemata'.

The ω-rule was also formulated by Carnap (1934, Section 14, p. 38; 1935). He did not use any description of it, labeling it DC2 (one of the two rules of Direct Consequence for his language I). Rosser (1937) decided to call the ω-rule 'Carnap's rule'.

The name 'ω-rule' occured for the first time in Grzegorczyk *et al.* (1958) (in fact the phrase 'rule ω' is used there).

For the problem of motivations and reasons of introducing the ω-rule by various authors as well as the problem of (in)dependency see Isaacson (1992).

no connection between the proofs of the premises, i.e., the proofs of the premises need not be obtained uniformly, there need not be a uniform method generating a proof of $\varphi(\overline{n})$ when n is given, it follows that the ω-rule is ineffective. The answer to the question formulated above is affirmative, Peano arithmetic with ω-rule (this theory will be denoted by $(\mathrm{PA})_\omega$) is complete. This follows from the fact that the unique model of $(\mathrm{PA})_\omega$ up to isomorphism is the standard model $\mathfrak{N}_0$.

A next question one can ask is the following one: Assume that the theory $(\mathrm{PA})_\omega$ is complete. How many times must the ω-rule be applied to obtain a complete extension of PA? The answer depends on the way in which arithmetic has been axiomatized and on the way in which the degree of complexity of a proof has been defined. Let us explain this. First observe that if the usage of the ω-rule in a proof is admitted than a proof is no more a finite sequence of formulas such that each formula is either an axiom or can be obtained from previous ones by one of the rules of inference (cf. Section 2.1). Now a proof is a tree. So let us define first a proof tree. A proof tree is a graph whose nodes can be divided into finitely many disjoint levels (nodes of the ith level will be called nodes of the rank i) such that: (1) there exists exactly one node of the rank 0 (it is called the end node), (2) every node of the rank $i+1$ is connected to exactly one node of the rank i, (3) for every node P of rank i exactly one of the following cases holds: (a) the node P is connected to no node of rank $i+1$, (b) the node P is connected to one, two or countably many nodes of rank $i+1$ (nodes of rank $i+1$ to which the node P is connected are called predecessors of the node P), (4) every node of rank i can be connected only to nodes of rank $i-1$ and to nodes of rank $i+1$. A node of rank i which is not connected to any node of the rank $i+1$ is called initial. Some examples of proof trees are given by Diagrams 4.2 and 4.3 below.

Now we can say that a formula φ is a theorem of the theory $(\mathrm{PA})_\omega$ if and only if there exists a proof tree Δ and an assignment of formulas of the language L(PA) to nodes of δ such that

(1) formulas assigned to initial nodes are axioms of Peano arithmetic PA,

(2) if P is a noninitial node of the rank i then formulas assigned to nodes of Δ of the rank $i+1$ to which P is connected are premises of one of the inference rules of PA or of the ω-rule and the formula assigned to P is a conclusion of this rule,

(3) the formula φ is assigned to the terminal node of the proof-tree Δ.

Observe that admitting the ω-rule one can consider the system of arithmetic without the axiom scheme of induction. Indeed, let PA^- denote the theory based on the axioms of Peano arithmetic without the axiom scheme of induction. Then this scheme can be proved in $(\mathrm{PA}^-)_\omega$. To prove that let φ be a formula and assume that

$$(\mathrm{PA}^-)_\omega \vdash \varphi(0) \wedge \forall x[\varphi(x) \rightarrow \varphi(x)].$$

Using the substitution rule and *modus ponens* together with laws of the predicate calculus we obtain that $\mathrm{PA}^- \vdash \varphi(\overline{n})$ for any $n \in \mathbb{N}$. Applying the ω-rule one obtains $(\mathrm{PA}^-)_\omega \vdash \forall x \varphi(x)$.

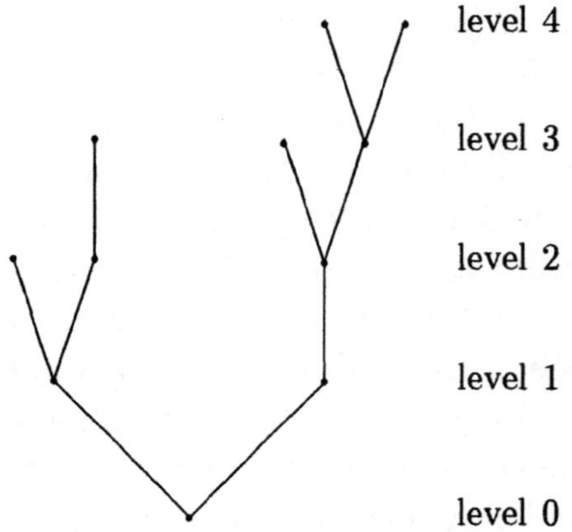

Diagram 4.2

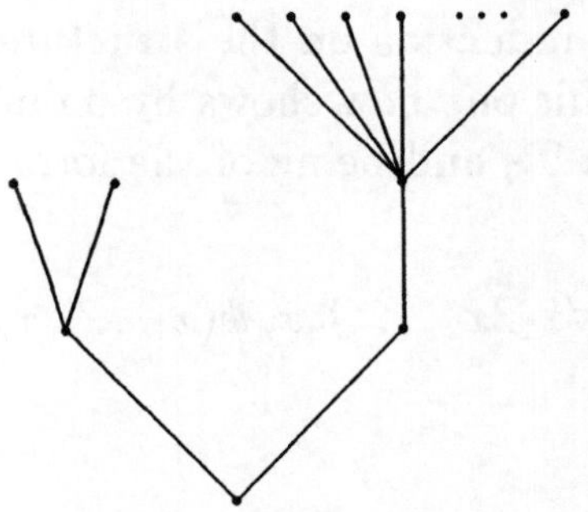

Diagram 4.3

Introduce now a notion of the rank of a theorem of $(\mathrm{PA})_\omega$. To do this define an increasing sequence PA^α (α being a countable ordinal) of sets of theorems of $(\mathrm{PA})_\omega$. Elements of PA^α will be called theorems of the rank α. We put

$$\mathrm{PA}^0 = \text{axioms of PA},$$

$$\mathrm{PA}^{\alpha+1} = \mathrm{PA}^\alpha \cup \{\varphi : \text{there exist formulas } \psi, \chi \in \mathrm{PA}^\alpha \text{ such that } \varphi \text{ is obtained from } \psi \text{ and } \chi \text{ by a single application of one of the inference rules of PA or there exists a formula } \psi \text{ such that for every } n \in \mathbb{N} \text{ the formula } \psi(\overline{n}) \text{ belongs to } \mathrm{PA}^\alpha \text{ and } \varphi \text{ is of the form } \forall x \psi(x)\},$$

$$\mathrm{PA}^\lambda = \bigcup_{\alpha<\lambda} \mathrm{PA}^\alpha \text{ for } \lambda \text{ limit.}$$

It can be easily seen that the set of theorems of $(\mathrm{PA})_\omega$ is equal to $\bigcup_{\alpha<\omega_1} \mathrm{PA}^\alpha$. The following theorem holds.

Theorem 4.1.2 *If a sentence φ is true in the standard model $\mathfrak{N}_0$ then $\varphi \in (\mathrm{PA})^{\omega\times 3}$, hence the set of theorems of $(\mathrm{PA})_\omega$ is equal to $\bigcup_{\alpha<\omega\times 3} \mathrm{PA}^\alpha$.*

Sketch of the proof. Observe that for any sentence φ which belongs to the set BQ (cf. Definition 2.4.23) and is true in the standard model $\mathfrak{N}_0$ there exists a $k \in \omega$ such that $\varphi \in \mathrm{PA}^k$. One can prove this by induction on the structure of formulas of the class BQ. Using this one now shows by induction on n that any sentence φ true in $\mathfrak{N}_0$ and being of the form

$$\forall x_1 \exists x_2 \ldots Q_n x_n \psi(x_1, \ldots, x_n)$$

or

$$\exists x_1 \forall x_2 \ldots Q'_n x_n \psi(x_1, \ldots, x_n)$$

(where ψ is a formula of the class BQ and Q_n is the existential quantifier and Q'_n the universal one if n is even, and *vice versa* if n is odd[2]) belongs to $\mathrm{PA}^{\omega+n}$. To finish the proof it suffices to observe that the transformation of any given formula to the prenex form can be done using only finitely many usual inference rules (without the ω-rule). Consequently, any sentence φ true in $\mathfrak{N}_0$ belongs to $\mathrm{PA}^{\omega+\omega+\omega}$, i.e., to $\mathrm{PA}^{\omega\times 3}$. Q.E.D.

Note that if one took as axioms not the axioms of Peano arithmetic PA but all sentences of the class BQ true in the standad model $\mathfrak{N}_0$ (denote such a theory by T), and if the rank of a theorem were defined as above, then the set of theorems of $(\mathrm{PA})_\omega$ would be equal to $\mathrm{T}^{\omega\times 2}$. Indeed, for any sentence φ of the class BQ true in $\mathfrak{N}_0$ one would now have $\varphi \in \mathrm{T}^0$.

Remark that the notion of the rank of a theorem can be defined in a different way, for instance as follows (cf. Kotlarski, 1986):

---

[2]Formulas of the form $\forall x_1 \exists x_2 \ldots Q_n x_n \psi(x_1, \ldots, x_n)$ where $\psi \in \mathrm{BQ}$ are said to be formulas of the class Π^0_n and formulas of the form $\exists x_1 \forall x_2 \ldots Q'_n x_n \psi(x_1, \ldots, x_n)$ where $\psi \in \mathrm{BQ}$ are said to be formulas of the class Σ^0_n.

$$\begin{aligned} T^0 &= T, \\ T^{\alpha+\frac{1}{2}} &= T^\alpha \cup \{\varphi : \varphi \text{ is of the form } \forall x\psi(x) \text{ and } \psi(\overline{n}) \in \\ &\quad T^\alpha \text{ for every } n \in \mathbb{N}\}, \\ T^{\alpha+1} &= \text{the smallest set of formulas containing } T^{\alpha+\frac{1}{2}} \\ &\quad \text{and closed under the rules of inference of PA}, \\ T^\lambda &= \bigcup_{\alpha<\lambda} T^\alpha \text{ for } \lambda \text{ limit}. \end{aligned}$$

By this definition of rank only the application of the ω-rule increases the rank of a theorem while an application of any other rule (from PA) does not increase it (by the previous definition any application of any rule increased the rank). One can now show that the set of theorems of $(\mathrm{PA})_\omega$ is equal to T^ω — independently from which set of sentences was adopted as the set of axioms: the axioms of Peano arithmetic PA or the set of all sentences of the class BQ true in $\mathfrak{N}_0$. Indeed, one can now prove that for any sentence φ of the class Σ^0_{2n+1} true in the model $\mathfrak{N}_0$, $\varphi \in T^n$.

Theorem 4.1.2 and the above remarks answer our question: how many times should the ω-rule be applied to obtain a complete extension of Peano arithmetic PA? They show that the answer to this question depends on the way in which arithmetic has been formalized and in which the notion of the complexity of a proof has been defined. Another illustration of that dependency are Rosser's results from 1937. He considered the system of arithmetic to be based on Whitehead and Russell's theory of types introduced in *Principia Mathematica* (it was also used by Gödel in 1931). Denote it by P_0. Next Rosser defined systems P_α for ordinals $\alpha < \omega^2$ in the following way:

1. if the system P_α has been defined then let P'_α be an extension of P_α obtained by adding all formulas of the form $\forall x\psi(x)$ such that $\psi(\overline{n})$ is a theorem of P_α for every n. The system $P_{\alpha+1}$ is now defined as the smallest set of formulas containing P'_α and closed under the usual inference rules (cf. Section 2.1),

2. if α is a limit ordinal then $P_\alpha = \bigcup_{\beta<\alpha} P_\beta$.

The following theorems hold.

Theorem 4.1.3 *Let $n \in \mathbb{N}$ and $n \neq 0$. If P_n is consistent then*
(i) *there exist sentences undecidable in P_n,*
(ii) *the system P_n is not closed under the ω-rule,*
(iii) *the sentence Con_{P_n} is not provable in P_n.*

Theorem 4.1.4 *If $n \in \mathbb{N}$ and $P_{\omega+n}$ is consistent then*
(i) *there exist sentences undecidable in $P_{\omega+n}$,*
(ii) *the system $P_{\omega+n}$ is not closed under the ω-rule,*
(iii) *the sentence $\mathrm{Con}_{P_{\omega+n}}$ is not provable in $P_{\omega+n}$.*

Similar theorems can also be proved for each of the systems $P_{m\omega+n}$ for $m, n \in \mathbb{N}$.

As noted above the ω-rule is ineffective. Various suggestions have been proposed for modifying it to make it more effective. One of such suggestions is to require the existence of an effective method of obtaining a proof of $\varphi(\overline{n})$ when n is given. It turns out — as shown by Shoenfield (1959) — that this rule is no less powerful than the ω-rule, i.e., Peano arithmetic with this rule is complete. Add that this modification of the ω-rule was suggested by Novikoff in a somewhat different context; a similar rule was also investigated by Parsons (1958).

We sketch the result of Shoenfield following his paper (1959). First one must precisely define the restricted effective version of the ω-rule. To do this assume that an arithmetization of Peano arithmetic is given (cf. Section 2.3). Let φ be a formula of L(PA). Denote by $E_n(\varphi)$ the set of n-tuples $(a_1, \ldots, a_n)$ of elements of $\mathbb{N}$ such that

$$\mathrm{PA} \vdash \varphi(\overline{a_1}, \ldots, \overline{a_n}).$$

By Representability Theorem 2.2.12 every n-ary recursive relation R is $E_n(\varphi)$ for some formula φ of the language L(PA). Define now a recursive function S_n by

$$S_n(e, a) = Sub(e, \langle 2n \rangle, Num(a)).$$

Then

$$S_n(\ulcorner\varphi\urcorner, a) = \ulcorner\varphi(x_n/\overline{a})\urcorner.$$

Now define a recursive function Sb_n such that

$$Sb_n(\ulcorner\varphi\urcorner, a_1, \ldots, a_n) = \ulcorner\varphi(\overline{a_1}, \ldots, \overline{a_n})\urcorner.$$

Precisely we define by induction on n:

$$Sb_0(e) = e$$

and

$$Sb_{n+1}(e, a_1, \ldots, a_n, a_{n+1}) = Sb_n(S_n(e, a_{n+1}), a_1, \ldots, , a_n).$$

We have now

$$(a_1, \ldots, a_n) \in E_n(\varphi) \equiv \exists x Prf(x, Sb_n(\ulcorner\varphi\urcorner, a_1, \ldots, a_n)). \quad (4.1)$$

By Theorem on Normal Form 1.4.17 for any n-ary recursive function f there exists an elementary recursive relation R such that

$$f(\overline{a}) = b \equiv \exists y R(\overline{a}, y, b).$$

Choose φ such that R is $E_{n+2}(\varphi)$ and let $e = \ulcorner\varphi\urcorner$. Then by (4.1):

$$\begin{aligned} f(\overline{a}) = b \ &\equiv\ \exists x \exists y Prf(x, Sb_{n+2}(e, \overline{a}, y, b)) \\ &\equiv\ \exists x Prf((x)_0, Sb_{n+2}(e, \overline{a}, (x)_1, b)). \end{aligned}$$

Hence by Theorem 1.4.17

$$f(\overline{a}) = (\mu z Prf((z)_{1,0}, Sb_{n+2}(e, \overline{a}, (z)_{1,1}, (z)_0)))_0,$$

where $(z)_{1,0} = ((z)_1)_0$ and similarly $(z)_{1,1} = ((z)_1)_1$. We define a recursive relation T_n and a recursive function U by

$$T_n(e, \overline{a}, z) \equiv Prf((z)_{1,0}, Sb_{n+2}(e, \overline{a}, (z)_{1,1}, (z)_0))),$$

$$U(z) = (z)_0.$$

We then have

$$f(\overline{a}) = U(\mu z T_n(e, \overline{a}, z)). \tag{4.2}$$

A number e is called an index of an n-ary function f if (4.2) holds for all $\overline{a}$. The function f is then denoted by $\{e\}$. In this way we showed that every recursive function has an index. Conversely, if e is an index of a function f, then f has the explicite definition (4.2) and is recursive. So a partial function is recursive if and only if it has an index.

Having defined the notion of the index we can now precisely describe the concept of an effective ω-rule. We define the notion of a number of a proof in $(\mathrm{PA})^*_\omega$ ($(\mathrm{PA})^*_\omega$ denotes Peano arithmetic with the effective ω-rule). For $0 \leqslant i \leqslant k$, let $s_i = \langle a_i, b_i \rangle$, where $a_i = \ulcorner \varphi_i \urcorner$ and b_i is 0 or the index of a recursive function. We say that $s = \langle s_0, s_1, \ldots s_k \rangle$ is a number of a proof in $(\mathrm{PA})^*_\omega$ of φ_k if and only if the following conditions hold for $i < k$: (1) if $b_i = 0$ then φ_i is an axiom of PA or a logical consequence of certain φ_j with $j < i$, (2) if $b_i \neq 0$ then φ_i is of the form $\forall x \varphi(x)$ and for each n, $\{b_i\}(n)$ is the number of a proof in $(\mathrm{PA})^*_\omega$ of $\varphi(\overline{n})$. We say that φ is provable in $(\mathrm{PA})^*_\omega$ if there exists a number of a proof in $(\mathrm{PA})^*_\omega$ of φ.

From the above definitions follows that an application of the ω-rule is allowed in a proof only if the proofs of the premises $\varphi(\overline{n})$ can be obtained in a uniform way, i.e., if there exists a recursive function f such that for every $n \in \mathbb{N}$, $f(n)$ is a (code of a) proof of $\varphi(\overline{n})$.

Using several technical facts one can show that if φ is true in the standard model $\mathfrak{N}_0$ then φ is provable in $(\mathrm{PA})^*_\omega$, i.e., the theory $(\mathrm{PA})^*_\omega$ is complete.

A natural question arises (similar to the one formulated in connection with the ω-rule): how many times must the restricted effective ω-rule be applied to obtain a complete extension of PA? To answer this define a sequence of subtheories of $(\mathrm{PA})^*_\omega$.

$$
\begin{aligned}
(\mathrm{PA})_\omega^{*,0} &= \mathrm{PA},\\
(\mathrm{PA})_\omega^{*,\alpha+1} &= \text{the class of sentences which are provable from sentences of } (\mathrm{PA})_\omega^{*,\alpha} \text{ by the restricted effective } \omega\text{-rule},\\
(\mathrm{PA})_\omega^{*,\lambda} &= \bigcup_{\alpha<\lambda} (\mathrm{PA})_\omega^{*,\alpha} \text{ for } \lambda \text{ limit.}
\end{aligned}
$$

It can be shown now that the theory $(\mathrm{PA})_\omega^{*,\omega^\omega}$ is complete, i.e., every sentence φ true in the standard model $\mathfrak{N}_0$ is an element of $(\mathrm{PA})_\omega^{*,\omega^\omega}$. Hence ω^ω applications (in the above defined sense) of the effective ω-rule suffice to obtain the complete extension of Peano arithmetic.[3]

It was shown in Corrollary 2.4.22 that for every recursive extension T of Peano arithmetic PA the set Thm_{T} is not recursive (in this case the theory T is said to be undecidable). This means that there is no mechanical effective method of deciding in a finite number of steps whether a given sentence of the language L(T) is provable in T or not. Most mathematical theories are undecidable. This fact should not be astonishing. Indeed, problems formulated in decidable theories are not any longer scientific problems — they can be solved in a mechanical way (the only trouble here can be the time needed to obtain a solution — cf. remarks on the complexity of decision procedures in Section 3.2). So if richer and interesting mathematical theories were decidable then one should wonder why there still are so many open problems in them.

The negative solution to the decision problem for most theories in mathematics also has practical consequences — it implies that there are and always will be enough problems to be solved. Tarski put it in an amusing way:

> I have no doubt that many mathematicians experienced
> a profound feeling of relief when they heard of this result.
> Perhaps sometimes in their sleepless nights they thought

[3]For current discussions of the ω-rule see, e.g., Isaacson (1992) and Ignjatovic (1994).

> with horror of the moment when some wicked metamathematician would find a positive solution of the problem, and design a machine which would enable us to solve any mathematical problems in a purely mechanical way, so that any further creative mathematical thought would become a worthless hobby. The danger is now over, that such a robot will ever be created; mathematicians have regained their *raison d'être* and can sleep quietly. (1995, p. 166)

The non-recursiveness of the set Thm_{T} implies that it is not representable in PA (in T). This corollary indicates certain limits in defining notions in formal systems. An interesting comment to this was given by W.V.O. Quine who said that those systems wanted to swallow a greater piece of ontology than they were able to digest. On the other hand, the cardinality argument shows that this phenomenon is quite normal: in fact there are uncountably many subsets of the set $\mathbb{N}$ of natural numbers while the set of formulas of the language of PA (T) is countable. What is surprising here is that among sets undefinable (non strongly representable) in the given theory is just the set of Gödel numbers of its theorems.

In Section 2.4 Tarski's theorem on the undefinability of truth was proved (cf. Theorem 2.4.11). It states that in no recursive extension of Peano arithmetic PA the notion of truth for formulas of its language can be expressed within the system, in fact methods stronger than those available in a given system must be used. It seems that this result contributes to the discussions on the philosophically adequate, general and universal definition of truth.

Gödel's Second Incompleteness Theorem on the unprovability of consistency of a given theory in itself struck Hilbert's program (the problem of whether Gödel's results rejected it will be discussed below), in particular his proposal to prove consistency of the whole of mathematics by finitistic methods. The theorem showed that the infinite cannot be justified by the finite. To prove consistency of a given theory (containing Peano arithmetic) one needs stronger methods than those available in it (hence methods which are less certain and more 'suspicious' and questionable from the point of view of consistency).

The philosophical meaning and significance of Gödel's Second Theorem is weakened in a sense by the fact that it holds only if the formula expressing in the language of the considered system the metamathematical property of consistency fulfills appropriate conditions, i.e., if its structure reflects the real metamathematical properties of provability and is not only numerically (i.e., in the sense of strong representability) compatible with the notion of demonstrability.

The significance of Gödel's First Incompleteness Theorem was diminished by the fact that it gave an example of an undecidable sentence of a metamathematical contents which had no clear mathematical (arithmetical) meaning. Though it was about natural numbers, it arose from certain metamathematical considerations (about arithmetic as a formalized system) and one could doubt whether it would appear in studies of 'normal' number-theorists. Other undecidable sentences found later had the same character (cf. sentences of Rosser, Kreisel and Levy, Kent, Mostowski, Shepherdson — see Smoryński, 1981). Interesting for logicians they were rather artificial from a mathematical point of view. Hence a problem arose: is it possible to indicate examples of undecidable sentences of mathematical, in particular number-theoretical, contents? The problem was solved by Paris and Harrington in 1977 and by Paris and Kirby in 1982 — their results were presented in Section 2.5. Undecidable sentences constructed by them are of mathematically interesting contents (whatever it means) and give us more information on incompleteness. They show that Peano arithmetic and its extensions are incomplete in the domain of sentences strongly connected with the research practice of mathematicians, i.e., with the really developed combinatorics or number theory. Furhermore, the new undecidable sentences are examples of arithmetical propositions which have no 'pure', i.e., no arithmetical proof. More precisely, they are examples of sentences about certain combinatorial properties of finite sets (Paris-Harrington sentence) or about properties of sequences of natural numbers (Paris-Kirby sentence) such that their natural proofs use infinite sets. On the other hand, they cannot be proved in Peano arithmetic, so any proof of them must use something that goes beyond the domain of finite objects. Observe that 'impure' proofs appear rather often in mathematics — one often uses complicated objects to prove properties of objects of a rather simple type. But here we have examples of propositions having 'impure' but no 'pure' proofs.

One could say that Gödel's undecidable sentence was in fact a sentence about natural numbers and hence a proposition of a mathematical contents. If one had enough patience (and paper!) one would be able to write it down as a sentence in the language of PA (without using any metamathematical abbreviations). Hence there is no difference between it and the new Paris-Harrington or Paris-Kirby sentences. This argument seems to be inappropriate. One should doubt whether Gödel's sentence would ever have been considered if it had no metamathematical source (and contents). Gödel obtained it translating (by the method of aritmetization) certain metamathematical sentence. Sentences of a similar type are usually not considered by number-theorists, not because they do not like long propositions and prefer statements whose contents can be easily grasped, but because such sentences are not interesting from a mathematical point of view and their contents becomes clear only by metamathematical interpretation.

It turns out that Paris-Harrington and Paris-Kirby undecidable sentences, though of mathematically interesting contents, are in fact equivalent to a certain metamathematical sentence. Indeed, one can show that they are provably equivalent (in Peano arithmetic PA) to the reflection principle for Σ^0_1 formulas, i.e., to the following sentence:

> for any sentence φ of the language L(PA) of Peano arithmetic such that it is in the prenex form and all

quantifiers occurring in it are existential:

if φ is provable then φ is true.

Using the arithmetization of syntax one can write this principle as a sentence of L(PA). Denote it by $\mathrm{Rfn}_{\Sigma_1}(\mathrm{PA})$. McAloon (1979) proved that

$$\mathrm{PA} \vdash \varphi_0 \equiv \mathrm{Rfn}_{\Sigma_1}(\mathrm{PA}),$$

where φ_0 is a Paris-Harrington sentence. Similar equivalence can be also shown for a Paris-Kirby sentence. This phenomenon is really interesting from the point of view of philosophy of mathematics.

New undecidable sentences turn out to be stronger than Gödel's one. As shown above (cf. Theorem 2.4.7 and its proof) Gödel's undecidable sentence is provably (in PA) equivalent to the sentence $\mathrm{Con}_{\mathrm{PA}}$. Paris-Harrington and Paris-Kirby sentences are undecidable not only in PA but also in Peano arithmetic PA with an additional axiom: $\mathrm{Con}_{\mathrm{PA}}$.

What are the reasons for the undecidability of the new sentences? They are different from the ones in the case of Gödel's proposition (see above). Remind that they are of the form $\forall x \exists y \psi(x, y)$ where $\psi \in \mathrm{BQ}$. If such a sentence were provable in PA, then there would exist a provably recursive function $f(x)$ such that $\mathrm{PA} \vdash \forall x \psi(x, F(x))$ where F is a term representing the function f. Such a function f indicating examples of objects fulfilling the formula ψ is called 'choice function'. Recall that a function $f : \mathbb{N} \to \mathbb{N}$ is said to be provably recursive if and only if f is recursive and there exists a term F of the language L(PA) representing f in PA and such that

$$\mathrm{PA} \vdash \forall x \exists! y [F(x) = y],$$

where $\exists! y$ means "there exists exactly one y". The notion of representability used here is a bit different from the one used in Section 2.2 but they are equivalent. If it is said that a term F represents a function f in PA, then we mean that for any $n \in \mathbb{N}$,

PA $\vdash F(\overline{n}) = \overline{k}$ where $k = f(n)$. Hence if the term F represents the function f in PA then for every $n \in \mathbb{N}$

$$\text{PA} \vdash \exists! y[F(\overline{n}) = y]$$

but it can happen that PA $\nvdash \forall x \exists! y[F(x) = y]$. Hence not every recursive function (which is of course representable in PA — cf. Theorem 2.2.12) is provably recursive.

If PA $\nvdash \forall x \exists y \psi(x, y)$ then the reason for the undecidability of this sentence is the fact that for any provably recursive function f, if F represents f then PA $\nvdash \forall x \psi(x, F(x))$, i.e., no provably recursive function is a choice function for the formula ψ.

To explain this problem better we will introduce a hierarchy of functions. Let $f_0(x) = x + 1$, $f_{n+1}(x) = f_n^{x+1}(x)$ where $f_n^{x+1}(x) = \underbrace{f_n \circ \ldots \circ f_n}_{x+1}(x) = \underbrace{f_n(\ldots(f_n}_{x+1}(x))\ldots)$. So in particular

$$f_1(x) = 2x + 1, \quad f_2(x) \approx 2^x \text{ and } f_3(x) \approx \left.2^{2^{\cdot^{\cdot^{2}}}}\right\} x \text{ times}.$$

A function h_1 is said to be similar to a function h_2 if and only if $\mathcal{F}(h_1) = \mathcal{F}(h_2)$, where $\mathcal{F}(h)$ is the smallest family of functions containing the functions h, S, $+$ and $\cdot$ and closed under composition. One can now prove that for any $n \in \mathbb{N}$, the function f_{n+1} is not similar to f_n and that f_{n+1} majorizes all the functions belonging to the family $\mathcal{F}(f_n)$, i.e., for any function $f \in \mathcal{F}(f_n)$ there exists a number k such that for every $l > k$, $f_{n+1}(l) > f(l)$. The hierarchy of functions f_n $(n \in \mathbb{N})$ can be extended into transfinite. This can be done, for example, in the following way: if an ordinal α is a successor, i.e., $\alpha = \beta + 1$ for some ordinal β, then we put

$$f_\alpha(x) = f_{\beta+1}(x) = f_\beta^{x+1}(x),$$

and if α is a limit ordinal then f_α is defined by diagonalization, so in particular

$$f_\omega(x) = f_x(x).$$

Recall that the ordinal ε_0 is the smallest ordinal α such that $\omega^\alpha = \alpha$. It can also be defined as the limit of the following sequence of ordinals:

$$\omega, \omega^\omega, \omega^{\omega^\omega}, \omega^{\omega^{\omega^\omega}}, \ldots$$

Now the following theorem holds.

Theorem 4.1.5 *The class of functions* $\bigcup_{\alpha<\varepsilon_0} \mathcal{F}(f_\alpha)$ *is equal to the class of provably recursive functions.*

After those (long and a bit tedious) introductory explanations we can return to our problem. Consider, e.g., the Paris-Harrington undecidable sentence φ_0. If φ_0 were provable in PA, i.e., if PA $\vdash \forall x \exists y \psi(x, y)$ then according to Theorem 4.1.5 there would exist an ordinal $\alpha < \varepsilon_0$ such that PA $\vdash \forall x \exists y < f_\alpha(x) \psi(x, y)$. But PA $\nvdash \forall x \exists y \psi(x, y)$. So for any $\alpha < \varepsilon_0$:

$$\text{PA} \nvdash \forall x \exists y < f_\alpha(x) \psi(x, y).$$

Consequently, the choice function f for the formula φ_0 increases faster than the functions f_α for $\alpha < \varepsilon_0$. Indeed, one can prove that the function $H(x + 1, x, x)$ (H is here the function from Paris-Harrington sentence φ_0) is similar to the function f_{ε_0}, i.e., the choice function for the sentence φ_0 majorizes all provably recursive functions and therefore Peano arithmetic is unable to express it or its properties. And this is exactly the source of the undecidability of φ_0 in PA. A similar phenomenon holds for the Paris-Kirby undecidable sentence.

4.2 Gödel's Theorems vs. Hilbert's Program

It was stated above that Gödel's incompleteness theorems struck Hilbert's program. The question now is: did they reject it? This question cannot be answered definitely for the simple reason that

Hilbert's program was not formulated precisely enough, it used vague terms as 'finitistic', 'real', 'ideal' which were never precisely defined. In this situation various opinions are formulated and defended. On the one hand it is claimed that Gödel's incompleteness results showed the failure of Hilbert's program (cf. e.g., Smoryński, 1977; 1985; 1988). On the other hand it is argued that it is not the case. Several reasons are given here. For example, Detlefsen (1979) says that the second Gödel's theorem does not imply the rejection of Hilbert's proposal, because it is possible that the unprovable formal sentence stating the consistency of the theory "does not really express consistency" in the sense used by Hilbert. In fact, he points out that some argument is needed to show that the properties of the provability predicate used in the proof of the considered theorem are necessary properties of any formula capable of expressing consistency. Detlefsen (1990) argued that Gödel's First Incompleteness Theorem does not refute the program, because Hilbert did not demand the conservation property but only a weak conservation, i.e., conservation with respect to real sentences which can be finitistically decided, and he did not claim that every real sentence may be decided in such a way.

As indicated above, Gödel's Second Incompleteness Theorem requires certain assumptions about the formal translation of some metamathematical notions (such as proof, provability, etc.). Detlefsen (1990) observes that they are not satisfied by various theories or theory-like arrangements of proofs and theorems which nonetheless do satisfy the conditions required by Gödel's First Theorem.[4] He considers so-called consistency-minded theories which incorporate consistency constraints into the very conditions on proof, provability, etc. Two types of such theories are distinguished: Rosser systems (studied in Rosser 1936;

[4]The First Incompleteness Theorem only requires that the set of theorems be representable, i.e., the formal system considered may be 'identified' with its set of theorems. Consequently, all we need to know about it is its set of theorems. Recall that the Second Incompleteness Theorem requires certain properties of the provability to be formally derivable.

Kreisel and Takeuti, 1974; Guaspari and Solovay, 1979; Visser, 1989; Arai, 1990) and Feferman-systems (introduced in Feferman, 1960, and studied in Jaroslov, 1975; Visser, 1989). Since some of those theories constitute plausible ways in which the Hilbertian might go about constructing his ideal theories, Detlefsen formulates the following open question: "whether G2 [i.e. Gödel's Second Incompleteness Theorem — R.M.] applies to Hilbert's Program *per se*, or only to those versions of it which needlessly restrict themselves to theory construction of the usual static variety" (p. 345).

Resnik (1974) argues that the incompleteness theorem "has less bearing upon the [Hilbert's] programme than is often credited to it!" This thesis is based on the claim that: "every formal system to which Gödel's theorems apply is complete with respect to its real sentences. Thus the undecidable sentences are ideal sentences."

Note also that the failure of Hilbert's program for a certain formalized system of arithmetic need not be a failure of Hilbert's program for elementary number theory in the informal sense. In fact one cannot exclude the possibility that the latter can be formalized in a system which can be justified on finitistic grounds.

And what were the reactions and opinions of Hilbert and Gödel? First note that Hilbert somewhat reformulated his original program. Indeed, at the beginning it was demanded that the main goal of *Beweistheorie* is to prove the consistency of axioms of mathematics (cf. Hilbert, 1923). On the other hand, in the first volume of (Hilbert and Bernays, 1934–1939) Hilbert wrote (p. v) that "the final aim [of *Beweistheorie* — R.M.] is to show that the usual methods of mathematics are consistent". A shift from stressing the need of proving the consistency of axioms to stressing the necessity of proving the consistency of the methods of mathematics can be observed here.

Hilbert proposed to add to the rules of inference a simple form of the ω-rule considered above which allows the derivation of all true arithmetical sentences. In the Preface to the first volume of Hilbert and Bernays (1934–1939) he wrote:

> ... the occasionally held opinion, that from the results of Gödel follows the non-executability of my Proof Theory, is shown to be erroneous. This result shows indeed only that for more advanced consistency proofs one must use the finite standpoint in a deeper way than is necessary for the consideration of elementary formalisms.

Hence Hilbert now allowed to certain extent infinitistic rules in his program.

To better understand Hilbert's new position and Gödel's position (which will be discussed below) it is worth digressing from the proper subject and considering some views of Leibniz. It will turn out that the reflection of some Leibnizian elements can be found in opinions of Hilbert and Gödel on the perspectives of realizing Hilbert's program.[5]

According to Leibniz, theorems are either primitive truths, i.e., axioms of a given theory, or propositions (called 'derived truths') which can be reduced to the primitive ones by means of a proof. A proof consists either of a finite number of steps (in this case one talks about finitistic truth or about finitely analytic sentences) or of an infinite number of steps (one deals with infinitistic truths or with infinitely analytic sentences). Primitive truths are necessary and finitely (directly) analytic, derived truths are either necessary (though not identical), i.e., finitely analytic, or contingent, i.e., infinitely analytic. On the other hand, Leibniz claimed that necessary truths possess finite proofs while contingent ones have only infinite proofs. So the difference between them is a rather practical one and not one of a substantial character. A being with unbounded calculating possibilities would be able to decide all the truths directly, hence any truth would be necessary for him. Here is the source and reason for Leibniz's idea of a real logical calculus, *ars combinatoria* (only

[5]Cf. Murawski (199?a). Note also that the original Hilbert's program beared the Kantian character — cf. Introduction.

its fragments are known to us in the form of finitistic systems of logic). This calculus must be infinitistic. So Leibniz allowed infinite elements in reasonings and proofs. Consequently Hilbert's proposal to allow the ω-rule is compatible with Leibniz's ideas and has in fact the Leibnizian character.

An explicit influence of Leibniz on Gödel's reactions on the new situation in logic and the foundations of mathematics after incompleteness results can also be seen. Before we present and discuss some details note that after 1945 Gödel's interests were concentrated almost exclusively on the philosophy of mathematics and on philosophy in general. He studied works of Kant and Leibniz as well as phenomenological works of Husserl (especially in the fifties). In his *Nachlaß* several notes on the works of those philosophers and on their views were found. Gödel claimed that it was Leibniz who had mostly influenced his own scientific thinking and activity (Gödel studied Leibniz's works already in the thirties). Hao Wang writes that "Gödel's major results and projects can be viewed as developments of Leibniz's conceptions along several directions" (cf. Wang Hao, 1987, p. 261). Gödel accepted the main ideas of Leibniz's monadology, he was interested in the realization of a modified form of *characteristica universalis* (Gödel's incompleteness theorems indicated that the idea cannot be fully realized). Both Leibniz and Gödel were convinced of the meaning and significance of the axiomatic method. Gödel's results indicated the necessity of some changes and modifications in Leibniz's program, though Gödel was still looking for axiomatic principles for metaphysics from which the whole of knowledge could be deduced (or which at least would be a base of any knowledge). Among other things he was searching for a method of analyzing concepts which would induce methods allowing to obtain new results. Add also that Gödel first got the idea of his proof of the existence of God in reading Leibniz (cf. Wang Hao, 1987, p. 195).

Having presented the connections between Leibniz and Gödel in general, let us turn now to problems related to the incompleteness theorems. First observe that in his first philosophical paper "Russell's Mathematical Logic" (1944) one of the questions Gödel has turned to was the question whether (and in which sense) the axioms of the *Principia Mathematica* can be considered to be analytic. And he answered that if analyticity is understood as reducibility by explicit or contextual definitions to instances of the law of identity, then even arithmetic is not analytic because of its undecidability. The inspiration of Leibniz can be easily seen here (compare Leibniz's finitely analytic truths). But if infinite reduction, with intermediary sentences of infinite length, is allowed (as would be suggested by Leibniz's theory of contingent propositions) then all axioms of the *Principia* can be proved analytic, but the proof would require "the whole of mathematics ..., e.g., the axiom of choice can be proved to be analytic only if it is assumed to be true" (Gödel, 1944, p. 151).

Already in 1931, Gödel expicitly wrote about problems directly connected with the incompleteness results:

> I wish to note expressly that Theorem XI (and the corresponding results for M and A) do not contradict Hilbert's formalistic viewpoint. For this viewpoint presupposes only the existence of a consistency proof in which nothing but finitary means of proof is used, and it is conceivable that there exist finitary proofs that cannot be expressed in the formalism of P (or M or A).[6]

And in footnote 48[a] (evidently an afterthought) to (1931) Gödel wrote:

> As will be shown in Part II of this paper, the true reason for the incompleteness inherent in all formal systems of mathematics is that the formation of ever higher types can be continued into the transfinite ... while in any formal system at most denumerably many of them are avail-

[6]Theorem XI states that if P (the system of arithmetic of natural numbers used by Gödel in (1931) and based on the system of *Principia*) is consistent then its consistency is not provable in P; M is set theory and A is classical analysis. English translation according to (Heijenoort, 1967, p. 615).

able. For it can be shown that the undecidable propositions constructed here become decidable whenever appropriate higher types are added (for example, the type ω to the system P). An analogous situation prevails for the axiom system of set theory.[7]

At the Vienna Circle meeting on 15th January 1931 Gödel argued that it is doubtful, "whether all intuitionistically correct proofs can be captured in a *single* formal system. That is the weak spot in Neumann's argumentation" (quotation taken from Sieg, 1988).

Gödel suggested that Hilbert's program may be continued by allowing two principles which can be treated as finitistic, namely (1) the principle of transfinite induction on certain primitive recursive well-orderings, and (2) a notion of computable functions of a finite type (i.e., of computable functionals), to which the process of primitive recursion can be extended in a natural way.

The first principle (more exactly, the induction up to the ordinal ε_0) was later applied by Gerhard Gentzen to prove the consistency of the arithmetic of natural numbers (we discuss the proof below).

[7]English translation taken from (Heijenoort, 1967, p. 610).

Compare the above remarks of Gödel with Turing's ideas from his paper (1939). In the introduction to this paper Turing wrote:

> The well-known theorem of Gödel (1931) shows that every system of logic is in a certain sense incomplete, but at the same time it indicates means whereby from a system L of logic a more complete system L' may be obtained. By repeating the process we get a sequence L, $L_1 = L'$, $L_2 = L_1'$, ... each more complete than the proceeding. A logic L_ω may then be constructed in which the provable theorems are the totality of theorems provable with the help of logics L, L_1, L_2, ... Proceeding in this way we can associate a system of logic with any constructive ordinal. It may be asked whether such a sequence of logics of this kind is complete in the sense that to any problem A there corresponds an ordinal α such that A is solvable by means of the logic L_α.

See also Feferman (1988) where Turing's idea and its development are discussed.

The second principle was applied by Gödel (1958). He considered the question of how far finitary reasoning might reach. The problem was considered by him also in 1972 (this paper was a rivised and expanded English version of (1958)). Here Gödel claimed that concrete finitary methods are insufficient to prove the consistency of elementary number theory and some abstract concepts must be used in addition. He wrote (footnote b):

> There is nothing in the term 'finitary' which would suggest a restriction to concrete knowledge. Only Hilbert's special interpretation of it makes this restriction.

However, he was convinced that a precise definition of concrete finitary method would have to be given in order to establish with certainty the necessity of using abstract concepts. In 1946 he explicitely called for an effort to use progressively more powerful transfinite theories to derive new arithmetical theorems (Gödel, 1946).

A similar postulate was also formulated by R. Carnap who wrote:

> ... *alles Mathematische ist formalisierbar; aber die Mathematik ist nicht durch Ein System erschöpfbar*, sondern erfordert eine Reihe immer reicherer Sprachen. (... *all that is mathematical can be formalized; yet the whole of mathematics cannot be grasped by one system* but an infinite series of still richer and richer languages is necessary.) (1934, p. 274)

To end this section let us add that also Zermelo (cf. Section 2.6 for his reaction on incompleteness results) proposed to allow infinitary methods to overcome restrictions revealed by Gödel. According to Zermelo the existence of undecidable propositions was a consequence of the restriction of the notion of proof to finitistic methods (he said here about "finitistic prejudice"). This situation could be changed if one used a more general "scheme" of proof. Zermelo had here in mind an infinitary logic, in which there were infinitely long sentences and rules of inference with infinitely many premises. In such a logic, he insisted, "*all* propositions are decidable!" Note that that time was not yet ripe for such an infinitary logic. Systems of such a logic, though in a more restricted form than demanded by Zermelo, and without escaping incompleteness, were constructed in the mid-fifties in works of Henkin, Karp and Tarski (cf. Barwise, 1980 and Moore, 1980).

4.3 Generalized Hilbert's Program

The natural consequence of the situation which had arosen after Gödel's incompleteness results was — as indicated above — the idea of extending the admissible methods and allowing stronger ones. In particular it was proposed to allow general constructive methods. It seems that Bernays was among the first to recognize this need. He wrote (cf. Bernays, 1967, p. 502):

> It thus became apparent that the 'finite Standpunkt' is not the only alternative to classical ways of reasoning and is not necessary implied by the idea of proof theory. An enlarging of the methods of proof theory was therefore suggested: instead of a restriction to finitist methods of reasoning, it was required only that the arguments be of a constructive character, allowing us to deal with more general forms of inference.

One of the motivations of this shift from the original Hilbert's program can be sought in Gödel's reduction (independently found also by Gentzen) of Peano arithmetic PA to the intuitionistic system HA of Heyting's arithmetic — it was shown that PA is consistent relative to HA (cf. Gödel, 1933; Gentzen, 1969).[8] Another reason may be Gentzen's proof of consistency

[8]Gentzen's paper "Über das Verältniss zwischen intuitionistische und klassische Arithmetik" was submitted and accepted by *Mathematische Annalen* in 1933 but withdrawn on account of Gödel's publication. An English translation was published in 1969.

of PA by transfinite induction up to ε_0 (cf. Gentzen, 1936; 1938) which was apparently accepted by Hilbert and Bernays in the second volume of *Grundlagen der Mathematik.*

Gentzen showed that in the system of Peano arithmetic PA extended by admitting the principle of transfinite induction up to the ordinal ε_0 one can prove that PA is consistent. Note that there exists a primitive recursive relation $\prec$ well-ordering the set $\mathbb{N}$ of natural numbers in the type ε_0. Hence it is strongly representable in PA. Let $\dot{\prec}$ be a formula of L(PA) strongly representing $\prec$ in PA. The principle of ε_0-induction is the following axiom scheme

$$\varphi(0) \wedge \forall x[\forall z(z \dot{\prec} x \rightarrow \varphi(z)) \rightarrow \varphi(x)] \rightarrow \forall x \varphi(x).$$

In Peano arithmetic PA extended by this axiom scheme one can prove that PA is consistent. It is impossible to present here this technically complicated proof in detail. So let us only say that its idea is as follows: to every formal proof in PA one assigns an ordinal $\alpha < \varepsilon_0$ called 'the height of the proof'; next one shows that if a proof ended by a formula "$0 = 1$" then there would exists a proof of a smaller height whose end formula would be "$0 = 1$" as well. Thus the existence of a proof of the formula "$0 = 1$" would contradict the ε_0-induction. Consequently, there is no proof of "$0 = 1$" in PA and PA is consistent.[9]

It is worth noting here that the first version of Gentzen's proof was submitted in 1935 but withdrawn after criticism directed against the means used in the proof which were considered to be too strong.[10]

Of course, a problem now arises: what is meant by constructivity? In general this concept is much less clear than that of finitism. Nevertheless, the broadening of the original Hilbert's proof theory postulated by Bernays has become an accepted

---

[9]In Kotlarski and Ratajczyk (1990, Section 5) a semantical proof of Gentzen's theorem can be found.

[10]This version became publically known because of a paper Bernays (1970) and was recently published in the name of Gentzen (cf. Gentzen, 1974).

paradigm. Investigations were carried out in this direction and several results were obtained. Studies following the idea of Gentzen of using a transfinite induction on a certain recursive well-ordering for some ordinal form a part of them. Gentzen-style consistency proofs have been given for a variety of subsystems of analysis and set theory by the use of effective transfinite induction on larger and larger ordinals (cf. Schütte, 1977; Takeuti, 1987).

Another program of reductionism was elaborated by Feferman, namely the program of predicative reductionism (cf. Feferman, 1964–68; see also Simpson, 1985). Gödel has proposed (see Section 4.2) an "extension" of the "finite Standpunkt" by way of primitive recursive functionals of higher type (cf. Gödel, 1958).

A further refinement of the original Hilbert's program was suggested in Kreisel (1958) and elaborated in Kreisel (1968) where a call for a hierarchy of Hilbert's programs can be found. Besides reduction to finitary and constructive conceptions, Kreisel also considers the nonconstructive predicative conception and within constructivity itself he analyzes which specific principles would be needed for various pieces of reductive work.

It is impossible to report here on the particular results obtained along those lines (it would require a lot of technical notions). One can find a survey of them in Feferman (1988a). Let us note only that the researches have led to two surprising insights: (i) classical analysis can be formally developed in conservative extensions of elementary number theory and (ii) strong impredicative subsystems of analysis can be reduced to constructively meaningful theories, i.e., relative consistency proofs can be given by constructive means for impredicative parts of second-order arithmetic.

At the end of this section we want to stress that all generalizations discussed above are very different from the original Hilbert's program. Hilbert's postulate was the validation and justification of classical mathematics by a reduction to finitistic mathematics. The latter was important here for philosophical and, say, ideological reasons: finitistic objects and reasonings have clear physical meaning and are indispensable for all scientific thought. None of the proposed generalizations can be viewed as finitistic (whatever it means). Hence they have another value and meaning from a methodological and generally philosophical point of view. They are not contributing directly to Hilbert's program but, on the other hand, they are in our opinion compatible with Hilbert's reductionist philosophy.

4.4 Relativized Hilbert's Program vs. Reverse Mathematics

Another consequence of Gödel's incompleteness results (besides those described above) is the so-called 'relativized Hilbert's program'. If the entire infinitistic mathematics cannot be reduced to and justified by finitistic mathematics then one can ask: for which part of it is that possible? In other words: how much of infinitistic mathematics can be developed within formal systems which are conservative over finitistic mathematics with respect to real sentences? This constitutes the relativised version of the program of Hilbert. In what follows we would like to show how the so-called 'reverse mathematics' of Friedman and Simpson contributes to it providing us with a partial realization of Hilbert's original program.

To be able to consider the problem one should specify what is meant by finitistic mathematics and by real sentences. We follow here Tait (1981) who claimed that Hilbert's finitism is captured by the formal system PRA of primitive recursive arithmetic (also called 'Skolem arithmetic') (cf. Introduction). Its language contains the constant 0, successor function S and a function symbol **f** for each primitive recursive function f. Its nonlogical axioms are: some trivial axioms concerning the constant terms 0, $S0$, $SS0$, ... (i.e., names of particular natural numbers) and the successor function as well as the defining equations of the primitive recursive functions and induction on quantifier-free formulas. The theory PRA is certainly finitistic and 'logic-free'. On the other hand, it is powerful enough to accommodate all elementary reasonings about natural numbers and manipulations of finite strings of symbols. By 'real' sentences we shall understand Π^0_1 sentences, i.e., sentences of the language of Peano arithmetic of the form $\forall x \varphi(x)$ where φ contains only atomic formulas, connectives and bounded quantifiers.

It turns out not only that one can formalize classical mathematics in set theory but also that most of its parts (such as geometry, number theory, analysis, differential equations, complex analysis etc.) can be formalized in a weaker system called 'second-order arithmetic A_2^-' (denoted sometimes as Z_2). We described this system in Section 2.6.

Possible models of A_2^- are structures of the form $\langle \mathcal{X}, \mathfrak{M}, \in \rangle$ where $\mathfrak{M}$ is a model of PA and $\mathcal{X}$ is a family of subsets of M (the universe of $\mathfrak{M}$). Observe that for the standard model $\mathfrak{N}_0$ of PA the structure $\langle \mathcal{P}(\mathbb{N}), \mathfrak{N}_0, \in \rangle$ is a model of A_2^- but this is not the case for a nonstandard model $\mathfrak{M}$ of PA, i.e., $\langle \mathcal{P}(M), \mathfrak{M}, \in \rangle$ is never a model of A_2^-. (For standard models of A_2^- see, e.g., Apt and Marek (1974) and for nonstandard models of A_2^- see Murawski (1976–77) and Murawski (1984) where a bibliography can also be found).

Second-order arithmetic is a nice system because one avoids troubles connected with set theory (for example paradoxical consequences of the axiom of choice or philosophical problems connected with the adoption of these or those axioms) and on the other hand, it is strong enough to prove many important theorems of classical mathematics. There is only the problem of impredicativity of A_2^-: a formula φ in the comprehension axiom may be *any* formula of the language of A_2^- , so it may be of the form $\forall Y \psi(Y, x, \ldots)$, i.e., defining one particular set it may refer to the family of all sets (recall Poincaré's objections against such definitions mentioned in the Introduction). But it turns out that in many cases certain fragments of A_2^- suffice, i.e. only particular special forms of the comprehension axiom are needed. And in this way we come to reverse mathematics.

At the Congress of Mathematicians in Vancouver in 1974 Harvey Friedman formulated a program of foundations of mathematics today called 'reverse mathematics' (cf. Friedman, 1975). Its aim is to study the role of set existence axioms, i.e., comprehension axioms in ordinary mathematics. The main problem is: Given a specific theorem τ of ordinary mathematics, which set existence axioms are needed in order to prove τ? This research program turned out to be very fruitful and led to many interesting results[11] and ... showed that Hilbert's program can be partially realized! (There is much literature about reverse mathematics. The most comprehensive report on it will be the book Simpson (1998), one can consult also Simpson (1985a, 1987 and 1988) where a bibliography can be found).

The procedure used in the reverse mathematics (it reveals the inspiration for its name) is the following: assume we know that a given theorem τ can be proved in a particular fragment $S(\tau)$ of A_2^-. Is $S(\tau)$ the weakest fragment with this property? To answer this question positively one shows that the principal set existence axiom of $S(\tau)$ is equivalent to τ, the equivalence being provable in some weaker system in which τ itself is not provable. In a proof of this equivalence the most difficult part is usually the proof of the implication: if τ then the set existence axiom of $S(\tau)$. This explains the term 'reverse mathematics'—indeed, one derives here the axiom from the theorem while in ordinary mathematics we deduce theorems from axioms.

Some specific systems — fragments of A_2^- — arose in this context; the most important are: RCA_0, WKL_0, ACA_0, ATR_0 and $\Pi_1^1 - CA_0$. We shall describe only the first three of them. To do this we need a hierarchy of formulas of the language of A_2^- (cf. Section 1.5). By an arithmetical formula we mean a formula containing no quantifiers bounding set-variables. Bounded formulas are arithmetical formulas with only bounded quantifiers, i.e., quantifiers of the form: $\forall x < y$, $\exists x < y$. The class of all

[11]It is even claimed that the implications of the results of reverse mathematics "make much of what was written in the past on the philosophy of mathematics obsolete" (cf. Drake, 1989).

such formulas is denoted by Δ_0^0 (or BQ) (cf. Definition 2.4.23). Formulas of the class Σ_1^0 are formulas equivalent in PA to formulas of the form $\exists x_1 \ldots \exists x_n \psi(x_1, \ldots, x_n, y_1, \ldots, y_k)$ where $\psi \in \Delta_0^0$ and formulas of the class Π_1^0 are negations of Σ_1^0 formulas.

The system RCA_0 is a theory in the language of A_2^- based on the following axioms:

(i) PA^- (i.e. axioms of Peano arithmetic PA without the axiom scheme of induction),

(ii) extensionality axiom,

(iii) scheme of induction for Σ_1^0 formulas, i.e.:

$$\varphi(0) \wedge \forall x[\varphi(x) \rightarrow \varphi(Sx)] \rightarrow \forall x \varphi(x),$$

where φ is a Σ_1^0 formula,

(iv) (recursive comprehension axiom)

$$\forall x[\varphi(x) \equiv \psi(x)] \rightarrow \exists X \forall x[x \in X \equiv \varphi(x)],$$

where φ is Σ_1^0 and ψ is Π_1^0.

The name of axiom (iii) explains the name RCA_0 of the theory. It can be shown that $\langle Rec, \mathfrak{N}_0, \in \rangle$, where Rec is the family of all recursive sets, is a model of RCA_0.

The theory WKL_0 consists of RCA_0 plus a further axiom known as 'weak König's lemma' (therefore the name WKL_0) which states that every infinite binary tree has an infinite path (this can be formulated in the language of A_2^- using coding). It is stronger than RCA_0 what follows for example from the fact that $\langle Rec, \mathfrak{N}_0, \in \rangle$ is not a model of WKL_0 (this is a consequence of Gödel's theorem on the essential undecidability of Peano arithmetic).

The theory ACA_0 is PA^- plus induction axiom plus arithmetical comprehension, i.e., comprehension scheme for any arithmetical formula (possibly containing set parameters) (this explains the name ACA_0). This theory is not weaker than WKL_0

because it proves weak König's lemma. It is in fact stronger than WKL_0 what follows from the fact that there are models of WKL_0 consisting of sets definable in $\mathfrak{N}_0$ by formulas of a given class[12] whereas for any model $\langle \mathcal{X}, \mathfrak{N}_0, \in \rangle$ of ACA_0 the family $\mathcal{X}$ must be closed with respect to arithmetical definality.

The specified subsystems of A_2^- are appropriate for particular parts of classical mathematics. In RCA_0 one can construct real numbers, define notions of the convergence of a sequence, of a continuous function, of Riemann's integrability, etc., and prove positive results of recursive analysis and recursive algebra. For example, one can prove in RCA_0 that every countable field has an algebraic closure, that every countable ordered field has an extension to a real closed field as well as the intermediate value theorem for continuous functions (cf. Simpson, 1998).

The theory WKL_0 turns out to be a quite strong theory, in particular one can prove in it the following theorems:

— the Heine–Borel covering theorem (every covering of $[0,1]$ by a countable sequence of open intervals has a finite subcovering) (cf. Friedman, 1976),

— every continuous function on $[0,1]$ is uniformly continuous (cf. Simpson, 1998),

— every continuous function on $[0,1]$ is bounded (cf. Simpson, 1998),

— every continuous function on $[0,1]$ has a supremum (cf. Simpson, 1998),

— every uniformly continuous function on $[0,1]$, which has a supremum, attains it (cf. Simpson, 1998),

— every continuous function on $[0,1]$ attains a maximum value (cf. Simpson, 1998),

[12] A family of Δ_2^0 sets is an example of a model of WKL_0, where Δ_2^0 sets are sets definable in the standard model of Peano arithmetic PA simultaneously by Σ_2^0 and Π_2^0 formulas; recall that Σ_2^0 formulas are formulas equivalent in PA to formulas of the form $\exists x_1 \ldots x_n \psi(x_1, \ldots, x_n, y_1, \ldots, y_k)$, $\psi \in \Pi_1^0$ and Π_2^0 formulas are negations of Σ_2^0 formulas.

— the Hahn-Banach theorem (cf. Brown and Simpson, 1986 as well as Brown, 1987),

— the Cauchy-Peano theorem on the existence of solutions of ordinary differential equations (cf. Simpson, 1984),

— every countable commutative ring has a prime ideal (cf. Friedman *et al.*, 1983),

— every countable formally real field can be ordered (cf. Friedman *et al.*, 1983),

— every countable formally real field has a real closure (cf. Friedman *et al.*, 1983),

— Gödel's completeness theorem for the predicate calculus (cf. Friedman, 1976 and Simpson, 1998).

Even more: if S is one of the above-stated theorems then $\mathrm{RCA}_0 + S$ is equivalent to WKL_0. This means that $\mathrm{RCA}_0 + S$ and WKL_0 have the same theorems. Hence, in particular, weak König's lemma is necessary and sufficient to prove (on the base of RCA_0) each of the indicated above theorems, and, on the other hand, one can deduce (in RCA_0) Peano's theorem from Hahn-Banach theorem and *vice versa* (similarly for other pairs of the indicated theorems).

To indicate the strength of ACA_0 let us mention that the following theorems can be proved in it:

— the Bolzano-Weierstrass theorem (every bounded sequence of real numbers has a convergent subsequence) (cf. Friedman, 1976),

— every Cauchy sequence of reals is convergent (cf. Simpson, 1998),

— every bounded sequence of reals has a supremum (cf. Friedman, 1976),

— every bounded increasing sequence of real numbers is convergent (cf. Friedman, 1976),

— the Arzela-Ascoli lemma (any bounded equicontinuous sequence of functions on $[0,1]$ has a uniformly convergent subsequence) (cf. Simpson, 1998),

— every countable vector space has a basis (cf. Friedman *et al.*, 1983),

— every countable commutative ring has a maximal ideal (cf. Friedman *et al.*, 1983),

— every countable Abelian group has a unique divisible closure (cf. Friedman *et al.*, 1983).

And again, if S is any of those theorems then $\mathrm{RCA}_0 + S$ is equivalent to ACA_0. Hence the arithmetical comprehension axiom is necessary and sufficient to prove (on the base of RCA_0) each of the indicated theorems. Furthermore, one can deduce (in RCA_0) the theorem on the existence of a base in a countable vector space from Bolzano-Weierstrass theorem and *vice versa* (and similarly for other pairs of the indicated theorems).

What is the meaning of those results? First of all they indicate how much of A_2^- we need in fact to prove various particular theorems of classical mathematics. And it is interesting from the philosophical point of view that certain particular fragments of A_2^- (described above) turn out to be adequate with respect to ordinary mathematical practice. But note that proofs (in the formalized subsystems of the second-order arithmetic) of uniqueness are usually more difficult and more complicated than proofs of the existence (in mathematical practice the former are usually simple consequences of the latter). There is also no direct connection between the complexity of a classical proof of a theorem and the level in the hierarchy of subsystems of A_2^- in which a formalized version of it can be proved (as an example can serve here the theorem that every Abelian group has a torsion subgroup which is trivial in classical algebra but RCA_0 + *this theorem* is equivalent to ACA_0, hence is not a theorem of, say, WKL_0).

In mathematical practice we often encounter the following situation: Assume that a certain theorem τ can be proved in set theory. The natural question is now: can τ be proved in an elementary way? Observe that if τ can be classified in the hierarchy of subsystems of A_2^- at a level higher than RCA_0 then the answer is negative, i.e., an abstract, infinitistic proof of τ is indispensible and necessary.

Results of reverse mathematics have also interesting mathematical, and not only logical, applications. As an example can serve here the Cauchy-Peano theorem on the existence of solutions of ordinary differential equations. Since it is equivalent to WKL_0 over RCA_0 and since the structure $\langle Rec, \mathfrak{N}_0, \in\rangle$ is not a model of WKL_0 we get that there exists a differential equation $y' = f(x, y)$, where f is a recursive continuous function, such that it has no recursive solution.

Observe that not every mathematical theorem can be classified in Friedman's hierarchy of subsystems of A_2^-. As an example one can give here Hilbert's basis theorem (cf. Simpson, 1988a) which states that for any field K and any natural number n all ideals in the ring $K[x_1, \ldots, x_n]$ can be finitely generated. This theorem is provable in ACA_0 but RCA_0 + "Hilbert's basis theorem" is not equivalent to any of the considered systems. It it worth mentioning here the following phenomenon interesting from a methodological point of view. Denote by $WO(\alpha)$ the sentence stating that the ordinal α is a well-ordered set. One can prove in RCA_0 that $WO(\omega^\omega)$ is equivalent to Hilbert's basis theorem. But the sentence $WO(\omega^\omega)$ is incomparable with WKL_0. On the other hand, for any given natural number n one can prove in RCA_0 that $WO(\omega^n)$ is equivalent to Hilbert's basis theorem for n. What more, $WO(\omega^n)$ is provable in RCA_0. So here we have a certain analogy with the ω-incompleteness of Peano arithmetic (cf. Gödel, 1931).

Add also that there are sentences which are unprovable in the full system A_2^- of second-order arithmetic but can be proved in Zermelo-Fraenkel set theory (cf. Friedman, 1981).

To indicate the connections between the reverse mathematics and Hilbert's program we must recall some results. In the early eighties, Harrington and Ratajczyk proved a theorem on conservativeness of WKL_0 (none of them published it, the proof can be found in Simpson, 1998).

Theorem 4.4.1 *If* $\langle \mathcal{X}, M, \in \rangle$ *is a countable model of* RCA_0 *and* $A \in \mathcal{X}$ *then there exists a family* $\mathcal{Y} \subseteq \mathcal{P}(M)$ *such that* $A \in \mathcal{Y}$ *and* $\langle \mathcal{Y}, M, \in \rangle$ *is a model of* WKL_0.

Corollary 4.4.2 *The theory* WKL_0 *is conservative over* RCA_0 *with respect to* Π^1_1 *sentences, i.e., every* Π^1_1 *sentence*[13] *provable in* WKL_0 *can be proved in* RCA_0.

What more, in 1977 Friedman proved (this result was not published; it can be found also in Kirby and Paris, 1977) that WKL_0 is a conservative extension of Skolem arithmetic PRA with respect to Π^0_2 sentences. His proof used modeltheoretical methods. Sieg improved this result giving an alternative proof which uses Gentzen-style methods and exhibiting a primitive recursive proof tranformation. Hence the reducibility of WKL_0 to PRA is itself provable in PRA.

Combining those results with the fact that WKL_0 is a fairly strong theory (what was indicated above) one comes to the conclusion that a large and significant part of classical mathematics is finitistically reducible. This means in fact that Hilbert's program can be partially realized!

Add that all this has also some 'practical' consequences. First observe that the class of Π^0_2 sentences is rather broad — many theorems of number theory can be formulated as sentences belonging to that class. Since one can formalize within WKL_0 the technique of contour integration, we conclude that any Π^0_2 number-theoretic theorem which is provable with the help of it can also be proved 'elementarily', i.e., within PRA and, even more, one can effectively (at least theoretically) find such an 'elementary' proof. To give one more example consider Artin's theorem (being a solution to Hilbert's seventeenth problem). In

[13] Π^1_1 formulas are formulas of the language of A_2^- which are equivalent in PA to formulas of the form $\forall X \psi$ where ψ is an arithmetical formula.

his seventeenth problem Hilbert asked "whether every definite form [of any number of variables with real coefficients – R.M.] may not be expressed as a quotient of sums of squares of forms" (cf. Hilbert, 1901). Recall that a form is called 'definite' if it becomes negative for no real values of the variables. In 1926, Artin answered this question positively. His theorem can be written as a Π^0_2 sentence. Since all results of the theory of real closed fields needed in its proof are provable in WKL_0, it follows by Friedman's and Sieg's theorems that Artin's theorem can be proved in PRA, i.e., in an elementary way.

It seems that Hilbert would be satisfied by such results.

4.5 Hilbert's Tenth Problem

Gödel's incompleteness theorems are also connected with Hilbert's tenth problem (cf. Hilbert, 1901). It concerned the solvability of diophantine equations. More precisely, Hilbert asked for a general methods which would allow us to decide in the case of any given diophantine equation whether it has roots or not. Recall the appropriate definition.

Definition 4.5.1 *A diophantine equation is an equation of the form $f(x_1, \ldots, x_n) = 0$ where $f(x_1, \ldots, x_n)$ is a polynomial with integer coefficients and the solutions are sought in integers only (this is the meaning of "diophantine").*

The original Hilbert's formulation was as follows (cf. English translation of (Hilbert, 1901), p. 18):

> Given a diophanitine equation with any number of unknown quantities and with rational integral numerical coefficients: *To devise a process according to which it can be determined by a finite number of operations whether the equation is solvable in rational integers.*

It can be easily seen that the essence of Hilbert's problem will not be changed if the solutions will be sought in natural numbers only. Indeed, both problems, i.e., the problem of deciding whether a given diophantine equation has solutions in integers, and the problem of solvability in natural numbers, are equivalent. It follows from the fact that any integer is a difference of two natural numbers and from Lagrange's theorem stating that any natural number is a sum of four squares of integers.

Definition 4.5.2 *A relation* $R \subseteq \mathbb{N}^n$ *is said to be diophantine if and only if there exists a polynomial* $p(x_1, \ldots, x_n, y_1, \ldots, y_k)$ *with integer coefficients such that*

$$R(x_1, \ldots, x_n) \equiv \exists y_1 \ldots \exists y_k [p(x_1, \ldots, x_n, y_1, \ldots, y_k) = 0].$$

The following theorem holds.

Theorem 4.5.3 (Matiyasevich) *A relation R is recursively enumerable if and only if R is diophantine.*

This theorem was proved by Matiyasevich (1970). He used some earlier results of Davis, Putnam and J. Robinson (cf. Robinson, 1952 and Davis *et al.*, 1961; see also Davis *et al.*, 1976).[14] A full account of Matiyasevich's theorem can be found in, e.g., Bell and Machover (1977).

In Corollary 2.4.22 examples of recursively enumerable but not recursive sets were given. It implies that not every recursively enumerable set is recursive. By Matiyasevich's theorem not every diophantine relation is recursive. Observe that the fact that a diophantine relation R holds for $a_1, \ldots, a_n$, i.e.,

$$\exists y_1 \ldots \exists y_k [p(a_1, \ldots, a_n, y_1, \ldots, y_k) = 0],$$

can be interpreted as a statement that the polynomial p (with given parameters $a_1, \ldots, a_n$) has solutions $y_1, \ldots, y_k$. Hence, if Hilbert's tenth problem had a positive solution then every diophantine equation would be recursive. But this is not true. Consequently Matiyasevich's theorem and the corollary from Gödel's theorem imply the negative solution to Hilbert's tenth problem.

[14]A history of researches which led to the proof of Matiyasevich's theorem can be found in Reid (1996). The volume contains also papers by Davis and Matiyasevich.

Note also that by Matiyasevich's theorem for any recursively axiomatizable extension T of Peano arithmetic PA there exists a polynomial p_T with integer coefficients (p_T depends on T) such that the sentence

there exists no solution of the diophantine equation $p_T = 0$

is true but undecidable in T, provided T is consistent.

4.6 Gödel's Theorems and Computer Science

Gödel's incompleteness theorems were also used as arguments in discussions of problems related to computer science, especially artificial intelligence. The origins of the latter are connected with Turing's paper "Computing Machinary and Intelligence" (1950). One of the main problems investigated in this domain is the question of whether machines can act in an intelligent way. Note that this problem is in fact very old and was considered already by René Descartes (1596–1650) and Julien Offray de La Mettrie (1709–1751) who discussed the question of whether a human being is a mechanical machine. Descartes answered this question negatively in *Discours de la méthode pour bien conduire sa raison et chercher la vérité dans les sciences* (1637) arguing that it is impossible to construct a (mechanical) machine which could perform acts that are characteristic for a human being. On the contrary, La Mattrie claimed in *L'homme-machine* (1748) that thinking is a function of the body and that a human being can be treated as a machine.

The perspective of the eighteenth-century thinkers was naturally restricted to mechanical machines. New aspects of the problem appeared with the introduction of symbolical machines (such as Turing machines) and modern computers. It can now be formulated in the form: does the human mind work as a formalized system, can a computer program be treated as a model of the human mind? At the beginning of the sixties two papers appeared in which an attempt was made to answer this question using Gödel's theorems: (Smart, 1961) and (Lucas, 1961). They caused a long and interesting debate.[15]

By a machine Smart understands a Turing machine and is of opinion that (cf. Smart, 1961, p. 105):

> Recent tendencies in biology and psychology have made plausible, to some of us at any rate, the view that men are complicated mechanisms.

In his argumentation he uses Gödel's theorem and writes (pp. 105–106):[16]

> Consider a formal language L_0 adequate for elementary number theory. Gödel has shown that if elementary number theory is consistent there is some closed arithmetical sentence expressible in the symbolism of L_0 which can not be proved or disproved in L_0. By reasoning which makes use of the syntax language of L_0, however, we can show that this proposition which is undecidable within L_0 is in fact true. Corresponding to any formal language L_0 (with constructive rules of proof) we can specify some Turing machine (or computer) T_0 which if given enough time will produce a proof or disproof of any decidable closed arithmetical sentence of L_0. The machine will not of course ever produce for us a proof of a Gödelian undecidable proposition of L_0, for no such proof exists. A proof does exist in the language L_1, which is the language got by adjoining the syntax language of L_0 to L_0. To L_1 will correspond a machine T_1, which if given enough time will produce a proof of the original sentence undecidable in L_0. However in L_1 there will be another proposition undecidable within it but which we by using the syntax

[15]The arguments of Lucas were criticized in particular by Good in (1967 and 1969) and Benacerraf in (1967). Lucas defended his position in (1967 and 1968). See also (Hofstadter, 1979) and (Penrose, 1989 and 1994).

[16]Observe that Smart uses here the notion of the syntax language introduced by Carnap; Gödel spoke about metamathematical considerations.

> language of L_1 can show to be true. And so on. Consider the sequence of languages L_0, L_1, L_2, ... and the sequence of machines T_0, T_1, T_2, ... Whatever you like to mention there will always be some proposition which we can prove but it can not.

Smart's argument is now based on the claim that a machine can transform itself into another machine with greater logical power and in this way he comes to the conclusion (p. 106):

> If it [the human nervous system — R.M.] is a machine there must be some Turing machine which can do anything it can do.

Let us turn now to Lucas who claims at the very beginning of his paper (1961) that (p. 112):

> Gödel's Theorem seems to me to prove that Mechanism is false, that is, that minds cannot be explained as machines.

His argumentation is as follows.[17] Lucas wants to show that the mathematical abilities of a human being, e.g., his own abilities, are bigger than those of a machine, even if restricted to arithmetical propositions. If a Mechanist says that a machine M is equivalent to Lucas, or rather to the mathematical powers of Lucas, then Lucas produces the Gödel formula φ_M corresponding to the theory $T(M)$ consisting of arithmetical statements provable by M. It is inessential how we define the theory $T(M)$ — the only important thing is that $T(M)$ is semidecidable (i.e., recursively enumerable). It is always the case if one assumes that M is equivalent to a Turing machine (or other standard notion). There are two possibilities now:

Case 1: the theory $T(M)$ is consistent.

Case 2: the theory $T(M)$ is inconsistent.

---

[17]A nice presentation and discussion of Lucas' argument can be found, e.g., in Krajewski (1993). See also Rucker (1982).

In the first case the sentence φ_M is unprovable by M, i.e., it is unprovable in the theory $T(M)$, but Lucas can prove it by showing that it is true (Lucas uses here Gödel's theorem together with the assumption that $T(M)$ is consistent). In the second case every sentence is provable by M. But such a provability is useless and Lucas has certainly another notion of provability and consequently is different from the machine M. So in any case Lucas has shown that the machine M cannot be equivalent to himself.

In this way Lucas comes in (1961) to the following conclusion (p. 116):

> We are trying to produce a model of the mind which is mechanical — which is essentially "dead" — but the mind, being in fact "alive", can always go on better than any formal, ossified, dead, system can. Thanks to Gödel's theorem, the mind always has the last word.

Observe that the above procedure can be mechanized. Let us assume that all machines are listed in a sequence M_1, M_2, M_3, ... It is easy to show that there is a recursive function g such that for every n

(∗) if $T(M_n)$ is consistent then $g(n)$ is the Gödel number of a (Gödel) sentence which is true and unprovable in $T(M_n)$.

An objection to the above argumentation can be made on the basis of (∗). Indeed, if a machine corresponding to the function g can simulate the procedure described above (sometimes called 'out-Gödeling of M') then machines are not really worse than we are. It can be replied that this machine can be out-Gödeled too. In fact, the aim is not to dominate all machines at once but each machine proposed by the Mechanist. Therefore Lucas describes the procedure as dialectical and he writes:

> The argument is dialectical. It is an argument between two persons, not a proof sequence constructed by one. (1968, p. 154)

Hence it is a game in which Lucas wins in every move.

Another objection that can be raised is that the procedure does not work if the program of a machine is not known. Hence the mathematical powers of Lucas might be equal to a machine but he would not be able to find an n in order to produce $g(n)$. This leads to the following possibility: perhaps we are a machine but we do not know which one. This can be answered as follows: the appropriate Gödel sentence exists in an objective way and can be used by us at any time. On the other hand, if a Mechanicist presents a number n then Lucas can produce $g(n)$. The sentence corresponding to it will be true provided the theory $T(M_n)$ is consistent.

But now the next problem appears: how do we know that $T(M_n)$ is consistent? The set $C = \{n : T(M_n) \text{ is consistent}\}$ is not recursive,[18] so the problem "is a given machine consistent" is undecidable. Consequently, it is theoretically impossible to distinguish Case 1 and Case 2. To distinguish them requires nonrecursive skills, i.e., the procedure of out-Gödeling assumes a non-mechanical nature of Lucas. But this is the thesis that was supposed to be proved!

In this situation one can argue that either Case 1 or Case 2 hold and hence the whole argument still works. But now the procedure becomes nonconstructive and its dialectical character is doubtful.

Since Gödel sentences for inconsistent theories are false and contradict even the most elementary arithmetical truths, i.e., truths that can be formulated as formulas with bounded quantifiers only,[19] Lucas is not allowed to commit even a single mistake. Indeed, his argument in Case 2 depends entirely on his consistency and if he presented Gödel sentence as a sentence

[18]Assume that C is recursive. Then $C' = \{g(n) : n \in C\}$ is recursively enumerable. Hence $C' = T(M_k)$ for some k. Since all $g(n)$ for $n \in C$ are true, it follows that $k \in C$. By the definition $g(k)$ is unprovable in $T(M_k)$. But on the other hand $g(k) \in T(M_k)$, a contradiction.

[19]Compare remarks on the form of Gödel sentences and the fact that every formula of the form $\exists x \varphi(x)$ (φ of the class BQ) true in the standard model of PA, is provable in PA — cf. Theorem 2.4.27.

proving his superiority over machines, he would fall into inconsistency.

So the question if we are consistent turns out to be significant. Lucas (1961) claims that the answer is negative and writes (p. 120):

> ... are not men inconsistent too? Certainly women are, and politicians; and even male non-politicians contradict themselves sometimes, and a single inconsistency is enough to make a system inconsistent.

And he adds (p. 123):

> ... if we find that no system containing simple arithmetic can be free of contradictions, we shall have to abandon not merely the whole of mathematics and the mathematical sciences, but the whole of thought.

The assumption that mathematics (and we ourselves as well!) are consistent is necessary to do mathematics. But it should be noted that it may happen that proofs of contradictions are too long to matter in practice and that it is possible to remain in safe contradiction-free areas all the time.[20] Gödel's Second Theorem seems to imply (compare restrictions indicated in Section 2.4) that our consistency cannot be proved in a mathematical, formal way even if we are consistent. In practice we believe that we are in principle consistent or rather that we can correct and remove any inconsistencies when they appear. Lucas puts it in the following way (1961, p. 122):

> ... the mind does indeed try out dubious axioms and rules of inference; but if they are found to lead to contradiction, they are rejected altogether. We try out axioms and rules of inference provisionally-true: but we do not keep them, once they are found to lead to contradictions.

[20]This is the case with large computer programs. Note also that the infinitesimal calculus was developed for centuries on inconsistent foundations.

The belief that we are in principal consistent does not, however, assure that Lucas' argument is right. Indeed, G. Lee Bowie (1982) remarked that whatever Lucas may claim about his own consistency, we know enough to prove that he is actually inconsistent. Lucas' procedure can be expressed as the recursiveness of the function g satisfying the condition $(*)$ formulated above. But the range of g, i.e., the set

$$A = \{\text{the sentence with Gödel number } g(n) : n \in \mathbb{N}\}$$

is inconsistent. This can be proved in the following way: Since g is recursive, the set A is recursively enumerable. Hence one can assume that it is generated by a machine M_k for some $k \in \mathbb{N}$. If A were consistent then by $(*)$ the sentence with Gödel number $g(k)$ would be unprovable in $T(M_k)$. Yet it is an element of A and consequently it belongs to $T(M_k)$. This is a contradiction.

It seems impossible to modify Lucas' argument to avoid the above criticism (cf. Krajewski, 1983 and 1993).

Though no convincing arguments in favour of or against the mechanistic thesis were proposed so far, it seems that Gödel's results support the moderate opinion of Nagel and Newman (1958, pp. 100-101) formulated in the following way:[21]

> Given a definite problem, a machine ... might be built for solving it; but no one such machine can be built for solving every problem. The human brain may, to be sure, have built-in limitations of its own, and there may be mathematical problems it is incapable of solving. But, even so, the brain appears to embody a structure of rules of operation which is far more powerful than the structure of currently conceived artificial machines. There is no immediate prospect of replacing the human mind by robots.

[21]Though formulated almost forty years ago, when computers and artifical intelligence were at the beginnings of their development, it still seems to be valid.

What was the opinion of Gödel? He was looking for arguments that "laws of thought are not mechanical" (cf. Kreisel, 1980, p. 216). In the 25th Josiah Willard Gibbs Lecture "Some basic theorems on the foundations of mathematics and their implications"[22] Gödel considered implications of his incompleteness theorems for the problem of relations between mind and machine. One of the implications is that the human mind is incapable of formulating or mechanizing all its mathematical intuitions. Therefore one can speak about certain incompletability of mathematics. Gödel wrote (p. 309):

> ... *it* [Second Incompleteness Theorem — R.M.] *makes it impossible that someone should set up a certain well-defined system of axioms and rules and consistently make the following assertion about it: All of these axioms and rules I perceive (with mathematical certitude) to be correct, and moreover I believe that they contain all of mathematics.* If someone makes such a statement he contradicts himself. For if he perceives the axioms under consideration to be correct, he also perceives (with the same certainty) that they are consistent. Hence he has a mathematical insight not derivable from his axioms.

On the other, hand Gödel noted that on the basis of what has been proved so far, it remains possible that (pp. 309–310):

> the human mind (in the realm of pure mathematics) *is* equivalent to a finite machine that, however, is unable to completely understand its own functioning.

This means that such a machine would be equivalent to mathematical intuition, but, on the other hand, it would be impossible to prove that fact, or even to prove that the machine yields

---

[22]Gödel delivered this lecture at the meeting of the American Mathematical Society at Brown University on 26th December 1951. It was first published in the volume III of his *Collected Works* (cf. Gödel, 1951).

only correct theorems. This possibility follows from Gödel's Second Theorem under the assumption that our mind is consistent. Gödel treated the latter as granted.

The second important implication is formulated by Gödel in the following way (p. 310):

> *Either mathematics is incompletable in this sense, that its evident axioms can never be comprised in a finite rule, that is to say, the human mind (even within the realm of pure mathematics) infinitely surpasses the powers of any finite machine, or else there exist absolutely unsolvable diophantine problems* ... (where the case that both terms of the disjunction are true is not excluded, so that there are, strictly speaking, three alternatives).

Gödel claimed hovever that the history of mathematics forced us to reject the second alternative. His argumentation is summerized by Hao Wang in the following way (1974, pp. 324–325):

> If it were true it would mean that human reason is utterly irrational by asking questions it cannot answer, while asserting emphatically that only reason can answer them. Human reason would then be very imperfect and, in some sense, even inconsistent, in glaring contradiction to the fact that those parts of mathematics which have been systematically and completely developed (such as, e.g. the theory of 1st and 2nd degree Diophantine equations, the latter with two unknowns) show an amazing degree of beauty and perfection.

Certain 'rationalistic optimism' can be seen here. In this way Gödel shared Hilbert's belief expressed in 1926 in the words:

> In mathematics there is no *ignorabimus*,

and repeated at the end of his speech over the local radio station in Königsberg in September 1930, just after the conference during which Gödel announced his incompleteness result (cf. Section 2.6):

Wir müssen wissen. Wir werden wissen.
(We must know. We shall know.)

In Gödel's opinion the attempted proofs for the equivalence of the mind and machines were fallacious. This follows, e.g., from the following unpublished paragraph about Turing's alleged proof that every mental procedure for producing an infinite series of integers is equivalent to a mechanical procedure:[23]

> Turing, in *Proc. Lond. Math. Soc.* 42 (1936), p. 250, gives an argument which is supposed to show that mental procedures cannot carry any further than mechanical procedures. However, this argument is inconclusive, because it depends on the supposition that a finite mind is capable of only a finite number of distinguished states. What Turing disregards completely is the fact that *mind, in its use, is not static, but constantly developing.* This is seen, e.g., from the infinite series of ever stronger axioms of infinity in set theory, each of which expresses a new idea or insight. A similar process takes place with regard to the primitive terms. E.g., the iterative concept of set became clear only in the past few decades. Several more primitive ideas now appear on the horizon, e.g., the self-reflexive concept of proper class. Therefore, although at each stage of the mind's development the number of its possible states is finite, there is no reason why this number should not converge to infinity in the course of its development. Now there may exist systematic methods of accelerating, specializing, and uniquely determining this development, e.g., by asking the right questions on the basis of a mechanical procedure. But it must be admitted that the precise definition of a procedure of this kind would require a substantial deepening of our understanding of the basic operations of the mind. Vaguely

[23]This paragraph was to be added as a footnote at the word 'mathematics' on p. 73, line 3 of the English translation of Gödel (1931) in Davis (1965). It is quoted here according to Wang Hao (1974, pp. 326–327).

> defined procedures of this kind, however, are known, e.g., the process of defining recursive well-orderings of integers representing larger and larger ordinals or the process of forming stronger and stronger axioms of infinity in set theory.

In discussions with Hao Wang (cf. 1974, p. 326) Gödel added that Turing's argument became valid under two additional assumptions: (1) there is no mind separated from matter, and (2) the brain functions basically like a digital computer (or (2') the physical laws, in their observable consequences, have a finite limit of precision). Gödel treated (2) as very likely and (2') as practically certain. On the other hand he believed that (1) was (cf. Wang Hao, 1974, p. 326):

> a prejudice of our time, which will be disproved scientifically (perhaps by the fact that there aren't enough nerve cells to perform the observable operations of the mind).

Gödel's theorems were investigated and interpreted also from the point of view of the information theory. We cannot present here the appropriate technical considerations (they can be found, e.g., in Chaitin (1974 and 1982)). The information theory suggests that the phenomenon revealed by Gödel's incompleteness theorems is natural and widespread, not pathological and unusual. The results obtained in it lead to the following corollaries: if a certain effective (recursive) set of methods of reasoning has been precized beforehand then there is an upper bound to the complexity of theorems that can be proved in such a system. Consequently, if one wishes to obtain more complex theorems (in the sense of the information theory, i.e., theorems containing more information) then one will have to continually introduce new axioms and new methods. Neither the admissible methods and rules can be fixed and codified nor the concept of a correct mathematical proof can be defined once and for ever. Hence, progress in mathematics seems to be much like progress in the natural sciences than hitherto expected. Note that such a thesis was already stated by Gödel in (1944) where he wrote on pp. 127–128:

> The analogy between mathematics and a natural science is enlarged upon by Russell also in another respect ... axioms need not be evident in themselves, but rather their justification lies (exactly as in physics) in the fact that they make it possible for these "sense perceptions" to be deduced ... I think that ... this view has been largely justified by subsequent developments, and it is to be expected that it will be still more so in the future. It has turned out that solution of certain arithmetical problems requires the use of assumptions essentially transcending arithmetic ... Furthermore it seems likely that for deciding certain questions of abstract set theory and even for certain related questions of the theory of real numbers new axioms based on some hitherto unknown idea will be necessary. Perhaps also the apparently unsurmountable difficulties which some other mathematical problems have been presenting for many years are due to the fact that the necessary axioms have not yet been found. Of course, under these circumstances mathematics may lose a good deal of its "absolute certainty"; but, under the influence of the modern criticism of the foundations, this has already happened to a large extent.

All such arguments are used by quasi-empiricism (cf., e.g., Putnam, 1975) which claims that mathematical knowledge is not *a priori*, absolute and certain but it is rather quasi-empirical, fallible and probable, it is in fact much like natural sciences.

4.7 Gödel's Theorems and General Philosophy

Several attempts were (and still are) made to apply Gödel's incompleteness results to problems of the general philosophy and other domains such as biology, psychology, linguistics, theology, etc. (cf., e.g., papers collected in the volume Wolkowski (1993)). The problem arises of whether it is legitimate to use Gödel's results in such contexts and to apply it in domains far from the mathematical world and objects.

It seems that first of all a distinction should be made between considerations inspired by the theorems or by the method of Gödel on the one hand, and genuine applications of them on the other. A nice example of distinguishing those two aspects is provided by the book *Gödel, Escher, Bach* by D. Hofstadter (1979). He claims that (p. 969):

> If one uses Gödel's Theorem as a metaphor, as a source of inspiration, rather than trying to translate it literally into the language of psychology or of any other discipline, then perhaps it can suggest new truths in psychology or other areas. But it is quite unjustifiable to translate it directly into a statement of another discipline and take that as equally valid. It would be a large mistake to think that what has been worked out with the utmost delicacy in mathematical logic should hold without modification in a completely different area.
>
> I think it can have suggestive value to translate Gödel's theorem into other domain, provided one specifies in advance that the translations are metaphorical and are not intended to be taken literally.

There are natural obstacles and difficulties in applying Gödel's results in fields other than mathematics. First, there is no first-order theory of the whole of reality, or even of given fragments of the reality. And Gödel's theorems apply only to such theories. Second, philosophical interpretations of mathematical theorems are a subtle issue. Indeed, since the terminology of mathematics and philosophy are different, a certain translation or interpretation of mathematical terms and concepts is necessary and all applications depend on them. And this task is rather difficult, if not hopeless, since in mathematics notions are precisely defined whereas philosophical concepts are rather vague. Maybe it was just this what Wittgenstein had in mind when he claimed that Gödel's results had no meaning and significance for the philosophy.

4.8 Conclusions

Gödel's incompleteness theorems struck Hilbert's program and indicated several difficulties in attempts to realize it. On the other hand they inspired various investigations in the foundations and philosophy of mathematics. It was shown in particular that Hilbert's program can be partially realized.

Gödel's results also indicated certain weaknesses of the axiomatic-deductive method and of the formalization based on it. But it does not mean that the method itself should be rejected and not used any more in mathematics. Just the opposite! The axiomatic-deductive method is the best method we have. It enables us to systematize mathematics and to present it in an intersubjectively understandable way. It also makes it possible to specify the notion of a proof and of the correctness of methods admissible in mathematics. If we resigned from it then the notion of a proof and consequently that of a theorem would become entirely subjective.

Gödel's results showed that the hopes and expectations with regard to the application of the axiomatic-deductive method in mathematics were not fully justified. Indeed, more was expected than could be realized. In particular, we now know that the whole of mathematics cannot be formalized as one first-order system, that it is impossible to construct a complete first-order theory even of natural numbers, that in order to formalize something, something else must remain non-formalized. We understand better the axiomatic-deductive method and mathematics itself. We know that such categories as contents, sense, meaning, intuition cannot be fully eliminated even from mathematics (or rather from doing mathematics).[24]

Together with subtle and sophisticated methods of the metamathematics the axiomatic-deductive method makes the precise proofs possible that certain statements do or do not follow from given axioms. This should be considered in fact as a virtue of this method and not as a weakness.

Gödel's incompleteness theorems indicate that, as Post has put it, "a mathematical proof is in fact a result of a creative activity of our mind", that one cannot restrict the invention of mathematicians. A problem unsolvable by a certain given methods (i.e., in a given theory) can turn out to be solvable by other, richer and stronger methods (in a stronger theory). In their research practice mathematicians do not work in one given formal theory. They use various methods and means that could help to find a solution to a considered problem. But the results of their investigations can always be presented and ordered by using the axiomatic-deductive method. This makes them objective and verifiable in an intersubjective way.

Gödel's results, as shown above, substantially shaped mathematical logic as well as the foundations and the philosophy of mathematics after 1931. There remains the problem of the influence of those results on 'normal' mathematicians (i.e., those working in other fields than the mentioned foundational ones) and their research practice. As Kreisel remarked (1980, p. 149) incompleteness theorems have not changed the conception of mathematics held by the majority of mathematicians. Only in set theory and general topology, and to some extent in algebra, they have played a major role — indeed, they helped to clarify the need for new axioms. Moore (1991, p. 10) formulates the following hypothesis:

---

[24]It is reflected in new directions in the philosophy of mathematics – cf., e.g., papers collected in the anthology (Tymoczko, 1985).

Perhaps the incompleteness theorems will only be truly appreciated by non-logicians when some central problem of analysis, such as the Riemann Hypothesis, is shown to be undecidable on the basis of the accepted axioms. But that day has not yet come.

Bibliography

Ackermann, W.: 1924-25, 'Begründung des tertium non datur mittels der Hilbertschen Theorie der Widerspruchsfreiheit', *Mathematische Annalen* **93**, 1–36.

Ackermann, W.: 1928, 'Zum Hilbertschen Aufbau der reellen Zahlen', *Mathematische Annalen* **99**, 118–133; English translation: 'On Hilbert's Construction of Real Numbers', in: Heijenoort, J. van (Ed.) *From Frege to Gödel. A Source Book in Mathematical Logic, 1879–1931*, Harvard University Press, Cambridge, Mass., 1967, pp. 495–507.

Ackermann, W.: 1940, 'Zur Widerspruchsfreiheit der Zahlentheorie', *Mathematische Annalen* **117**, 162–194.

Adamowicz, Z.: 1995, 'Diagonal Reasonings in Mathematical Logic'. In: Jakubczyk, B. *et al.* (Eds.) *Panoramas of Mathematics. Colloquia 93–94*, Banach Center Publications, Polish Academy of Sciences, Institute of Mathematics, Warszawa, pp. 9–18.

Adamowicz, Z. and Zbierski, P.: 1997, *Logic of Mathematics*, John Wiley & Sons, Inc., New York.

Ajdukiewicz, K.: 1928, *Główne zasady metodologii nauk i logiki formalnej* (Main Principles of the Methodology of Science and Formal Logic), Nakładem Komisji Wydawniczej Koła Matematyczno-Fizycznego Słuchaczów Uniwersytetu Warszawskiego, Warszawa.

Apt, K. and Marek, W.: 1974, 'Second Order Arithmetic and Related Topics', *Annals of Mathematical Logic* **6**, 177–239.

Arai, T.: 1990, 'Derivability Conditions on Rosser's Provability Predicates', *Notre Dame Journal of Formal Logic* **31**, 487–497.

Ax, J.: 1968, 'The Elementary Theory of Finite Fields', *Annals of Mathematics* **88**, 239–271.

Ax, J. and Kochen, S.: 1965, 'Diophantine Problems Over Local Fields', *American Journal of Mathematics* **87**, 605–630.

Ax, J. and Kochen, S.: 1965a, 'Diophantine Problems Over Local Fields. II: A Complete Set of Axioms For *p*-adic Number Theory', *American Journal of Mathematics* **87**, 631–648.

Ax, J. and Kochen, S.: 1966, 'Diophantine Problems Over Local Fields. III: Decidable Fields', *Annals of Mathematics* **83**, 437–456.

Barwise, J.: 1980, 'Infinitary Logics'. In: Agazzi, E. (Ed.) *Modern Logic — A Survey*, D. Reidel Publ. Comp., Dordrecht, pp. 3–112.

Barzin, M.: 1940, 'Sur la Portée du Théorème de M. Gödel', *Académie Royale de Belgique, Bulletin de la Classe des Sciences*, Series 5, **26**, 230–239.

Behmann, H.: 1922, 'Beiträge zur Algebra und Logik, insbesondere zum Entscheidungsproblem', *Mathematische Annalen* **86**, 163–229.

Bell, J. and Machover, M.: 1977, *A Course in Mathematical Logic*, North-Holland Publ. Comp., Amsterdam.

Benacerraf, P.: 1967, 'God, the Devil, and Gödel', *The Monist* **51**, 9–32.

Bernays, P.: 1967, 'Hilbert David'. In: Edwards, P. (Ed.) *Encyclopedia of Philosophy*, vol. 3, Macmillan and Free Press, New York, pp. 496–504.

Bernays, P.: 1970, 'On the Original Gentzen Consistency Proof for Number Theory'. In: Myhill, J. *et al.* (Eds.) *Intuitionism and Proof Theory*, North-Holland Publ. Comp., Amsterdam, pp. 409–417.

Bernays, P. and Schönfinkel, M.: 1928, 'Zum Entscheidungsproblem der mathematischen Logik', *Mathematische Annalen*, **99**, 342–372.

Boolos, G.: 1979, *The Unprovability of Consistency. An Essay In Modal Logic*, Cambridge University Press, Cambridge.

Boolos, G.: 1989, 'A New Proof of the Gödel Incompleteness Theorem', *Notices of the American Mathematical Society* **36**, 388–390.

Boolos, G.: 1989a, 'A Letter from George Boolos', *Notices of the American Mathematical Society* **36**, p. 676.

Boolos, G.: 1993, *The Logic of Provability*, Cambridge University Press, Cambridge.

Boolos, G. and Sambin, G.: 1991, 'Provability: the Emergence of a Mathematical Modality', *Studia Logica* **50**, 1–23.

Bowie, G.L.: 1973, 'An Argument Against Church's Thesis', *Journal of Philosophy* **70**, 66–76.

Bowie, G.L.: 1982, 'Lucas' Number Is Finally Up', *Journal of Philosophical Logic* **11**, 279–285.

Börger, E., Grädel, E. and Gurevich, Y.: 1997, *The Classical Decision Problem*, Springer-Verlag, Berlin-Heidelberg.

Bridgman, P.W.: 1934, 'A Physicist's Second Reaction to Mengenlehre', *Scripta Mathematica* **2**, 101–117 and 234.

Brown, D.K.: 1987, *Functional Analysis in Weak Subsystems of Second-Order Arithmetic*, Ph.D. Thesis, Pennsylvania State University, University Park, Pa.

Brown, D.K., and Simpson, S.G.: 1986, 'Which Set Existence Axioms Are Needed to Prove the Separable Hahn-Banach Theorem?', *Annals of Pure and Applied Logic* **31**, 123–144.

Buldt, B: 1991, *Die Sätze von Gödel. Logische und philosophische Perspektiven*, Ph.D. Thesis, Ruhr-Universität Bochum.

Büchi, J.R.: 1962, 'On A Decision Method In Restricted Second Order Arithmetic'. In: *Logic, Methodology and Philosophy of Science. Proceedings of the 1960 Congress*, Stanford, pp. 1–11.

Calude, Ch., Marcus, S. and Tevy, I.: 1979, 'The First Example of a Recursive Function Which Is Not Primitive Recursive', *Historia Mathematica* **6**, 380–384.

Cantor, G.: 1874, 'Über eine Eigenschaft des Inbegriffs aller reellen algebraischen Zahlen', *Journal für die reine und angewandte Mathematik* **77**, 258–262.

Cantor, G.: 1892, 'Über eine elementare Frage der Mannigfaltigkeitslehre', *Jahresberichte der Deutschen Mathematiker Vereinigung* **1**, 75–78.

Cantor, G.: 1932, *Gesammelte Abhandlungen mathematischen und philosophischen Inhalts mit erläuternden Anmerkungen sowie mit Ergänzungen aus dem Briefwechsel Cantor-Dedekind*, hrsg. von E. Zermelo, Verlag von Julius Springer, Berlin. Reprinted by Springer-Verlag, Berlin-Heidelberg-New York 1980.

Carnap, R.: 1934, 'Die Antinomien und die Unvollständigkeit der Mathematik', *Monatshefte für Mathematik und Physik* **41**, 263–284.

Carnap, R.: 1934a, *Logische Syntax der Sprache*, Springer-Verlag, Wien.

Carnap, R.: 1935, 'Ein Gültigkeitskriterium für die Sätze der klassischen Mathematik', *Monatshefte für Mathematik und Physik*, **42**, 163–190.

Cegielski, P.: 1981, 'Théorie élémentaire de la multiplication des entiers naturels'. In: Berline, C., McAloon, K., Ressayre, J.P. (Eds.) *Model Theory and Arithmetic*, Lecture Notes in Mathematics 890, Springer-Verlag, Berlin-Heidelberg-New York, pp. 44–89.

Chaitin, G.: 1974, 'Information-Theoretic Computational Complexity', *IEEE Transactions on Information Theory* **IT-20**, 10–15. Reprinted in: Tymoczko, Th. (Ed.) *New Directions in the Philosophy of Mathematics. An Anthology*, Birkhäuser, Boston-Basel-Stuttgart 1985 (second revised and expanded edition: Princeton University Press, Princeton, NJ, 1998), pp. 289–299.

Chaitin, G.: 1982, 'Gödel's Theorem and Information', *International Journal of Theoretical Physics* **21**, 941–954. Reprinted in: Tymoczko, Th. (Ed.) *New Directions in the Philosophy of Mathematics. An Anthology*, Birkhäuser, Boston-Basel-Stuttgart 1985 (second revised and expanded edition: Princeton University Press, Princeton, NJ, 1998), pp. 300–311.

Church, A.: 1936, 'A Note on the Entscheidungsproblem', *Journal of Symbolic Logic* **1**, 40–41. Reprinted in: Davis, M. (Ed.) *The Undecidable: Basic Papers on Undecidable Propositions, Unsolv-*

able Problems, and Computable Functions, Raven Press, Hewlett, N.Y., 1965, pp. 108–115.

Church, A.: 1936a, 'An Unsolvable Problem of Elementary Number Theory', *American Journal of Mathematics* **58**, 345–363. Reprinted in: Davis, M. (Ed.) *The Undecidable: Basic Papers on Undecidable Propositions, Unsolvable Problems, and Computable Functions*, Raven Press, Hewlett, N.Y., 1965, pp. 88–107.

Church, A.: 1937, Review of Turing (1936), *Journal of Symbolic Logic*, **2**, 42–43.

Church, A.: 1941, *The Calculi of Lambda-Conversion*, Princeton University Press, Princeton.

Church, A. and Quine, W.V.O.: 1952, 'Some Theorems on Definability and Decidability', *Journal of Symbolic Logic* **17**, 179–187.

Cobham, A.: 1964, 'The Intrinsic Computational Difficulty of Functions'. In: Bar-Hillel, Y. (Ed.) *Logic, Methodology and Philosophy of Science. Proceedings of the 1964 International Congress*, North-Holland Publ. Comp., Amsterdam, pp. 24–30.

Cohen, P.J.: 1969, 'Decision Procedures For Real and *p*-adic Fields', *Comm. Appl. Math.* **22**, 131–151.

Cook, S.A.: 1971, 'The Complexity of Theorem Proving Procedures'. In: *Proceedings of the Third Annual ACM Symposium on the Theory of Computing*, Association for Computing Machinery, New York, pp. 151–158.

Crossley, J.N.: 1974, 'Reminiscences of Logicians (Reported by J.N. Crossley)'. In: Crossley, J.N. (Ed.) *Algebra and Logic*, Springer-Verlag, Berlin, pp. 1–62.

Davis, M.: 1958, *Computability and Unsolvability*, McGraw-Hill, New York.

Davis, M. (Ed.): 1965, *The Undecidable: Basic Papers on Undecidable Propositions, Unsolvable Problems, and Computable Functions*, Raven Press, Hewlett, N.Y.

Davis, M.: 1982, 'Why Gödel Didn't Have Church's Thesis', *Information and Control* **54**, 3–24.

Davis, M., Matiyasevich, Y. and Robinson, J.: 1976, 'Hilbert's Tenth Problem. Diophantine Equations: Positive Aspects of a Negative

Solution'. In: Browder, F. (Ed.) *Mathematical Developments Arising from Hilbert's Problems*, Proceedings of the Symposia in Pure Mathematics 28, American Mathematical Society, Providence, RI, 1976, pp. 323–378. Reprinted in: *The Collected Works of Julia Robinson*, ed. by Feferman, S., American Mathematical Society, Providence, RI, 1996, pp. 269–324.

Davis, M., Putnam, H. and Robinson, J.: 1961, 'The Decision Problem for Exponential Diophantine Equations', *Annals of Mathematics* **74**, 425–436. Reprinted in: *The Collected Works of Julia Robinson*, ed. by Feferman, S., American Mathematical Society, Providence, RI, 1996, pp. 77–88.

Dawson, J.W., Jr.: 1984, 'Discussion on the Foundation of Mathematics', *History and Philosophy of Logic* **5**, 111–129.

Dawson, J.W., Jr.: 1985, 'The Reception of Gödel's Incompleteness Theorems'. In: Asquith, P.D. and Kitcher, Ph. (Eds.) *PSA 1984: Proceedings of the 1984 Biennial Meeting of the Philosophy of Science Association*, vol. 2, pp. 253–271. Reprinted in: Shanker, S.G. (Ed.) *Gödel's Theorem in Focus*, Croom Helm, London (reprinted by Routledge, London and New York 1989), pp. 74–95.

Dawson, J.W., Jr.: 1985a, 'Completing the Gödel-Zermelo Correspondence', *Historia Mathematica* **12**, 66–70.

Dawson, J.W., Jr.: 1993, 'Prelude to Recursion Theory: The Gödel-Herbrand Correspondence'. In: Wolkowski, Z.W. (Ed.) *First International Symposium on Gödel's Theorems*, World Scientific, Singapore-New Jersey-London-Hong Kong, pp. 1–13.

Dedekind, R.: 1888, *Was sind und was sollen die Zahlen?*, Friedrich Vieweg und Sohn, Braunschweig.

Dehornoy, P.: 1993, *Complexité et Décidabilité*, Springer-Verlag, Paris-Berlin.

Descartes R.: 1637, *Discours de la méthode pour bien conduire sa raison et chercher la vérité dans les sciences*. Reprinted in: *Oeuvres complètes*, Adam, Ch. and Tannery, P. (Eds.), Paris 1897–1913, second edition: Paris 1956–1957.

Detlefsen, M.: 1979, 'On Interpreting Gödel's Second Theorem', *Journal of Philosophical Logic* **8**, 297–313. Reprinted in: Shanker,

S.G. (Ed.) *Gödel's Theorem in Focus*, Croom Helm, London (reprinted by Routledge, London and New York 1989), pp. 131–154.

Detlefsen, M.: 1986, *Hilbert's Program. An Essay on Mathematical Instrumentalism*, D. Reidel Publ. Comp., Dordrecht-Boston-Lancaster-Tokyo.

Detlefsen, M.: 1990, 'On the Alleged Refutation of Hilbert's Program Using Gödel's First Incompleteness Theorem', *Journal of Philosophical Logic* **19**, 343–377.

Detlefsen, M.: 1993, 'The Kantian Character of Hilbert's Formalism'. In: Czermak, J. (Ed.) *Philosophy of Mathematics. Proceedings of the 15th International Wittgenstein-Symposium*, Part 1, Verlag Hölder-Pilcher-Tempsky, Wien 1993, 195–205.

Doner, J. and Hodges, W.: 1988, 'Alfred Tarski and Decidability Theory', *Journal of Symbolic Logic* **53**, 20–35.

Doner, J., Mostowski, A. and Tarski, A.: 1978, 'The Elementary Theory of Well-Ordering: A Metamathematical Study'. In: MacIntyre, A. *et al.* (Eds.) *Logic Colloquium'77*, North-Holland Publ. Comp., Amsterdam, pp. 1–54.

Drake, F.R.: 1989, 'On the Foundations of Mathematics in 1987'. In: Ebbinghaus, H.-D. *et al.* (Eds.) *Logic Colloquium'87*, Elsevier Science Publishers, Amsterdam, pp. 11–25.

Dummett, M.: 1978, 'Wittgenstein's Philosophy of Mathematics'. In: Dummett M.: *Truth and Other Enigmas*, Harvard University Press, Cambridge, Mass., pp. 166–185.

Edmonds, J.: 1965, 'Paths, Trees and Flowers', *Canadian Journal of Mathematics* **3**, 449–467.

Ehrenfeucht, A.: 1959, 'Decidability of the Theory of Linear Ordering Relation', *Notices of the American Mathematical Society* **6**, 268–269.

Ehrenfeucht, A. and Feferman, S.: 1960, 'Representability of Recursively Enumerable Sets in Formal Theories', *Archiv f. Math. Logik und Grundlagenforschung* **5**, 37–41.

Ershov, Yu.L., Lavrov, I.A., Taimanov, A.D. and Taitslin, M.A.: 1965, 'Elementary Theories', *Russian Mathematical Surveys*, **20**, 35–105.

Feferman, S.: 1960, 'Arithmetization of Metamathematics in a General Setting', *Fundamenta Mathematicae* **49**, 35–92.

Feferman, S.: 1964–68, 'Systems of Predicative Analysis', Part I: *Journal of Symbolic Logic* **29**, (1964), 1–30; Part II: *Journal of Symbolic Logic* **33**, (1968), 193–220.

Feferman, S.: 1984, 'Kurt Gödel: Conviction and Causation', *Philosophia naturalis.* A separate issue in: Weingartner, P. *et al.* (Eds.) *Philosophy of Science — History of Science. A Selection of Contributed Papers of the 7th International Congress of Logic, Methodology and Philosophy of Science, Salzburg, 1983*, Verlag Anton Hain, Meisenheim/Glan. Reprinted in: Shanker, S.G. (Ed.) *Gödel's Theorem in Focus*, Croom Helm, London (reprinted by Routledge, London and New York 1989), pp. 96–114.

Feferman, S.: 1986, Introductory note to: "Review of Hilbert, 'Die Grundlegung der elementaren Zahlentheorie' ". In: Gödel, K. *Collected Works*, vol. I, ed. by Feferman, S. *et al.*, Oxford University Press, New York and Clarendon Press, Oxford, 1986, pp. 208–213.

Feferman, S.: 1988, 'Turing and the Land of $O(z)$'. In: Herken, R. (Ed.) *The Universal Turing Machine — A Half-Century Survey*, Oxford University Press, New York and Oxford, pp. 113–147.

Feferman, S.: 1988a, 'Hilbert's Program Revisited: Proof-Theoretical and Foundational Reductions', *Journal of Symbolic Logic* **53**, 364–384.

Finsler, P.: 1926, 'Formale Beweise und die Entscheidbarkeit', *Mathematische Zeitschrift* **25**, 676–682. English translation: 'Formal Proofs and Undecidability', in: Heijenoort, J. van (Ed.) *From Frege to Gödel. A Source Book in Mathematical Logic, 1879–1931*, Harvard University Press, Cambridge, Mass., 1967, pp. 440–445; also in: P. Finsler, *Set Theory: Platonism and Circularity*, ed. by D. Booth and R. Ziegler, Birkhäuser, Basel 1966.

Fisher, M.J. and Rabin, M.O.: 1974, 'Super Exponential Complexity of Presburger's Arithmetic'. In: Karp, R. (Ed.) *Complexity of Computation, Proc. SIAM-AMS Sympos. Appl. Math.* **7**, pp. 27–41.

Floyd, J.: 1995, 'On Saying What You Really Want to Say: Wittgenstein, Gödel, and the Trisection of the Angle'. In: Hintikka J. (Ed.) *From Dedekind to Gödel. Essays on the Development of the Foundations of Mathematics*, Kluwer Academic Publishers, Dordrecht, pp. 373–425.

Folina, J.: 1998, 'Church's Thesis: Prelude to a Proof', *Philosophia Mathematica* (3) **6**, 302–323.

Fraenkel, A.: 1935, 'Zum Diagonalverfahren Cantors', *Fundamenta Mathematicae* **25**, 45–50.

Friedman, H.: 1975, 'Some Subsystems of Second Order Arithmetic and Their Use'. In: *Proceedings of the International Congress of Mathematicians*, Vancouver 1974, vol. 1, Canadian Mathematical Congress, pp. 235–242.

Friedman, H.: 1976, 'Systems of Second Order Arithmetic with Restricted Induction' (abstract), *Journal of Symbolic Logic* **41**, 557–559.

Friedman, H.: 1981, 'On the Necessary Use of Abstract Set Theory', *Advances in Mathematics* **41**, 209–280.

Friedman, H., Simpson, S.G. and Smith, R.L.: 1983, 'Countable Algebra and Set Existence Axioms', *Annals of Pure and Applied Logic* **25**, 141–181; addendum: *Annals of Pure and Applied Logic* **28**, 319–320.

Gandy, R.: 1988, 'The Confluence of Ideas in 1936'. In: Herken, R. (Ed.) *The Universal Turing Machine*, Oxford University Press, New York and Oxford, pp. 123–148.

Garey, M.R. and Johnson, D.S.: 1979, *Computers and Intractability: A Guide to the Theory of NP-completeness*, Freeman, San Francisco.

Gentzen, G.: 1936, 'Die Widerspruchsfreiheit der reinen Zahlentheorie', *Mathematische Annalen* **112**, 493–565.

Gentzen, G.: 1938, 'Neue Fassung des Widerspruchsfreiheitsbeweises für die reine Zahlentheorie', *Forschung zur Logik und Grundlagen der exakten Wissenschaften, New Series*, **4**, 19–44.

Gentzen, G.: 1969, 'On the Relation Between Intuitionistic and Classical Arithmetic'. In: Szabo, M.E. (Ed.) *The Collected Papers of*

Gerhard Gentzen, North-Holland Publ. Comp., Amsterdam, pp. 53–67.

Gentzen, G.: 1974, 'Der erste Widerspruchsfreiheitsbeweis für die klassische Zahlentheorie', *Archiv für math. Logik und Grundlagenforschungen* **16**, 97–112. English translation in: Szabo, M.E. (Ed.) *The Collected Papers of Gerhard Gentzen*, North-Holland Publ. Comp., Amsterdam 1969.

Gödel, K.: 1929, 'Über die Vollständigkeit des Logikkalküls', doctoral dissertation, submitted in 1929; published and translated in: Gödel, K. *Collected Works*, vol. I, ed. by Feferman, S. *et al.*, Oxford University Press, New York and Clarendon Press, Oxford, 1986, pp. 60–101.

Gödel, K.: 1930, 'Die Vollständigkeit der Axiome des logischen Funktionenkalküls', *Monatshefte für Mathematik und Physik* **37**, 349–360. Reprinted with English translation 'The Completeness of the Axioms of the Functional Calculus of Logic' in: Gödel, K. *Collected Works*, vol. I, ed. by Feferman, S. *et al.*, Oxford University Press, New York and Clarendon Press, Oxford, 1986, pp. 102–123.

Gödel, K.: 1930a, 'Einige metamathematische Resultate über Entscheidungsdefinitheit und Widerspruchsfreiheit', *Anzeiger der Akademie der Wissenschaften in Wien* **67**, 214–215. Reprinted with the English translation 'Some Metamathematical Results on Completeness and Consistency' in: Gödel, K. *Collected Works*, vol. I, ed. by Feferman, S. *et al.*, Oxford University Press, New York and Clarendon Press, Oxford, 1986, pp. 140–143.

Gödel, K.: 1931, 'Über formal unentscheidbare Sätze der 'Principia Mathematica' und verwandter Systeme. I', *Monatshefte für Mathematik und Physik* **38**, 173–198. Reprinted with English translation 'On Formally Undecidable Propositions of Principia Mathematica and Related Systems' in: Gödel, K. *Collected Works*, vol. I, ed. by Feferman, S. *et al.*, Oxford University Press, New York and Clarendon Press, Oxford, 1986, pp. 144–195.

Gödel, K.: 1932, 'Über Vollständigkeit und Widerspruchsfreiheit', *Ergebnisse eines Mathematischen Kolloquiums* **3**, 12–13. Reprinted with English translation 'On Completeness and Consistency' in Gödel, K. *Collected Works*, vol. I, ed. by Feferman, S. *et al.*,

Oxford University Press, New York and Clarendon Press, Oxford, 1986, pp. 234–237.

Gödel, K.: 1933, 'Zur intuitionistischen Arithmetik und Zahlentheorie', *Ergebnisse eines Mathematischen Kolloquiums* **4**, 34–38. Reprinted with English translation 'On Intuitionistic Arithmetic and Number Theory' in: Gödel, K. *Collected Works*, vol. I, ed. by Feferman, S. *et al.*, Oxford University Press, New York and Clarendon Press, Oxford, 1986, pp. 286–295.

Gödel, K.: 1934, *On Undecidable Propositions of Formal Mathematical Systems* (mimeographed lecture notes, taken by S.C. Kleene and J.B. Rosser), Princeton; reprinted with revisions in: Davis, M. (Ed.) *The Undecidable: Basic Papers on Undecidable Propositions, Unsolvable Problems, and Computable Functions*, Raven Press, Hewlett, N.Y., 1965, pp. 39–74. Also in Gödel, K. *Collected Works*, vol. I, ed. by Feferman, S. *et al.*, Oxford University Press, New York and Clarendon Press, Oxford, 1986, pp. 346–371.

Gödel, K.: 1944, 'Russell's Mathematical Logic'. In: Schilpp, P.A. (Ed.) *The Philosophy of Bertrand Russell*, Northwestern University, Evanston, pp. 123–153. Reprinted in: Gödel, K. *Collected Works*, vol. II, ed. by Feferman, S. *et al.*, Oxford University Press, New York and Oxford, 1990, pp. 119–141.

Gödel, K.: 1946, 'Remarks Before the Princeton Bicentennial Conference on Problems in Mathematics', 1–4; first published in: Davis, M. (Ed.) *The Undecidable: Basic Papers on Undecidable Propositions, Unsolvable Problems, and Computable Functions*, Raven Press, Hewlett, N.Y., 1965, pp. 84–88. Reprinted in: Gödel, K. *Collected Works*, vol. II, ed. by Feferman, S. *et al.*, Oxford University Press, New York and Oxford, 1990, pp. 150–153.

Gödel, K.: 1951, 'Some Basic Theorems on the Foundations of Mathematics and Their Implications', first published in: Gödel, K. *Collected Works*, vol. III, ed. by Feferman, S. *et al.*, Oxford University Press, New York and Oxford 1995, pp. 304–323.

Gödel, K.: 1958, 'Über eine bisher noch nicht benützte Erweiterung des finiten Standpunktes', *Dialectica* **12**, 280–287. Reprinted with English translation 'On a Hitherto Unutilized Extension of the Finitary Standpoint' in: Gödel, K. *Collected Works*, vol. II, ed. by Feferman, S. *et al.*, Oxford University Press, New York and Oxford, 1990, pp. 240–251.

Gödel, K.: 1972, 'On an Extension of Finitary Mathematics Which Has Not Yet Been Used', revised and expanded English version of Gödel (1958), to have appeared in *Dialectica*, first published in: K. Gödel *Collected Works*, vol. II, ed. by Feferman, S. *et al.*, Oxford University Press, New York-Oxford 1990, pp. 271–280.

Good, I.J.: 1967, 'Human and Machine Logic', *British Journal for the Philosophy of Science* **18**, 145–146.

Good, I.J.: 1969, 'Gödel's Theorem is a Red Herring', *British Journal for the Philosophy of Science* **19**, 357–358.

Gordeev, L.: 1994, 'A Modified Sentence Unprovable in **PA**', *Journal of Symbolic Logic*, **59**, 1154–1157.

Grattan-Guinness, I.: 1979, 'In Memoriam Kurt Gödel: His 1931 Correspondence with Zermelo on his Incompletability Theorem', *Historia Mathematica* **6**, 294–304.

Grzegorczyk, A.: 1951, Undecidability of Some Topological Theories', *Fundamenta Mathematicae* **38**, 137–152.

Grzegorczyk, A.: 1953, 'Some Classes of Recursive Functions', *Rozprawy Matematyczne* **4**, 1–46.

Grzegorczyk, A.: 1962, 'An example of two weak essentially undecidable theories F and F^*', *Bulletin de l'Académie Polonaise des Sciences, Série des sciences math., astr. et phys.* **10**, 5–9.

Grzegorczyk, A., Mostowski, A. and Ryll-Nardzewski, C.: 1958, 'The Classical and the ω-Complete Arithmetic', *Journal of Symbolic Logic*, **23**, 188–206.

Guaspari, D. and Solovay, R.: 1979, 'Rosser Sentences', *Annals of Mathematical Logic* **16**, 81–99.

Hájek, P. and Pudlák, P.: 1993, *Metamathematics of First-Order Arithmetic*, Springer-Verlag, Berlin-Heidelberg-New York.

Heijenoort, J. van (Ed.): 1967, *From Frege to Gödel. A Source Book in Mathematical Logic, 1879–1931*, Harvard University Press, Cambridge, Mass.

Heintz, J., Roy, M.-F. and Solerno, P.: 1989, 'Complexité du principe de Tarski-Seidenberg', *C. R. Acad. Sci. Paris*, Sér. Math. **309**, 825–830.

Henkin, L.: 1952, 'A Problem Concerning Provability', *Journal of Symbolic Logic* **17**, 160.

Herbrand, J.: 1928, 'Sur la théorie de la démonstration', *Comptes rendus hebdomadaires des séances de l'Académie des sciences (Paris)* **186**, 1274–1276.

Herbrand, J.: 1930, 'Reserches sur la théorie de la démonstration', *Travaux de la Société des Sciences et des Lettres de Varsovie*. English translation in: J. Herbrand, *Logical Writings*, Harvard University Press, Harvard, Cambridge, Mass., 1971, pp. 46–188, 272–276.

Herbrand, J.: 1931, 'Sur la non-contradiction de l'arithmétique', *Journal für die reine und angewandte Mathematik* **166**, 1–8; English translation: 'On the Consistency of Arithmetic', in: Heijenoort, J. van (Ed.) *From Frege to Gödel. A Source Book in Mathematical Logic, 1879–1931*, Harvard University Press, Cambridge, Mass., 1967, pp. 618–628.

Herbrand, J.: 1931a, 'Sur le Problème fondamental de la logique mathématique, *Sprawozdania z posiedzeń Towarzystwa Naukowego Warszawskiego, Wydzia/l III*, **24**, 12–56. English translation in: J. Herbrand, *Logical Writings*, Harvard University Press, Harvard, Cambridge, Mass., 1971, pp. 215–259.

Hilbert, D.: 1901, 'Mathematische Probleme', *Archiv der Mathematik und Physik* **1**, 44–63 and 213–237. Reprinted in: Hilbert, D. *Gesammelte Abhandlungen*, Verlag von Julius Springer, Berlin, Bd. 3, pp. 290–329. English translation: 'Mathematical Problems', in: Browder, F. (Ed.) *Mathematical Developments Arising from Hilbert's Problems*, Proceedings of the Symposia in Pure Mathematics 28, American Mathematical Society, Providence, RI, 1976, pp. 1–34.

Hilbert, D.: 1918, 'Axiomatisches Denken', *Mathematische Annalen* **78**, 405–415. Reprinted in: Hilbert, D. *Gesammelte Abhandlungen*, Verlag von Julius Springer, Berlin 1935, Bd. 3., pp. 146–156.

Hilbert, D.: 1923, 'Die logischen Grundlagen der Mathematik', *Mathematische Annalen* **88**, 151–165. Reprinted in: Hilbert, D. *Gesammelte Abhandlungen*, Verlag von Julius Springer, Berlin 1935, Bd. 3., pp. 178–191.

Hilbert, D.: 1926, 'Über das Unendliche', *Mathematische Annalen* **95**, 161–190; English translation: 'On the Infinite', in: Heijenoort, J. van (Ed.) *From Frege to Gödel. A Source Book in Mathematical Logic, 1879–1931*, Harvard University Press, Cambridge, Mass., 1967, pp. 367–392.

Hilbert, D.: 1927, 'Die Grundlagen der Mathematik', *Abhandlungen aus dem mathematischen Seminar der Hamburgischen Universität* **6**, 65–85. Reprinted in: Hilbert, D. *Grundlagen der Geometrie*, 7th edition, Leipzig 1930. English translation: 'The Foundations of Mathematics', in: Heijenoort, J. van (Ed.) *From Frege to Gödel. A Source Book in Mathematical Logic, 1879–1931*, Harvard University Press, Cambridge, Mass., 1967, pp. 464–479.

Hilbert, D.: 1929, 'Probleme der Grundlegung der Mathematik'. In: *Atti del Congresso Internationale dei Mathematici, Bologna 3–10 September 1928*, Bologna, vol. I, pp. 135–141. Also in: *Mathematische Annalen* **102**, 1–9.

Hilbert, D.: 1930, 'Naturerkennen und Logik', *Die Naturwissenschaften* **18**, 959–963; reprinted in: Hilbert, D., *Gesammelte Abhandlungen*, Bd. 3, Verlag von Julius Springer, Berlin 1935, 378–387.

Hilbert, D.: 1931, 'Die Grundlegung der elementaren Zahlentheorie', *Mathematische Annalen*, **104**, 485–194.

Hilbert, D.: 1931a, 'Beweis des Tertium non datur', *Nachrichten der Gesellschaft der Wissenschaften zu Göttingen*, 120–125.

Hilbert, D.: 1971, *Gedenkband*, ed. by K. Reidemeister, Springer-Verlag, Berlin-Heidelberg.

Hilbert, D. and Ackermann, W.: 1928, *Grundzüge der theoretischen Logik*, Verlag von Julius Springer, Berlin. English translation of the second edition: *Principles of Mathematical Logic*, Chelsea Publishing Company, New York 1950.

Hilbert, D. and Bernays, P.: 1934–1939, *Grundlagen der Mathematik*, Bd. I 1934, Bd. II 1939, Springer-Verlag, Berlin.

Hofstadter, D.R.: 1979, *Gödel, Escher, Bach: An Eternal Golden Briad*, Basic Books, New York.

Hopcroft, J.E. and Ullman, J.D., *Introduction to Automata Theory, Languages, and Computation*, Addison-Wesley Publ. Comp., Reading, Massachusetts, 1979.

Husserl, E.: 1891, *Philosophie der Arithmetik*, Husserliana, 1970.

Ignjatovic, A.: 1994, 'Hilbert's Program and the Omega-Rule', *Journal of Formal Logic* **59**, 322–343.

Isaacson, D.: 1992, 'Some Considerations on Arithmetical Truth and the ω-Rule'. In: Detlefsen, M. (Ed.) *Proof, Logic and Formalization*, Routledge, London and New York, pp. 94–138.

Janiczak, A.: 1950, 'A Remark Concerning Decidability of Complete Theories', *Journal of Symbolic Logic* **15**, 277–279.

Janiczak, A.: 1953, 'Undecidability of Some Simple Formalized Theories', *Fundamenta Mathematicae* **40**, 131–139.

Janiczak, A.: 1955, 'On the Reducibility of Decision Problems', *Colloquium Mathematicum* **3**, 33–36.

Jaśkowski, S.: 1948, 'Sur le problème de décision de la topologie et de la théorie de groupes', *Colloquium Mathematicum* **1**, 176–179.

Jeroslov, R.G.: 1973, 'Redundances in the Hilbert-Bernays Derivability Conditions For Gödel's Second Incompleteness Theorem', *Journal of Symbolic Logic* **38**, 359–367.

Jeroslov, R.G.: 1975, 'Experimental Logics and Δ_2^0-Theories', *Journal of Philosophical Logic* **4**, 253–267.

Kalmár, L.: 1936, 'Zurückführung des Entscheidungsproblems auf den Fall von Formeln mit einer einzigen binären Funcktionsvariablen', *Comp. Mathematica* **4**, 137–144.

Kalmár, L.: 1943, 'Egyszerű példa eldönthetetlen aritmetikai problémára' (with a German summary: 'Ein einfaches Beispiel für ein unentscheidbares arithmetisches Problem'), *Matematikai és Fizikai Lapok* **50**, 1–23.

Kalmár, L.: 1959, 'An Argument Against the Plausibility of Church's Thesis'. In: Heyting, A. (Ed.) *Constructivity in Mathematics. Proceedings of the Colloquium Held at Amsterdam 1957*, North-Holland Publ. Comp., Amsterdam, pp. 72–80.

Kanamori, A. and McAloon, K.: 1987, 'On Gödel Incompleteness and Finite Combinatorics', *Annals of Pure and Applied Logic*, **33**, 23–41.

Kant, I.: 1787, *Kritik der reinen Vernunft*, J.H. Hartknoch, Riga, 2nd edition.

Karp, R.M.: 1972, 'Reducibility Among Combinatorial Problems'. In: Thatcher, J. (Ed.) *Complexity of Computer Computations*, Plenum Press, New York, pp. 85–104.

Kaye, R.: 1991, *Models of Peano Arithmetic*, Clarendon Press, Oxford.

Kennedy, H.C.: 1972, 'The Origins of Modern Axiomatics: Pasch to Peano', *American Mathematical Monthly* **79**, 133–136.

Kent, C.F. and Hodgson, B.R.: 1989, 'Extensions of Arithmetic for Proving Termination of Computations', *Journal of Symbolic Logic*, **54**, 779–794.

Ketelsen, Ch.: 1994, *Die Gödelschen Unvollständigkeitssätze. Zur Geschichte ihrer Entstehung und Rezeption*, Franz Steiner Verlag, Stuttgart.

Ketonen, J. and Solovay, R.: 1981, 'Rapidly Growing Ramsey Functions', *Annals of Mathematics* **113**, 267–314.

Kikutchi, M.: 1994, 'A Note on Boolos' Proof of the Incompleteness Theorem', *Mathematical Logic Quarterly* **40**, 528–532.

Kikutchi, M. and Tanaka, K.: 1994, 'On Formalization of Model-Theoretic Proofs of Gödel's Theorems', *Notre Dame Journal of Formal Logic* **35**, 403–412.

Kirby, L. and Paris, J.: 1977, 'Initial Segments of Models of Peano's Axioms'. In: Lachlan, A. *et al.* (Eds.) *Set Theory and Hierarchy Theory V. Bierutowice, Poland, 1976*, Springer-Verlag, Berlin-Heidelberg-New York, pp. 211–226.

Kirby, L. and Paris, J.: 1982, 'Accessible Independence Results for Peano Arithmetic', *Bulletin of London Mathematical Society* **14**, 285–293.

Kitcher, Ph.: 1976, 'Hilbert's Epistemology', *Philosophy of Science* **43**, 99–115.

Kleene, S.C.: 1936, 'General Recursive Functions of Natural Numbers', *Mathematische Annalen* **112**, 727–742. Reprinted in: Davis, M. (Ed.) *The Undecidable: Basic Papers on Undecidable Propositions, Unsolvable Problems, and Computable Functions*, Raven Press, Hewlett, N.Y., 1965, pp. 236–253.

Kleene, S.C.: 1936a, 'λ-definability and Recursiveness', *Duke Math. Journal* **2**, 340–353.

Kleene, S.C.: 1943, 'Recursive Predicates and Quantifiers', *Transactions of the American Mathematical Society* **53**, 41–73. Reprinted in: Davis, M. (Ed.) *The Undecidable: Basic Papers on Undecidable Propositions, Unsolvable Problems, and Computable Functions*, Raven Press, Hewlett, N.Y., 1965, pp. 254–287.

Kleene, S.C.: 1952, *Introduction to Metamathematics*, North-Holland Publ. Comp., Amsterdam and P. Noordhoff N.V., Groningen.

Kleene, S.C.: 1987, 'Reflections on Church's Thesis', *Notre Dame Journal of Formal Logic* **28**, 490–498.

Kleeve, J.P.: 1963, 'A Hierarchy of Primitive Recursive Functions', *Zeitschrift für Math. Logik und Grundlagen der Math.* **9**, 331–346.

Köhler, E.: 199?, 'Gödel und der Wiener Kreis'. In: Schimanovich, W. (Hrsg.) *Kurt Gödel — Wahrheit und Beweisbarkeit*, Hölder-Pichler-Tempsky, Wien (to appear).

Kossak, R.: 1985, 'A Note on Satisfaction Classes', *Notre Dame Journal of Formal Logic* **26**, 1–8.

Kotlarski, H.: 1986, 'Bounded Induction and Satisfaction Classes', *Zeitschrift für Mathematische Logik und Grundlagen der Mathematik* **32**, 531–544.

Kotlarski, H.: 1991, 'Full Satisfaction Classes: A Survey', *Notre Dame Journal of Formal Logic* **32**, 573–579.

Kotlarski, H.: 1994, 'On the Incompleteness Theorems', *Journal of Symbolic Logic* **59**, 1414–1419.

Kotlarski, H. and Ratajczyk, Z.: 1990, 'Inductive Full Satisfaction Classes', *Annals of Pure and Applied Logic* **47**, 199–223.

Kotlarski, H. and Ratajczyk, Z.: 1990a, 'More on Induction in the Language With a Full Satisfaction Class', *Zeitschrift für Mathematische Logik und Grundlagen der Mathematik* **36**, 441–454.

Krajewski, S.: 1976, 'Non-Standard Satisfaction Classes'. In: Marek, W. *et al.* (Eds.) *Set Theory and Hierarchy Theory*, Proc. Bierutowice Conf. 1975, Lecture Notes in Mathematics 537, Springer-Verlag, Berlin-Heidelberg-New York, pp. 121–144.

Krajewski, S.: 1983, 'Philosophical Consequences of Gödel's Theorems', *Bulletin of the Section of Logic* **12**, 157–164.

Krajewski, S.: 1993, 'Did Gödel Prove That we Are Not Machines?' In: Wolkowski, Z.W. (Ed.) *First International Symposium on Gödel's Theorems*, World Scientific, Singapore-New Jersey-London-Hong Kong, pp. 39–49.

Krajíček, J: 1995, *Bounded Arithmetic, Propositional Logic and Complexity Theory*, Cambridge University Press, Cambridge.

Krajíček, J.: 1996, 'A Fundamental Problem of Mathematical Logic', *Colloquium Logicum* **2**, 56–64.

Kreisel, G.: 1958, 'Hilbert's Programme', *Dialectica* **12**, 346–372. Revised, with Postscript in Benacerraf, P. and Putnam, H. (Eds.) *Philosophy of Mathematics. Selected Readings*, Prentice-Hall, Englewood Cliffs, New Jersey, 1964, pp. 157–180.

Kreisel, G.: 1968, 'A Survey of Proof Theory', *Journal of Symblic Logic* **33**, 321–388.

Kreisel, G.: 1976, 'What Have We Learned From Hilbert's Second Problem'. In: Browder, F. (Ed.) *Mathematical Developments Arising from Hilbert's Problems*, Proceedings of the Symposia in Pure Mathematics 28, American Mathematical Society, Providence, RI, 1976, pp. 93–130.

Kreisel, G.: 1980, 'Kurt Gödel 1906–1978', *Biographical Memoires of Fellows of the Royal Society* **26**, 149–224.

Kreisel, G., and Krivine, J.L.: 1967, *Éléments de logique mathématique*, Dunod, Paris. English translation: *Elements of Mathematical Logic*, North-Holland Publ. Comp., Amsterdam 1967.

Kreisel, G., and Takeuti, G.: 1974, 'Formally Self-Referential Propositions for Cut-Free Analysis and Related Systems', *Dissertationes Mathematicae* **118**, 4–50.

Kreisel, G., and Wang, Hao: 1955, 'Some Applications of Formalized Consistency Proofs', *Fundamenta Mathematicae* **42**, 101–110; II (Addition to the preceding) *Fundamenta Mathematicae* **45**, 334–335.

Kuczyński, J.: 1938, 'O twierdzeniu Gödla' (On Gödel's Theorem), *Kwartalnik Filozoficzny* **14**, 74–80.

Kuratowski, K.: 1948, *Topologie*, vol. I, Państwowe Wydawnictwo Naukowe, Warszawa.

Ladrière, J.: 1957, *Les limitations internes des formalismes. Étude sur la signification du théorème de Gödel et des théorèmes apparenté dans la théorie des fondements des mathématiques*, E. Nauwelaerts, Louvain and Gauthier-Villars, Paris.

La Mettrie, J.O. de: 1748, *L'homme-machine*. Critical edition, A. Vartanian (Ed.), Paris 1960.

Langford, C.H.: 1927, 'Some Theorems on Deducibility', *Annals of Mathematics*, 2nd ser., **28**, 16–40.

Langford, C.H.: 1927a, 'Theorems on Deducibility (Second paper)', *Annals of Mathematics*, 2nd ser., **28**, 459–471.

Läuchli, H. and Leonard, J.: 1966, 'On the Elementary Theory of Linear Order', *Fundamenta Mathematicae* **59**, 109–116.

Lewis, C.I. and Langford, C.H.: 1932, *Symbolic Logic*, The Century Co., New York.

Lindström, P.: 1997, *Aspects of Incompleteness*, Springer-Verlag, Berlin.

Löb, M.H.: 1955, 'Solution to a Problem of Leon Henkin', *Journal of Symbolic Logic* **20**, 115–118.

Löwenheim, L.: 1915, 'Über Möglichkeiten im Relativkalkül', *Mathematische Annalen* **76**, 447–470. English translation: 'On Possibilities in the Calculus of Relatives' in: Heijenoort, J. van (Ed.) *From Frege to Gödel. A Source Book in Mathematical Logic, 1879–1931*, Harvard University Press, Cambridge, Mass., 1967, pp. 228–251.

Lucas, J.R.: 1961, 'Minds, Machines and Gödel', *Philosophy* **36**, 112–127. Reprinted in: Anderson, A.R. (Ed.) *Minds and Machines*, Prentice Hall, Englewood Cliffs, 1964.

Lucas, J.R.: 1967, 'Human and Machine Logic. A Rejoinder', *British Journal for the Philosophy of Science* **19**, 155–156.

Lucas, J.R.: 1968, 'Satan Stultified: A Rejoinder to Paul Benacerraf', *The Monist* **52**, 145–158.

Łoś, J.: 1954, 'On the Categoricity In Power of Elementary Deductive Systems and Some Related Problems', *Colloquium Mathematicum* **3**, 58–62.

Marker, D.: 1996, 'Model Theory and Exponentiation', *Notices of the American Mathematical Society* **43**, 753–759.

Markov, A.A.: 1951, 'Teorya algorifmov', *Trudy Matem. Inst. AN SSSR im. W.A. Steklova* **38**, 176–189.

Markov, A.A.: 1954, 'Teorya algorifmov', *Trudy Matem. Inst. AN SSSR im. W.A. Steklova* **42**, Moscow, pp. 1–375; English translation: *The Theory of Algorithms*, Office of Technical Services, U.S. Department of Commerce, Washington, D.C., 1962.

Matiyasevich, Y.V.: 1970, 'Diofantovost' pieretchislimych množestv' (Recursively Enumerable Sets Are Diophantine), *Doklady Akademii Nauk SSSR* **191 (2)**, 279–282.

McAloon, K.: 1979, 'Formes combinatoires du théorème d'incomplétude (d'après J. Paris et d'autres)'. In: *Séminaire Bourbaki*, vol. 1977/78, Exposés 507–524, Springer-Verlag, Berlin-Heidelberg-New York, pp. 263–276.

McNulty, G.F.: 1986, 'Alfred Tarski and Undecidable Theories', *Journal of Symbolic Logic* **51**, 890–898.

Mendelson, E.: 1964, *Introduction to Mathematical Logic*, D. Van Nostrand Company, Inc., Princeton-Toronto-New York-London.

Mendelson, E.: 1990, 'Second Thoughts About Church's Thesis and Mathematical Proofs', *Journal of Philosophy*, **88**, 225–233.

Menger, K.: 1978, *Selected Papers in Logic and Foundations, Didactics, Economics*, D. Reidel Publ. Comp., Dordrecht-Boston-London.

Meyer, A.R.: 1975, 'The Inherent Complexity of Theories of Ordered Sets'. In: *Proceedings of the International Congress of Mathematicians, Vancouver 1974*, vol. 2, pp. 477–482.

Montague, R.: 1957, *Contributions to the Axiomatic Foundations of Set Theory*, dissertation, Berkeley.

Montague, R.: 1957a, 'Non-finite Axiomatizability', Summaries of talks presented at the Summer Institute of Symbolic Logic in 1957 at Cornell University, vol. 2, pp. 256–259 (mimeographed).

Montague, R.: 1962, 'Theories Incomparable With Respect to Relative Interpretability', *Journal of Symbolic Logic* **27**, 195–211.

Moore, G.H.: 1980, 'Beyond First-Order Logic: The Historical Interplay Between Mathematical Logic and Axiomatic Set Theory', *History and Philosophy of Logic* **1**, 95–137.

Moore, G.H.: 1991, 'Sixty Years After Gödel', *The Mathematical Intelligencer* **13**, 6–11.

Mostowski, A.: 1947, 'On Definable Sets of Positive Integers', *Fundamenta Mathematicae* **34**, 81–112. Reprinted in: Mostowski, A. *Foundational Studies. Selected Works*, vol. I, PWN – Polish Scientific Publishers, Warszawa and North-Holland Publ. Comp., Amsterdam-New York-Oxford, 1979, pp. 339–370.

Mostowski, A.: 1952, 'On Models of Axiomatic Systems', *Fundamenta Mathematicae* **39**, 133–158. Reprinted in: Mostowski, A. *Foundational Studies. Selected Works*, vol. II, PWN – Polish Scientific Publishers, Warszawa and North-Holland Publ. Comp., Amsterdam-New York-Oxford, 1979, pp. 198–217.

Mostowski, A.: 1955, 'Examples of Sets Definable by Means of Two or Three Quantifiers', *Fundamenta Mathematicae* **42**, 259–270. Reprinted in: Mostowski, A. *Foundational Studies. Selected Works*, vol. II, PWN – Polish Scientific Publishers, Warszawa and North-Holland Publ. Comp., Amsterdam-New York-Oxford, 1979, pp. 275–286.

Mostowski, A.: 1965, 'Thirty Years of Foundational Studies. Lectures on the Development of Mathematical Logic and the Study of the Foundations of Mathematics in 1930–1964', *Acta Philosophica Fennica* **XVII**, 1–180. Reprinted in: Mostowski, A. *Foundational Studies. Selected Works*, vol. I, PWN – Polish Scientific Publishers, Warszawa and North-Holland Publ. Comp., Amsterdam-New York-Oxford, 1979, pp. 1–176.

Mostowski, A., in collaboration with A. Grzegorczyk, S. Jaśkowski, J. Łoś, S. Mazur, H. Rasiowa and R. Sikorski: 1955, 'The Present State of Investigations On the Foundations of Mathematics', *Rozprawy Matematyczne* **9**, 1–48.

Mostowski, A., Robinson, R.M. and Tarski, A.: 1953, 'Undecidability and Essential Undecidability in Arithmetic'. In: Tarski, A., Mostowski, A. and Robinson, R.M., *Undecidable Theories*, North-Holland Publ. Comp., Amsterdam 1953, pp. 37–74.

Mostowski, A. and Tarski, A.: 1949, 'Undecidability in the Arithmetic of Integers and in the Theory of Rings', *Journal of Symbolic Logic* **14**, 76.

Murawski, R.: 1976–77, 'On Expandability of Models of Peano Arithmetic. I–III', *Studia Logica* **35** (1976), 409–419; **35** (1976), 421–431; **36** (1977), 181–188.

Murawski, R.: 1984, 'Expandability of Models of Arithmetic'. In: Wechsung, G. (Ed.) *Proceedings of Frege Conference 1984*, Akademie-Verlag, Berlin, pp. 87–93.

Murawski, R.: 1985, *Giuseppe Peano — Pioneer and Promoter of Symbolic Logic*, Komunikaty i Rozprawy Instytutu Matematyki Uniwersytetu im. Adama Mickiewicza, Poznań.

Murawski, R.: 1987, 'Giuseppe Peano a rozwój logiki symbolicznej' (Giuseppe Peano and The Development of Symbolic Logic), *Roczniki Polskiego Towarzystwa Matematycznego, Seria II: Wiadomości Matematyczne* **27**, 261–277.

Murawski, R.: 1988, 'Semantics for Nonstandard Languages', *Reports on Mathematical Logic* **22**, 105–114.

Murawski, R.: 1996, 'Contribution of Polish Logicians to Decidability Theory', *Modern Logic* **6**, 37–66.

Murawski, R.: 1997, 'Satisfaction Classes — A Survey'. In: Murawski, R. and Pogonowski, J. (Eds.) *Euphony and Logos*, Editions Rodopi, Amsterdam–Atlanta, GA, pp. 259–281.

Murawski, R.: 1998, 'Undefinability of Truth. The Problem of Priority: Tarski vs. Gödel', *History and Philosophy of Logic*, **19**, 153–160.

Murawski, R.: 199?, 'E.L. Post and the Development of Mathematical Logic and Recursion Theory', *Studies in Logic, Grammar and Retoric* (to appear).

Murawski, R.: 199?a, 'Leibniz's and Kant's Philosophical Ideas and the Development of Hilbert's Programme'. In: Perzanowski, J. (Ed.) *Being, Logos and Mathematics*, Editions Rodopi, Amsterdam-Atlanta, GA (to appear).

Mycielski, J.: 1983, 'The Meaning of the Conjecture $P \neq NP$ For Mathematical Logic', *American Mathematical Monthly* **90**, 129–130.

Mycielski, J.: 1985, 'What is a Mathematical Theory?', *Revista Colombiana de Matemáticas* **19**, 173–176.

Mycielski, J.: 1986, 'Locally Finite Theories', *Journal of Symbolic Logic* **51**, 59–62.

Mycielski, J., Pudlák, P. and Stern, A.S.: 1990, 'A Lattice of Chapters of Mathematics (Interpretations Between Theorems)', *Memoires of the American Mathematical Society* **84**, No 426, pp. 1–70.

Nagel, E. and Newman, J.R.: 1958, *Gödel's Proof*, Routledge & Kegan Paul, Ltd., London.

Neumann von, J.: 1927, 'Zur Hilbertschen Beweistheorie', *Mathematische Zeitschrift* **26**, 1–46. Reprinted in: *Collected Works*, vol. I: *Logic, Theory of Sets, and Quantum Mechanics*, ed. by A.H. Taub, Pergamon Press, Oxford 1961, pp. 256–300.

Odifreddi, P.: 1989, *Classical Recursion Theory*, North-Holland Publishing Company, Amsterdam.

Paris, J. and Harrington, L.: 1977, 'A Mathematical Incompleteness in Peano Arithmetic'. In: Barwise, J. (Ed.) *Handbook of Mathematical Logic*, North-Holland Publ. Comp., Amsterdam, pp. 1133–1142.

Parsons, C.: 1958, 'Elimination of a Restricted Infinite Induction', abstract, *Journal of Symbolic Logic* **23**, 106.

Peano, G.: 1889, *Arithmetices principia nova methodo exposita*, Bocca, Torino. English translation: 'The Principles of Arithmetic, Presented by a New Method', in: Heijenoort, J. van (Ed.) *From Frege to Gödel. A Source Book in Mathematical Logic, 1879–*

1931, Harvard University Press, Cambridge, Mass., 1967, pp. 83–97.

Peckhaus, V.: 1990, *Hilbertprogramm und Kritische Philosophie. Das Göttinger Model interdisziplinärer Zusammenarbeit zwischen Mathematik und Philosophie*, Vandenhoeck & Ruprecht, Göttingen.

Penrose, R.: 1989, *The Emperor's New Mind. Concerning Computers, Minds, and the Laws of Physics*, Oxford University Press, Oxford.

Penrose, R.: 1994, *Shadows of the Mind: a Search for the Missing Science of Consciousness*, Oxford University Press, Oxford-New York-Melbourne.

Pepis, J.: 1936, 'Beiträge zur Reduktionstheorie des logischen Entscheidungsproblems', *Acta Sc[ientiarum] Math[ematicarum]* (*Szeged*) **8**, Heft 1, 7–41.

Pepis, J.: 1937, 'O zagadnieniu rozstrzygalności w zakresie węższego rachunku funkcyjnego' (On the Decision Problem For Pure Functional Calculus), *Archiwum Towarzystwa Naukowego we Lwowie, Dział III* [*matematyczno-przyrodniczy*], tom VII, zeszyt 8, 1–172.

Pepis, J.: 1938, 'Untersuchungen über das Entscheidungsproblem der mathematischen Logik', *Fundamenta Mathematicae* **30**, 257–348.

Pepis, J.: 1938a, 'Ein Verfahren der mathematischen Logik', *Journal of Symbolic Logic* **3**, 61–76.

Perelman, Ch.: 1936, 'L'Antinomie de M. Gödel', *Academie Royale de Belgique, Bulletin de la Classe des Sciences*, Series 5, **22**, 730–736.

Péter, R.: 1934, 'Über den Zusammenhang der verschiedenen Begriffe der rekursiven Funktionen', *Mathematische Annalen* **110**, 612–632.

Péter, R.: 1935, 'Konstruktion nichtrekursiver Funktionen', *Mathematische Annalen* **111**, 42–60.

Péter, R.: 1937, 'Über die mehrfache Rekursion', *Mathematische Annalen* **113**, 489–527.

Péter, R.: 1951, *Rekursive Funktionen*, Akadémiai Kiadó, Budapest. English translation: *Recursive Functions*, Academic Press, New York 1967.

Post, E.: 1943, 'Formal Reductions of the General Combinatorial Decision Problem', *American Journal of Mathematics* **65**, 197–215.

Post, E.: 1944, 'Recursively Enumerable Sets of Positive Integers and Their Decision Problem', *Bulletin of the American Mathematical Society*, **50**, 284–316. Reprinted in: *Solvability, Provability, Definability. The Collected Works of Emil L. Post*, ed. by Davis, M., Birkhäuser, Boston-Basel-Berlin 1994, pp. 461–494.

Post, E.: 1965, 'Absolutely Unsolvable Problems and Relatively Undecidable Propositions. Account of an Anticipation'. In: Davis, M. (Ed.) *The Undecidable: Basic Papers on Undecidable Propositions, Unsolvable Problems, and Computable Functions*, Raven Press, Hewlett, N.Y., 1965, pp. 340–433.

Prawitz, D.: 1981, 'Philosophical Aspects of Proof Theory'. In: Fløistad, G. (Ed.) *Contemporary Philosophy. A New Survey*, Martinus Nijhoff Publishers, The Hague-Boston-London, pp. 235–277.

Presburger, M.: 1930, 'Über die Vollständigkeit eines gewissen Systems der Arithmetik ganzer Zahlen, in welchem die Addition als einzige Operation hervortritt'. In: *Sprawozdanie z I Kongresu matematyków krajów słowiańskich, Warszawa 1929 — Comptes-Rendus du I Congrès des Mathématiciens des Pays Slaves, Varsovie 1929*, Warszawa, 92–101, 395. English translation: 'On the Completeness of a Certain System of Arithmetic of Whole Numbers in Which Addition Occurs as the Only Operation', *History and Philosophy of Logic* **12**, 225–232.

Pudlák, P.: 1979, 'Another Combinatorial Sentence Independent of Peano's Axioms', unpublished notes.

Pudlák, P.: 1995, 'Logic and Complexity: Independence Results and the Complexity of Propositional Calculus. In: Chatterji, S.D. (Ed.) *Proceedings of the International Congress of Mathematicians, Zürich, Switzerland 1994*, Birkhäuser Verlag, Basel, pp. 288–297.

Putnam, H.: 1975, 'What Is Mathematical Truth?'. In: H. Putnam, *Mathematics, Matter and Method. Philosophical Papers*, vol. I, Cambridge University Press, Cambridge-London-New York-Melbourne, pp. 60–78. Reprinted in: Tymoczko, Th. (Ed.)

New Directions in the Philosophy of Mathematics. An Anthology, Birkhäuser, Boston-Basel-Stuttgart 1985 (second revised and expanded edition: Princeton University Press, Princeton, NJ, 1998), pp. 50–65.

Rabin, M.: 1965, 'A Simple Method for Undecidability Proofs and Some Applications'. In: Bar-Hillel, Y. (Ed.) *Logic, Methodology and Philosophy of Science. Proceedings of the 1964 International Congress*, North-Holland Publ. Comp., Amsterdam, pp. 58–68.

Rabin, M.: 1969, 'Decidability of Second Order Theories and Automata on Infinite Trees', *Transactions of the American Mathematical Society* **141**, 1–35.

Ramsey, F.R.: 1929, 'On a Problem of Formal Logic', *Proc. London Math. Soc.* **(2) 30**, 264–286.

Reid, C.: 1970, *Hilbert*, Springer-Verlag, Berlin-Heidelberg-New York (new edition as part of Reid, 1986).

Reid, C.: 1986, *Hilbert-Courant*, Springer-Verlag, New York.

Reid, C.: 1996, *Julia. A Life in Mathematics*, The Mathematical Association of America, Washington, DC.

Resnik, M.D.: 1974, 'On the Philosophical Significance of Consistency Proofs', *Journal of Philosophical Logic* **3**, 133–147. Reprinted in: Shanker, S.G. (Ed.) *Gödel's Theorem in Focus*, Croom Helm, London (reprinted by Routledge, London and New York 1989), pp. 115–130.

Ritchie, R.W.: 1963, 'Classes of Predictably Computable Functions', *Transactions of the American Mathematical Society* **106**, 139–173.

Robinson A.: 1956, *Complete Theories*, North-Holland Publ. Comp., Amsterdam.

Robinson A.: 1963, 'On Languages Based on Nonstandard Arithmetic', *Nagoya Mathematical Journal* **22**, 83–107.

Robinson, J.: 1949, 'Definability and Decision Problems in Arithmetic', *Journal of Symbolic Logic* **14**, 98–114. Reprinted in: *The Collected Works of Julia Robinson*, ed. by Feferman, S., American Mathematical Society, Providence, RI, 1996, pp. 7–23.

Robinson, J.: 1950, 'General Recursive Functions', *Proceedings of the American Mathematical Society* **1**, 703–718. Reprinted in: *The Collected Works of Julia Robinson*, ed. by Feferman, S., American Mathematical Society, Providence, RI, 1996, pp. 25–40.

Robinson, J.: 1952, 'Existential Definability in Arithmetic', *Transactions of the American Mathematical Society* **72**, 437–449. Reprinted in: *The Collected Works of Julia Robinson*, ed. by Feferman, S., American Mathematical Society, Providence, RI, 1996, pp. 47–59.

Robinson, R.M.: 1947, 'Primitive Recursive Functions', *Bulletin of the American Mathematical Society* **53**, 915–942.

Robinson, R.M.: 1948, 'Recursion and Double Recursion', *Bulletin of the American Mathematical Society* **54**, 987–993.

Robinson, R.M.: 1951, 'Undecidable Rings', *Transactions of the American Mathematical Society* **70**, 137–159.

Rodríguez-Consuegra, F.A.: 1993, 'Russell, Gödel and Logicism'. In: Czermak, J. (Ed.) *Philosophy of Mathematics. Proceedings of the 15th International Wittgenstein-Symposium*, Part 1, Verlag Hölder-Pichler-Tempsky, Wien, pp. 233–242.

Ross, D.: 1974, 'Church's Thesis: What Its Difficulties Are and Are Not', *Journal of Philosophy* **71**, 515–525.

Rosser, J.B.: 1936, 'Extensions of Some Theorems of Gödel and Church', *Journal of Symbolic Logic* **1**, 87–91. Reprinted in: Davis, M. (Ed.) *The Undecidable: Basic Papers on Undecidable Propositions, Unsolvable Problems, and Computable Functions*, Raven Press, Hewlett, N.Y., 1965, pp. 230–235.

Rosser, J.B.: 1937, 'Gödel Theorems for Non-Constructive Logics', *Journal of Symbolic Logic* **2**, 129–137.

Rosser, J.B.: 1939, 'An Informal Exposition of Proofs of Gödel's Theorem and Church's Theorem', *Journal of Symbolic Logic* 4, 53–60. Reprinted in: Davis, M. (Ed.) *The Undecidable: Basic Papers on Undecidable Propositions, Unsolvable Problems, and Computable Functions*, Raven Press, Hewlett, N.Y., 1965, pp. 223–230.

Rucker, R.: 1982, *Infinity and the Mind*, Birkhäuser, Boston-Basel-Stuttgart.

Russell, B.: 1901, 'Recent Work in the Philosophy of Mathematics', *The International Monthly* **4**, 83–101. Reprinted as 'Mathematics and the Metaphysicians' in: *Mysticism and Logic and Other Essays*, Doubleday, New York 1914, pp. 70–92.

Russell, B.: 1959, *My Philosophical Development*, George Allen & Unwin Ltd., London.

Ryll-Nardzewski, C.: 1953, 'The Role of the Axiom of Induction in Elementary Arithmetic', *Fundamenta Mathematicae* **39**, 239–263.

Schultze, F.: 1882, *Philosophie der Naturwissenschaften.*

Schütte, K.: 1977, *Proof Theory*, Springer-Verlag, Berlin-Heidelberg-New York.

Schütte, K. and Simpson, S.G.: 1985, 'Ein in der reinen Zahlentheorie unbeweisbarer Satz über endliche Folgen von natürlichen Zahlen', *Archiv für Mathematische Logik und Grundlagenforschung*, **25**, 75–89.

Shanker, S.G.: 1988, 'Wittgenstein's Remarks on the Significance of Gödel's Theorem'. In: Shanker, S.G. (Ed.) *Gödel's Theorem in Focus*, Croom Helm, London (reprinted by Routledge, London and New York 1989), pp. 155–256.

Shapiro, S.: 1981, 'Understanding Church's Thesis', *Journal of Philosophical Logic* **10**, 353–365.

Shapiro, S.: 1993, 'Understanding Church's Thesis, Again', *Acta Analytica*, **11**, 59–77.

Shoenfield, J.R.: 1959, 'On a Restricted ω-Rule', *Bulletin de l'Académie Polonaise des Sciences. Serie des sci. math., astr. et phys.* **7**, 405–407.

Sieg, W.: 1988, 'Hilbert's Program Sixty Years Later', *Journal of Symbolic Logic* **53**, 338–348.

Sieg, W.: 1994, 'Mechanical Procedures and Mathematical Experience'. In: George, A. (Ed.) *Mathematics and Mind*, Oxford University Pess, Oxford, pp. 71–117.

Sieg, W.: 1997, 'Step by Recursive Step: Church's Analsis of Effective Calculability', *Bulletin of Symbolic Logic*, **3**, 154–180.

Sigmund, K.: 1995, 'A Philosopher's Mathematician: Hans Hahn and the Vienna Circle', *The Mathematical Intelligencer* **17**, 16–29.

Simpson, S.G.: 1984, 'Which Set Existence Axioms Are Needed to Prove the Cauchy/Peano Theorem for Ordinary Differential Equations?', *Journal of Symbolic Logic* **49**, 783–802.

Simpson, S.G.: 1985, 'Friedman's Research on Subsystems of Second Order Arithmetic'. In: Harrington, L.A. *et al.* (Eds.) *Harvey Friedman's Research in the Foundations of Mathematics*, North-Holland Publ. Comp., Amsterdam, pp. 137–159.

Simpson, S.G.: 1985a, 'Reverse Mathematics', *Proceedings of Symposia in Pure Mathematics* **42**, 461–471.

Simpson, S.G.: 1985b, 'Nichtbeweisbarkeit von gewissen kombinatorischen Eigenschaften endlicher Bäume', *Archiv für Mathematische Logik und Grundlagenforschungen* **25**, 45–65.

Simpson, S.G.: 1985c, 'Nonprovability of certain combinatorial properties of finite trees'. In: Harrington, L.A. *et al.* (Eds.) *Harvey Friedman's Research on the Foundations of Mathematics*, North-Holland Publ. Comp., Amsterdam, pp. 87–117.

Simpson, S.G.: 1987, 'Subsystems of Z_2 and Reverse Mathematics', Appendix to Takeuti, G. *Proof Theory*, second edition, North-Holland Publ. Comp., Amsterdam 1987, pp. 432–446.

Simpson, S.G.: 1987a, 'Unprovable Theorems and Fast-Growing Functions'. In: Simpson, S.G. (Ed.) *Logic and Combinatorics. Proceedings of a Summer Research Conference held August 4–10, 1985*, Contemporary Mathematics, vol. 65, American Mathematical Society, Providence, Rhode Island, pp. 359–394.

Simpson, S.G.: 1988, 'Partial Realizations of Hilbert's Program', *Journal of Symbolic Logic* **53**, 349–363.

Simpson, S.G.: 1988a, 'Ordinal Numbers and the Hilbert's Basis Theorem', *Journal of Symbolic Logic* **53**, 961–974.

Simpson, S.G.: 1998, *Subsystems of Second Order Arithmetic*, Springer-Verlag, Berlin.

Skolem, T.: 1919, 'Untersuchungen über die Axiome des Klassenkalküls und über Produktions- und Summationsprobleme, welche gewisse Klassen von Aussagen betreffen', *Videnskapsselskapets skrifter, I. Matematisk-naturvidenskabelig Klasse* **3**, 30–37.

Skolem, T.: 1923, 'Begründung der elementaren Arithmetik durch die rekurrierende Denkweise ohne Anwendung scheinbarer Veränderlichen mit unendlichem Ausdehnungsbereich', *Videnskapsselskapets skrifter, I. Matematisk-naturvidenskabelig Klasse* **6**, 38 pp. English translation: 'The Foundations of Elementary Arithmetic Established by Means of the Recursive Mode of Thought, without the Use of Apparent Variables Ranging over Infinite Domains', in: Heijenoort, J. van (Ed.) *From Frege to Gödel. A Source Book in Mathematical Logic, 1879–1931*, Harvard University Press, Cambridge, Mass., 1967, pp. 302–333.

Skolem, T.: 1930, 'Über einige Satzfunktionen in der Arithmetik', *Skriften utgit av videnskasselskapet i Kristina*, I Klassse, No 7, Oslo. Also in: T. Skolem, *Selected Works in Logic*, ed. by Fenstad, J.E., Oslo 1970, pp. 281–306.

Smart, J.J.C.: 1961, 'Gödel's Theorem, Church's Thesis, and Mechanism', *Synthese* **13**, 105–110.

Smoryński, C.: 1977, 'The Incompleteness Theorems'. In: Barwise, J. (Ed.) *Handbook of Mathematical Logic*, North-Holland Publ. Comp., Amsterdam, pp. 821–865.

Smoryński, C.: 1980, *Nonstandard Models of Arithmetic*, University of Utrecht, Preprint Nr 153, Utrecht.

Smoryński, C.: 1980a, 'Some Rapidly Growing Functions', *The Mathematical Intelligencer* **2**, 149–154.

Smoryński, C.: 1981, 'Fifty Years of Self-Reference in Arithmetic', *Notre Dame Journal of Formal Logic* **22**, 357–374.

Smoryński, C.: 1982, 'The Varieties of Arboreal Experience', *The Mathematical Intelligencer* **4**, 182–189.

Smoryński, C.: 1985, *Self-Reference and Modal Logic*, Springer-Verlag, New York-Berlin-Heidelberg-Tokyo.

Smoryński, C.: 1988, 'Hilbert's Programme', *CWI Quarterly* **1**, 3–59.

Smullyan, R.M.: 1992, *Gödel's Incompleteness Theorems*, Oxford University Press, New York-Oxford.

Smullyan, R.M.: 1996, *Diagonatization and Self-Reference*, Clarendon Press, Oxford.

Soare, R.I.: 1996, 'Computability and Recursion', *Bulletin of Symbolic Logic*, **2**, 284–321.

Sudan, G.: 1927, 'Sur le nombre transfini ω^{ω}', *Bulletin Mathématique de la Société Roumaine des Sciences* **30**, 11–30.

Sudkamp, Th.A.: 1988, *Languages and Machines. An Introduction to the Theory of Computer Science*, Addison-Wesley Publ. Comp., Reading, Massachusetts.

Szmielew, W.: 1949, 'Decision Problems in Group Theory', In: *Proceedings of the Xth International Congress of Philosophy*, vol. 1, Fascicule 2, Amsterdam, pp. 763–766.

Szmielew, W.: 1949a, 'Arithmetical Classes and Types of Abelian Groups', *Bulletin of the American Mathematical Society* **55**, 65; errata, 1192.

Szmielew, W.: 1950, *Arithmetical Properties of Abelian Groups*, doctoral dissertation, University of California, Berkeley.

Szmielew, W.: 1955, 'Elementary Properties of Abelian Groups', *Fundamenta Mathematicae* **41**, 203–271.

Szmielew, W. and Tarski, A.: 1952, 'Mutual Interpretability of Some Essentially Undecidable Theories'. In: *Proceedings of the International Congress of Mathematicians, Cambridge, Massachusetts, August 30–September 6, 1950*, vol. 1, American Mathematical Society, Providence, RI, p. 734.

Tait, W.W.: 1981, 'Finitism', *Journal of Philosophy* **78**, 524–546.

Takeuti, G.: 1987, *Proof Theory*, second edition, North-Holland Publ. Comp., Amsterdam.

Tarski, A.: 1933, *Pojęcie prawdy w językach nauk dedukcyjnych* (The Notion of Truth in Languages of Deductive Sciences), Nakładem Towarzystwa Naukowego Warszawskiego, Warszawa. English translation: 'The Concept of Truth in Formalized Languages', in: *Logic, Semantics, Metamathematics: Papers From 1923 To 1938 by Alfred Tarski*, Oxford University Press, Oxford 1956; second edition with corrections and emendations, ed. by J. Corcoran, Hackett, Indianapolis, IN, 1983, pp. 152–278.

Tarski, A.: 1933a, 'Einige Betrachtungen über die Begriffe ω-Widerspruchsfreiheit und der ω-Vollständigkeit', *Monatshefte für Mathematik und Physik*, **40**, 97–112. English translation: 'Some Observations on the Concepts of ω-consistency and ω-completeness', in: *Logic, Semantics, Metamathematics: Papers From 1923 To 1938 by Alfred Tarski*, Oxford University Press, Oxford 1956; second edition with corrections and emendations, ed. by J. Corcoran, Hackett, Indianapolis, IN, 1983, pp. 279–295.

Tarski, A.: 1936, 'Der Wahrheitsbegriff in den formalisierten Sprachen', *Studia Philosophica* **1**, 261–405 (offprints dated 1935).

Tarski, A.: 1948, *A Decision Method for Elementary Algebra and Geometry*, prepared for publication by J.C.C. McKinsey, U.S. Air Force Project RAND, R-109, the RAND Corporation, Santa Monica, California.

Tarski, A.: 1949, 'On essential undecidability', *Journal of Symbolic Logic* **14**, 75.

Tarski, A.: 1949a, 'Arithmetical Classes and Types of Boolean Algebras (Preliminary Report)', *Bulletin of the American Mathematical Society* **55**, 64.

Tarski, A.: 1949b, 'Undecidability of Group Theory', *Journal of Symbolic Logic* **14**, 76–77.

Tarski, A.: 1949c, 'Undecidability of the Theories of Lattices and Projective Geometries', *Journal of Symbolic Logic* **14**, 77–78.

Tarski, A.: 1951, *A Decision Method for Elementary Algebra and Geometry*, second revised edition, University of California Press, Berkeley and Los Angeles.

Tarski, A.: 1953, 'A General Method in Proofs of Undecidability'. In: Tarski,A., Mostowski, A. and Robinson, R.M., *Undecidable Theories*, North-Holland Publ. Comp., Amsterdam 1953, pp. 1–35.

Tarski, A.: 1953a, 'Undecidability of the Elementary Theory of Groups'. In: Tarski,A., Mostowski, A. and Robinson, R.M., *Undecidable Theories*, North-Holland Publ. Comp., Amsterdam 1953, pp. 75–87.

Tarski, A.: 1956, 'The Concept of Truth in Formalized Languages'. In: *Logic, Semantics, Metamathematics. Papers From 1923 To 1938*, Clarendon Press, Oxford, pp. 152–278.

Tarski, A.: 1956a, 'Some Observations on the Concept of ω-Consistency and ω-Completeness'. In: *Logic, Semantics, Metamathe-*

matics. Papers From 1923 To 1938, Clarendon Press, Oxford, pp. 279–295.

Tarski, A.: 1959, 'What is Elementary Geometry?'. In Henkin, L. et al. (Eds.) *Axiomatic Method, with Special Reference to Geometry and Physics*, North-Holland Publ. Comp., Amsterdam, pp. 16–29.

Tarski, A.: 1967, *The Completeness of Elementary Algebra and Geometry*, Institut Blaise Pascal, Paris (a reprint from page proofs of a work scheduled to appear in 1940 in *Actualités Scientifiques et Industrielles*, Hermann & Cie, Paris, but which did not appear due to the wartime conditions).

Tarski, A.: 1995, 'Some Current Problems in Metamathematics', ed. by J. Tarski and J. Woleński, *History and Philosophy of Logic* **16**, 159–168.

Tarski, A. and Mostowski, A.: 1949, 'Arithmetical Classes and Types of Well Ordered Systems', *Bulletin of the American Mathematical Society* **55**, 65.

Thiel, Ch.: 1980, 'Cantorsches Diagonalverfahren'. In: Mittelstraß, J. (Hrsg.) *Enzyklopedie Philosophie und Wissenschaftstheorie*, Band 1: A–G, Bibliographisches Institut, Mannheim-Wien-Zürich, p. 374.

Thiel, Ch.: 1995, *Philosophie und Mathematik. Eine Einführung in ihre Wechselwirkung und in die Philosophie der Mathematik*, Wissenschaftliche Buchgesellschaft, Darmstadt.

Turing, A.: 1936, 'On Computable Numbers with an Application to the Entscheidungsproblem', *Proc. London Math. Soc.*, ser. 2, **42**, 230–265. Reprinted in: Davis, M. (Ed.) *The Undecidable: Basic Papers on Undecidable Propositions, Unsolvable Problems, and Computable Functions*, Raven Press, Hewlett, N.Y., 1965, pp. 115–154.

Turing, A.: 1939, 'Systems of Logic Based on Ordinals', *Proc. London Math. Soc.*, ser. 2, **45**, 161–228.

Turing, A.: 1950, 'Computing Machinery and Intelligence', *Mind* **59**, 433–460. Reprinted in: Anderson, A.R. (Ed.) *Minds and Machines*, Prentice Hall, Englewood Cliffs 1964, pp. 4–42.

Tymoczko, Th. (Ed.): 1985, *New Directions in the Philosophy of Mathematics. An Anthology*, Birkhäuser, Boston-Basel-Stuttgart (second revised and expanded edition: Princeton University Press, Princeton, NJ, 1998).

Vaught, R.L.: 1953, 'Applications of the Generalized Skolem-Löwenheim Theorem to Problems of Completeness and Decidability', *Bulletin of the American Mathematical Society* **59**, 396–397.

Vinnikov, V.: 1999, 'We Shall Know: Hilbert's Apology', *The Mathematical Intelligencer* **21**, 42–46.

Visser, A.: 1989, 'Peano's Smart Children: A Provability Logical Study of Systems with Built-In Consistency', *Notre Dame Journal of Formal Logic* **30**, 161–196.

Wang, Hao: 1957, 'The Axiomatization of Arithmetic', *Journal of Symbolic Logic* **22**, 145–158.

Wang, Hao: 1974, *From Mathematics to Philosophy*, Routledge and Kegan Paul, London.

Wang, Hao: 1987, *Reflections on Kurt Gödel*, M.I.T. Press, Cambridge, Mass.

Wang, Hao: 1991, 'Imagined Discussions with Gödel and with Wittgenstein', Address to the Kurt Gödel Society Meeting at Kirchberg am Wechsel, Austria.

Wang, Hao: 1991a, 'To and From Philosophy — Discussions with Gödel and Wittgenstein', *Synthese* **88**(2), 229–277.

Webb, J.C.: 1980, *Mechanism, Mentalism and Metamathematics: An Essay on Finitism*, D. Reidel Publ. Comp., Dordrecht-Boston-London.

Whitehead, A.N. and Russell, B.: 1910–1913, *Principia Mathematica*, Cambridge University Press, Cambridge, vol. I 1910, vol. II 1912, vol. III 1913.

Wittgenstein, L.: 1956, *Remarks on the Foundations of Mathematics*, Basil Blackwell, Oxford.

Woleński, J.: 1989, *Logic and Philosophy in the Lvov-Warsaw School*, Kluwer Academic Publishers, Dordrecht-Boston-London.

Wolkowski, Z.W. (Ed.): 1993, *First International Symposium on Gödel's Theorems*, World Scientific, Singapore-New Jersey-London-Hong Kong.

Zygmunt, J.: 1991, 'Mojżesz Presburger: Life and Work', *History and Philosophy of Logic* **12**, 211–223.

List of Symbols

Index

SYNTHESE LIBRARY

162. M. A. Stewart (ed.), *Law, Morality and Rights.* 1983 ISBN 90-277-1519-X
163. D. Mayr and G. Süssmann (eds.), *Space, Time, and Mechanics.* Basic Structures of a Physical Theory. 1983 ISBN 90-277-1525-4
164. D. Gabbay and F. Guenthner (eds.), *Handbook of Philosophical Logic.* Vol. I: Elements of Classical Logic. 1983 ISBN 90-277-1542-4
165. D. Gabbay and F. Guenthner (eds.), *Handbook of Philosophical Logic.* Vol. II: Extensions of Classical Logic. 1984 ISBN 90-277-1604-8
166. D. Gabbay and F. Guenthner (eds.), *Handbook of Philosophical Logic.* Vol. III: Alternative to Classical Logic. 1986 ISBN 90-277-1605-6
167. D. Gabbay and F. Guenthner (eds.), *Handbook of Philosophical Logic.* Vol. IV: Topics in the Philosophy of Language. 1989 ISBN 90-277-1606-4
168. A. J. I. Jones, *Communication and Meaning.* An Essay in Applied Modal Logic. 1983 ISBN 90-277-1543-2
169. M. Fitting, *Proof Methods for Modal and Intuitionistic Logics.* 1983 ISBN 90-277-1573-4
170. J. Margolis, *Culture and Cultural Entities.* Toward a New Unity of Science. 1984 ISBN 90-277-1574-2
171. R. Tuomela, *A Theory of Social Action.* 1984 ISBN 90-277-1703-6
172. J. J. E. Gracia, E. Rabossi, E. Villanueva and M. Dascal (eds.), *Philosophical Analysis in Latin America.* 1984 ISBN 90-277-1749-4
173. P. Ziff, *Epistemic Analysis.* A Coherence Theory of Knowledge. 1984 ISBN 90-277-1751-7
174. P. Ziff, *Antiaesthetics.* An Appreciation of the Cow with the Subtile Nose. 1984 ISBN 90-277-1773-7
175. W. Balzer, D. A. Pearce, and H.-J. Schmidt (eds.), *Reduction in Science.* Structure, Examples, Philosophical Problems. 1984 ISBN 90-277-1811-3
176. A. Peczenik, L. Lindahl and B. van Roermund (eds.), *Theory of Legal Science.* Proceedings of the Conference on Legal Theory and Philosophy of Science (Lund, Sweden, December 1983). 1984 ISBN 90-277-1834-2
177. I. Niiniluoto, *Is Science Progressive?* 1984 ISBN 90-277-1835-0
178. B. K. Matilal and J. L. Shaw (eds.), *Analytical Philosophy in Comparative Perspective.* Exploratory Essays in Current Theories and Classical Indian Theories of Meaning and Reference. 1985 ISBN 90-277-1870-9
179. P. Kroes, *Time: Its Structure and Role in Physical Theories.* 1985 ISBN 90-277-1894-6
180. J. H. Fetzer, *Sociobiology and Epistemology.* 1985 ISBN 90-277-2005-3; Pb 90-277-2006-1
181. L. Haaparanta and J. Hintikka (eds.), *Frege Synthesized.* Essays on the Philosophical and Foundational Work of Gottlob Frege. 1986 ISBN 90-277-2126-2
182. M. Detlefsen, *Hilbert's Program.* An Essay on Mathematical Instrumentalism. 1986 ISBN 90-277-2151-3
183. J. L. Golden and J. J. Pilotta (eds.), *Practical Reasoning in Human Affairs.* Studies in Honor of Chaim Perelman. 1986 ISBN 90-277-2255-2
184. H. Zandvoort, *Models of Scientific Development and the Case of Nuclear Magnetic Resonance.* 1986 ISBN 90-277-2351-6
185. I. Niiniluoto, *Truthlikeness.* 1987 ISBN 90-277-2354-0
186. W. Balzer, C. U. Moulines and J. D. Sneed, *An Architectonic for Science.* The Structuralist Program. 1987 ISBN 90-277-2403-2
187. D. Pearce, *Roads to Commensurability.* 1987 ISBN 90-277-2414-8
188. L. M. Vaina (ed.), *Matters of Intelligence.* Conceptual Structures in Cognitive Neuroscience. 1987 ISBN 90-277-2460-1

189. H. Siegel, *Relativism Refuted.* A Critique of Contemporary Epistemological Relativism. 1987 ISBN 90-277-2469-5
190. W. Callebaut and R. Pinxten, *Evolutionary Epistemology.* A Multiparadigm Program, with a Complete Evolutionary Epistemology Bibliograph. 1987 ISBN 90-277-2582-9
191. J. Kmita, *Problems in Historical Epistemology.* 1988 ISBN 90-277-2199-8
192. J. H. Fetzer (ed.), *Probability and Causality.* Essays in Honor of Wesley C. Salmon, with an Annotated Bibliography. 1988 ISBN 90-277-2607-8; Pb 1-5560-8052-2
193. A. Donovan, L. Laudan and R. Laudan (eds.), *Scrutinizing Science.* Empirical Studies of Scientific Change. 1988 ISBN 90-277-2608-6
194. H.R. Otto and J.A. Tuedio (eds.), *Perspectives on Mind.* 1988 ISBN 90-277-2640-X
195. D. Batens and J.P. van Bendegem (eds.), *Theory and Experiment.* Recent Insights and New Perspectives on Their Relation. 1988 ISBN 90-277-2645-0
196. J. Österberg, *Self and Others.* A Study of Ethical Egoism. 1988 ISBN 90-277-2648-5
197. D.H. Helman (ed.), *Analogical Reasoning.* Perspectives of Artificial Intelligence, Cognitive Science, and Philosophy. 1988 ISBN 90-277-2711-2
198. J. Woleński, *Logic and Philosophy in the Lvov-Warsaw School.* 1989 ISBN 90-277-2749-X
199. R. Wójcicki, *Theory of Logical Calculi.* Basic Theory of Consequence Operations. 1988 ISBN 90-277-2785-6
200. J. Hintikka and M.B. Hintikka, *The Logic of Epistemology and the Epistemology of Logic.* Selected Essays. 1989 ISBN 0-7923-0040-8; Pb 0-7923-0041-6
201. E. Agazzi (ed.), *Probability in the Sciences.* 1988 ISBN 90-277-2808-9
202. M. Meyer (ed.), *From Metaphysics to Rhetoric.* 1989 ISBN 90-277-2814-3
203. R.L. Tieszen, *Mathematical Intuition.* Phenomenology and Mathematical Knowledge. 1989 ISBN 0-7923-0131-5
204. A. Melnick, *Space, Time, and Thought in Kant.* 1989 ISBN 0-7923-0135-8
205. D.W. Smith, *The Circle of Acquaintance.* Perception, Consciousness, and Empathy. 1989 ISBN 0-7923-0252-4
206. M.H. Salmon (ed.), *The Philosophy of Logical Mechanism.* Essays in Honor of Arthur W. Burks. With his Responses, and with a Bibliography of Burk's Work. 1990 ISBN 0-7923-0325-3
207. M. Kusch, *Language as Calculus vs. Language as Universal Medium.* A Study in Husserl, Heidegger, and Gadamer. 1989 ISBN 0-7923-0333-4
208. T.C. Meyering, *Historical Roots of Cognitive Science.* The Rise of a Cognitive Theory of Perception from Antiquity to the Nineteenth Century. 1989 ISBN 0-7923-0349-0
209. P. Kosso, *Observability and Observation in Physical Science.* 1989 ISBN 0-7923-0389-X
210. J. Kmita, *Essays on the Theory of Scientific Cognition.* 1990 ISBN 0-7923-0441-1
211. W. Sieg (ed.), *Acting and Reflecting.* The Interdisciplinary Turn in Philosophy. 1990 ISBN 0-7923-0512-4
212. J. Karpiński, *Causality in Sociological Research.* 1990 ISBN 0-7923-0546-9
213. H.A. Lewis (ed.), *Peter Geach: Philosophical Encounters.* 1991 ISBN 0-7923-0823-9
214. M. Ter Hark, *Beyond the Inner and the Outer.* Wittgenstein's Philosophy of Psychology. 1990 ISBN 0-7923-0850-6
215. M. Gosselin, *Nominalism and Contemporary Nominalism.* Ontological and Epistemological Implications of the Work of W.V.O. Quine and of N. Goodman. 1990 ISBN 0-7923-0904-9
216. J.H. Fetzer, D. Shatz and G. Schlesinger (eds.), *Definitions and Definability.* Philosophical Perspectives. 1991 ISBN 0-7923-1046-2
217. E. Agazzi and A. Cordero (eds.), *Philosophy and the Origin and Evolution of the Universe.* 1991 ISBN 0-7923-1322-4

218. M. Kusch, *Foucault's Strata and Fields*. An Investigation into Archaeological and Genealogical Science Studies. 1991 ISBN 0-7923-1462-X
219. C.J. Posy, *Kant's Philosophy of Mathematics*. Modern Essays. 1992 ISBN 0-7923-1495-6
220. G. Van de Vijver, *New Perspectives on Cybernetics*. Self-Organization, Autonomy and Connectionism. 1992 ISBN 0-7923-1519-7
221. J.C. Nyíri, *Tradition and Individuality*. Essays. 1992 ISBN 0-7923-1566-9
222. R. Howell, *Kant's Transcendental Deduction*. An Analysis of Main Themes in His Critical Philosophy. 1992 ISBN 0-7923-1571-5
223. A. García de la Sienra, *The Logical Foundations of the Marxian Theory of Value*. 1992 ISBN 0-7923-1778-5
224. D.S. Shwayder, *Statement and Referent*. An Inquiry into the Foundations of Our Conceptual Order. 1992 ISBN 0-7923-1803-X
225. M. Rosen, *Problems of the Hegelian Dialectic*. Dialectic Reconstructed as a Logic of Human Reality. 1993 ISBN 0-7923-2047-6
226. P. Suppes, *Models and Methods in the Philosophy of Science: Selected Essays*. 1993 ISBN 0-7923-2211-8
227. R. M. Dancy (ed.), *Kant and Critique: New Essays in Honor of W. H. Werkmeister*. 1993 ISBN 0-7923-2244-4
228. J. Woleński (ed.), *Philosophical Logic in Poland*. 1993 ISBN 0-7923-2293-2
229. M. De Rijke (ed.), *Diamonds and Defaults*. Studies in Pure and Applied Intensional Logic. 1993 ISBN 0-7923-2342-4
230. B.K. Matilal and A. Chakrabarti (eds.), *Knowing from Words*. Western and Indian Philosophical Analysis of Understanding and Testimony. 1994 ISBN 0-7923-2345-9
231. S.A. Kleiner, *The Logic of Discovery*. A Theory of the Rationality of Scientific Research. 1993 ISBN 0-7923-2371-8
232. R. Festa, *Optimum Inductive Methods*. A Study in Inductive Probability, Bayesian Statistics, and Verisimilitude. 1993 ISBN 0-7923-2460-9
233. P. Humphreys (ed.), *Patrick Suppes: Scientific Philosopher*. Vol. 1: Probability and Probabilistic Causality. 1994 ISBN 0-7923-2552-4
234. P. Humphreys (ed.), *Patrick Suppes: Scientific Philosopher*. Vol. 2: Philosophy of Physics, Theory Structure, and Measurement Theory. 1994 ISBN 0-7923-2553-2
235. P. Humphreys (ed.), *Patrick Suppes: Scientific Philosopher*. Vol. 3: Language, Logic, and Psychology. 1994 ISBN 0-7923-2862-0
Set ISBN (Vols 233–235) 0-7923-2554-0
236. D. Prawitz and D. Westerståhl (eds.), *Logic and Philosophy of Science in Uppsala*. Papers from the 9th International Congress of Logic, Methodology, and Philosophy of Science. 1994 ISBN 0-7923-2702-0
237. L. Haaparanta (ed.), *Mind, Meaning and Mathematics*. Essays on the Philosophical Views of Husserl and Frege. 1994 ISBN 0-7923-2703-9
238. J. Hintikka (ed.), *Aspects of Metaphor*. 1994 ISBN 0-7923-2786-1
239. B. McGuinness and G. Oliveri (eds.), *The Philosophy of Michael Dummett*. With Replies from Michael Dummett. 1994 ISBN 0-7923-2804-3
240. D. Jamieson (ed.), *Language, Mind, and Art*. Essays in Appreciation and Analysis, In Honor of Paul Ziff. 1994 ISBN 0-7923-2810-8
241. G. Preyer, F. Siebelt and A. Ulfig (eds.), *Language, Mind and Epistemology*. On Donald Davidson's Philosophy. 1994 ISBN 0-7923-2811-6
242. P. Ehrlich (ed.), *Real Numbers, Generalizations of the Reals, and Theories of Continua*. 1994 ISBN 0-7923-2689-X

9 780792 359043

GPSR Compliance

The European Union's (EU) General Product Safety Regulation (GPSR) is a set of rules that requires consumer products to be safe and our obligations to ensure this.

If you have any concerns about our products, you can contact us on ProductSafety@springernature.com

In case Publisher is established outside the EU, the EU authorized representative is:

Springer Nature Customer Service Center GmbH
Europaplatz 3
69115 Heidelberg, Germany

Batch number: 10584518

Printed by Printforce, the Netherlands